Principles and Techniques of Horticulture

Principles and Techniques of Horticulture

Edited by Elijah Baxter

SYRAWOOD
PUBLISHING HOUSE
New York

Published by Syrawood Publishing House,
750 Third Avenue, 9th Floor,
New York, NY 10017, USA
www.syrawoodpublishinghouse.com

Principles and Techniques of Horticulture
Edited by Elijah Baxter

© 2018 Syrawood Publishing House

International Standard Book Number: 978-1-68286-586-6 (Hardback)

Cataloging-in-Publication Data

Principles and techniques of horticulture / edited by Elijah Baxter.
      p. cm.
Includes bibliographical references and index.
ISBN 978-1-68286-586-6
   1. Horticulture. 2. Horticulture--Technological innovations. 3. Agricultural innovations. I. Baxter, Elijah.
SB318 .P75 2018
635--dc23

TABLE OF CONTENTS

PREFACE

Horticulture as field of agricultural science is concerned with the technology, art, business and science of growing and maintaining plants. Some of the main areas of horticulture are garden design, arboriculture, plant conservation, plant propagation, etc. The plants cultivated under this field are seaweed, nuts, mushrooms, medicinal plants, sprouts, fruits, etc. This book strives to provide a fair idea about the discipline and to help develop a better understanding of the latest advances within this field. Some of the diverse topics covered in it address the varied branches that fall under this category. It will help new researchers by foregrounding their knowledge in the branch of horticulture.

This book is a result of research of several months to collate the most relevant data in the field.

When I was approached with the idea of this book and the proposal to edit it, I was overwhelmed. It gave me an opportunity to reach out to all those who share a common interest with me in this field. I had 3 main parameters for editing this text:

1. Accuracy – The data and information provided in this book should be up-to-date and valuable to the readers.

2. Structure – The data must be presented in a structured format for easy understanding and better grasping of the readers.

3. Universal Approach – This book not only targets students but also experts and innovators in the field, thus my aim was to present topics which are of use to all.

Thus, it took me a couple of months to finish the editing of this book.

I would like to make a special mention of my publisher who considered me worthy of this opportunity and also supported me throughout the editing process. I would also like to thank the editing team at the back-end who extended their help whenever required.

Editor

Bioherbicides in Organic Horticulture

Xiaoya Cai and Mengmeng Gu *

Department of Horticultural Sciences, Texas A&M AgriLife Extension Service, College Station, TX 77843, USA; xiaoyacai@tamu.edu

* Correspondence: mgu@tamu.edu

Abstract: Organic horticulture producers rank weeds as one of their most troublesome, time-consuming, and costly production problems. With the increasing significance of organic horticulture, the need for new bioherbicides to control weeds has grown. Potential bioherbicides may be developed from pathogens, natural products, and extracts of natural materials. Fungal and bacteria pathogens are two important types of microbial agents that have potential to be used as bioherbicides. The byproducts of natural sources such as dried distillers grains with solubles (DDGS), corn gluten meal (CGM), and mustard seed meals (MSMs) have shown herbicidal activities in controlling many weed species. Some essential oil extracts have shown bioherbicide potential as well. The efficacy of a bioherbicide is the main limiting factor for its application, and it may be affected by environmental factors such as humidity and moisture, the application method, the spectrum of the bioherbicide, and the type of formulation. In addition to efficacy, costs and concerns about potential human health threats are also limitations to bioherbicide use. As the integration of bioherbicide technology into current weed management systems may help manage herbicide resistance, reduce production costs, and increase crop yields, future research should involve the development of more cost-effective and efficient bioherbicides for control of weeds, as well as the optimization of production methods and cultural practices with use of candidate bioherbicides.

Keywords: bioherbicide; distillers grains with solubles; corn gluten meal; mustard seed meal; *essential oil*

1. The Problem of Weeds in Organic Horticulture

Weeds are the most costly category of agricultural pests, causing great yield loss and labor expense [1]. Agricultural weeds can emerge rapidly, resulting in reduction of crop plant growth and quality by competing for nutrients and water provided to crops and producing chemicals that suppress crop growth. Annual weeds reproduce through prolific seed production, and they germinate in responses to light, increased fluctuations in soil temperature and moisture, improved aeration, and accelerated nutrient release, while perennial weeds regenerate new plants from small fragments of roots, rhizomes, stolons, and other underground structures [1]. Severe weed problems present a serious threat to horticultural crop production with favorable environmental conditions, a susceptible crop, or a large weed seed bank in the soil.

Current weed control in horticultural production includes conventional herbicides (pre-emergent and post-emergent), organic herbicides, physical methods (hand-weeding and mulches), and bioherbicides. Pre-emergence herbicides are effective before and during weed seed germination. When germinating seeds are in contact with the herbicide, the growth of emerging roots and/or shoots is inhibited, but pre-emergent herbicides may not be effective without good contact with germinating weed seeds [2]. Post-emergent herbicides are effective after weeds have emerged from the soil, ideally at the seedling stage. Organic herbicides need to be applied either prior to crop seedling emergence or transplanting, or post-directed to established crop plantings assuring that the herbicides do not cause injury on the crop plants. Current organic herbicides include ammonium nonanoate, fatty acids, vinegar, clove oil, and D-limonene [3]. Broadcast application of vinegar and clove oil has been studied for potential use in weed management on young, actively growing sweet corn, onion, and potato [4]. Physical methods in weed control include hand-pulling and mulches (including weed discs), which is necessary with some high-value crops but it is labor intensive, time-consuming, and expensive [5]. In addition to the above methods, grazing by domestic goats has also resulted in significant control of many weed species [6]. Combining hand-weeding and spot treatment with post-emergent herbicides after pre-emergent herbicide application has provided complete weed control [5].

Organic horticulture is expanding worldwide, driven by consumer demand, resource conservation, and food security in North American and European markets [7]. Expanding organic production has implied production of nutritionally-improved food crops while using fewer external inputs and reducing environmental impacts [8]. Horticultural crops, especially fruits and vegetables, are critical components of a healthy diet. In some studies, organic foods contained more nutrients and vitamins compared to conventionally-produced ones, and have grown to play an important role in consumer purchases [9]. In 2013, the North American organic food and drink market was valued at 35 billion US dollars, and a healthy market growth rate was predicted [10]. Organic horticultural crops may be more difficult to grow than conventionally-produced crops due to organic production regulations governing the use of materials for control of insect, disease, and weed control challenges. With the high costs of pest and weed control, and the time, and labor in managing the system, organic horticulture relies on price premiums for economic viability, which may make it more profitable than conventional horticulture depending on management strengths and cultural practices.

Weed control in organic horticulture does not have simple or standard solutions. Organic farmers need to take long-term approaches to control weeds without causing yield loss. Successful organic weed control needs to begin with an ecological understanding of weeds and their roles in the farm or garden ecosystem [1]. In organic horticulture, hand-weeding and cultural methods should be integrated to prevent the occurrence of weed-induced yield losses and to keep down costs for weed control. Because organic horticulture excludes the uses of synthetic herbicides due to their potential contamination of crops and natural resources, the use of bioherbicides to control weeds through the use of natural products, extracts, and natural biological agents such as fungi and bacteria to attack weeds is becoming an effective tool [11].

2. Bioherbicide Approach

Biological controls have been developed for weed management using either living organisms, such as insects, nematodes, bacteria, or fungi, or natural products. Bioherbicides offer a sustainable, low cost, and environmentally-friendly approach to complement conventional methods, which helps meet the need for new weed management strategies. There are two main approaches to biological weed control: classical biological control and bioherbicide approach [12]. The classical biological approach introduces a natural enemy that spreads throughout the area where the target weed occurred [13]. However, this approach has the risk of attacking non-target plants after the introduction of the biocontrol agent in a new area [14]. The classical approach is subjected to strict regulations because of the introduction of potentially harmful pathogens to agricultural production. The bioherbicide approach relies on natural enemies present within the native range of the weed to cause significant damage to the weed and reduce the negative impact on crop yield [13]. The classical approach is based on the innate capacity of natural enemies to reproduce, while the bioherbicide approach is based on reproduction of natural enemies under controlled conditions and subsequent spread by man [13]. The bioherbicide approach is preferred over the classical approach, because it offers diverse possibilities for use in agricultural systems, lawns, and gardens. With the increasing importance of the role of bioherbicides in organic horticulture, the main objective of the following discussion is to review the effectiveness of various bioherbicide approaches.

2.1. Bioherbicides from Pathogens

There have been many microbial agents under evaluation for their potential as bioherbicides with horticultural crops, turf, and forest trees, including obligate fungal parasites, soil-borne fungal pathogens, non-phytopathogenic fungi, pathogenic and non-pathogenic bacteria, and nematodes [11]. One of the first bioherbicides registered was DeVine (Encore Technologies, Plymouth, MN, USA) with the active ingredient *Phytophthora palmivora*, which was developed to control strangler vine (*Morrenia odorata*) on citrus in Florida [15]. In the subsequent quarter century, several more pathogenic fungi and bacteria have been developed to control weeds [15]. Using plant pathogens as biocontrol agents can cause severe damage to target weed species. In order to become suitable pathogens, they must be mass-produced and their pathogenicity tested on weeds in a range of environmental conditions, followed by field efficacy and host range tests [16]. A variety of phytotoxins produced by plant pathogens can interfere with plant metabolism, ranging from subtle effects on gene expression to plant mortality [17].

Some fungal pathogens are toxic to a wide range of weed species. The early mycoherbicides ("DeVine", "Collego" with the active ingredient *Colletotrichum gloeosporioides* f. sp. *aeschynomene*, "Biomal" with the active ingredient *Colletotrichum gloeosporioides*) had highly virulent fungal plant pathogens that could be mass-cultured to produce large quantities of inoculum for inundative application to the weed host. These fungi infect the aerial portion of weed hosts, resulting in visible disease symptoms [11]. The rust fungus *Puccinia canaliculata* is a foliar pathogen of yellow nutsedge (*Cyperus esculentus*), and it can be mass-cultured on the weed host in small field plots or the greenhouse [18]. Applying the fungal pathogen *Chonrotereum purpureum* to wounded branches or stumps of weedy tree species inhibited re-sprouting and decayed the woody tissues [19]. Weidemann *et al.* (1992) [20] reported that the fungal pathogen *Microsphaeropsis amaranthi* controlled certain pigweed (*Amaranthus*) species, while *Phoma proboscis* controlled field bindweed (*Convolvulus arvensis*) and *Colletotrichum capsici* controlled morning glory (*Ipomoea* spp.). The naturally occurring fungus *Phoma macrostoma* has been studied for control of dandelion (*Taraxacum officinale*), Canada thistle (*Cirsium arvense*), chickweed (*Stellaria media*) and scentless chamomile (*Matricaria perforata*), and its effect is equivalent to the industry standard synthetic herbicide pendimethalin [21]. One group of important bioherbicide candidates, soilborne fungi, significantly reduced weed populations by causing seed decay prior to emergence or killing seedlings shortly after emergence [22]. In a study by [23], *Trichoderma virens* (*Gliocladium virens*) colonized composted chicken manure and significantly reduced the emergence and growth of redroot pigweed (*Amaranthus retroflexus*) and broadleaf weeds in fields of horticulture crops.

Bacteria have also been studied in order to cause diseases in weeds, such as *Xanthomonas campestris* that is registered to control annual bluegrass [24]. Pathogenic bacteria *Xanthomonas campestris* pv *poannua* and *P. syringae* pv *tagetis* have been developed as bioherbicides to control annual bluegrass (*Poa annua*) and Asteraceae weeds, respectively [25]. The phytotoxin produced from a crude extract of *Pseudomonas syringae* reduced root and shoot growth of weeds in newly-established "Stevens" cranberry bogs [26]. In greenhouse and field tomato studies, applying the fungus *Myrothecium verrucaria* as a bioherbicide did not affect tomato growth throughout the growing season but killed 90%–95% of purslane species and 85%–95% of spurge species, and the yield was the same as with conventional herbicide application [27]. Spore suspensions of *Microsphaeropsis amaranthi* and *Phomopsis amaranthicola* alone, or a mixture of both organisms, were used as potential bioherbicides and significantly reduced the weed biomass of waterhemp (*Amaranthus rudis*) and pigweed, thereby increasing the yield of pumpkin and soybean [28]. Two fungi isolated from the parasitic weed dodder (*Cuscuta* spp.), *Fusarium tricinctum* and *Alternaria conjuncta/infectoria*, significantly controlled dodder without affecting cranberry growth, and these two fungi have potential to be used as bioherbicides in organic horticulture [29].

2.2. Bioherbicides from Natural Products

The byproducts of natural sources have been developed as potential bioherbicides to control weeds. Dried distillers grains with solubles (DDGS) is a byproduct of ethanol production that is commonly used as cattle feed, and is a potential fertilizer supplement in horticultural production systems due to its high nitrogen content [30]. Applying DDGS on the surface of potting mix at 800–1600 $g·m^2$ significantly reduced the number of annual bluegrass seedlings by 40%–57%, and common chickweed (*Stellaria media*) by 33%–58%, respectively [30]. The DDGS applied on the soil surface at 225 $g·m^2$ reduced the number of emerging creeping wood sorrel (*Oxalis corniculata*) seedlings by 25% [31]. A byproduct from corn wet-milling showing herbicidal activity is corn gluten meal (CGM), which has the potential to be used as a natural herbicidal product to control many broadleaf and grass species [32]. The CGM suppressed 22 germinating weed species at rates of 300–1000 $g·m^2$, and it caused reductions in plant survival, shoot length, and root development of black nightshade (*Solanum nigrum*), common lambsquarters (*Chenopodium album*), creeping bentgrass (*Agrostis palustris*), curly dock (*Rumex crispus*), purslane *(Portulaca oleracea)* and redroot pigweed when applied on the soil surface in a greenhouse [33]. Mustard seed meal (MSM) (*Sinapis alba* "IdaGold", a member of the Brassicaceae) is a byproduct of the commercial mustard oil pressing process [34]. The MSM contains glucosinolates (GLS) that can be enzymatically hydrolyzed to isothiocyanates, thiocyanate (SCN^-), nitriles, and other compounds. These biologically active compounds are toxic to many weed species [35,36]. Applying MSM to the soil surface of containers at 113, 225, and 450 $g·m^2$ reduced the number of annual bluegrass seedlings by 60%, 86%, and 98%, respectively [31]. With a MSM application rate of 225 $g·m^2$, the number of emerged seedlings and fresh weight of creeping woodsorrel were reduced by 90% and 95%, respectively. Post-emergence application of MSM at these three rates controlled liverwort from 83% to 97% without negative effects on plant growth [31]. However, there is a limitation to MSM use, because its application rate is 10–20-fold higher than typical granular herbicides used in nurseries [31]. Compared to nontreated controls, MSM application decreased emergence rates of kochia (*Bassia scoparia*), common lambsquarters, and barnyardgrass (*Echinochloa* spp.) by 83%, 73%, and 66%, respectively [37].

Bioherbicides from natural sources have shown great potential in organic production systems. Handiseni *et al.* (2012) [38] found that tomato and pepper seedling emergence in *Pythium ultimum*-infested soils have been improved by canola (*B. napus*) and mustard greens (*B. juncea*) seed meals. Brassicaceae seed meals (BSMs) were used to increase soil inorganic nitrogen and the yields of carrot, which had high efficacy in controlling weeds in organic production [39]. In strawberry production, after applying canola-derived BSM and MSM, the weed biomass of shepherd's purse (*Capsella bursa-pastoris*), Italian ryegrass (*Lolium multiflorum*), desert rock purslane (*Calandrinia ciliata*), and annual bluegrass decreased and strawberry fruit yields increased with BSM treatment, which indicated that BSMs may have potential use in organic horticulture systems as combined bioherbicides and green fertilizers [40]. Fennimore *et al.* [41] found that the combination of steam-disinfestation treatment with soil amendments of MSM showed improved strawberry yield as well as weed and pathogen control. In a lettuce field study, the application of meadowfoam (*Limnanthes alba*) seed meal suppressed weeds and increased lettuce yield and leaf nitrogen content [42]. Onions are poor competitors with weeds, which makes weed management in organically-grown onions difficult [34]. In a greenhouse study, MSM significantly decreased redroot pigweed emergence and slightly reduced total yield of onion, indicating MSM has potential to be used as a weed suppressive amendment in an organic onion production system [34]. In organically-grown broccoli and spinach, the application rate of 4.48 t/ha MSM and soybean seed meal significantly increased spinach yield, but broccoli yield was similar in all treatments [43]. However, application rates of 2.5% MSM and mustard greens seed meal significantly reduced heights and biomass in sorghum, and there was a negative effect of MSM on cotton yield [44]. Therefore, it is evident that the type, rate, and timing of seed meal applications should be considered to successfully manage weeds while producing an organic crop. As evidence, the combination of CGM, clove oil, and sweep cultivation had little impact on weed management for organic peanut production [45]. Russo and Webber (2012) [46] also reported that application of CGM and vinegar did not produce peanut pod or oil yields at levels produced with conventional weed control. Therefore, additional alternative weed control techniques and materials should be investigated for organic peanut production, as one example. In container-grown ornamentals, weed emergence was significantly reduced with DDGS application at 800 and 1600 g/m^2 to the soil surface, with no injury on *Rosa hybrid* "Red Sunblaze", *Phlox paniculata* "Franz Schubert", and *Coreopsis auriculata* "Nana", indicating opportunities for use of DDGS for organically-grown ornamentals [30].

2.3. Bioherbicides from Extracts

Extracts from natural sources may also have potential as bioherbicides. Five dipeptides extracted from hydrolyzed CGM inhibited root growth of germinating weeds [47]. Secondary metabolite extracts from the leaves of *Ailanthus altissima* had inhibitory effects on seed germination and plant growth of *Medicago saltiva* [48]. Rice hull extracts demonstrated a significant allelopathic potential. [49] reported that increasing concentrations of warm water hull extracts from selected rice cultivars resulted in inhibition of barnyardgrass germination, seedling growth, and weight. Nieves *et al.* (2011) [50] also reported that methanolic extracts of *Everniastrum sorocheilum*, *Usnea roccellina*, and *Cladonia confusa* inhibited germination and root growth of red clover (*Trifolium pratense*). Phenolics extracted from the lichen *Cladonia verticillaris* caused changes in the ultrastructure of both roots and leaves of lettuce seedlings, suggesting potential as powerful bioherbicides [51]. Black walnut (*Juglans nigra*) has allelopathic effects, and extracts from walnut have been commercially formulated as a bioherbicide [52]. A black walnut extract-based commercial product (NatureCur®, Redox Chemicals, LLC, Burley, ID, USA) completely inhibited growth of horseweed (*Conyza canadensis*) and hairy fleabane (*Conyza bonariensis*) at a concentration of 33.3%, showing potential as a pre- and post-emergent bioherbicide [52].

Herbs are rich in essential oil content, and essential oil extracts with allelopathic effects can be used for weed management [53]. The essential oils from eucalyptus (*Eucalyptus* spp.), Lawson cypress (*Chamaecyparis lawsoniana*), rosemary (*Rosmarinus officinalis*), and white cedar (*Thuja occidentalis*) significantly inhibited the weed species amaranth (*Amaranthus retroflexus*), purslane (*Portulaca oleracea*), and knapweed (*Acroptilon repens*), and may be applied for biological control of weeds as pre-emergent weed seed germination inhibitor [53]. Onen *et al.* (2002) [54] reported that the essential oils extracted from leaves and flowers of five different plant species (*Artemisia vulgaris*, *Mentha spicata* subsp. *spicata*, *Ocimum basilicum*, *Salvia officinalis*, *Thymbra spicata* subsp. *spicata*) were highly phytotoxic to seed germination and seedling growth of eight weed species from different families (*Agrostemma githago*, amaranth, *Cardaria draba*, *Chenopodium album*, *Echinochloa crus-galli*, *Reseda lutea*, *Rumex crispus*, *Trifolium pratense*). Manuka oil, the essential oil distilled from the manuka tree (*Leptospermum scoparium*), exhibited good post-herbicidal activity for control of the emergence of large crabgrass (*Digitaria* spp.) seedlings, which may be used as a potential bridge between traditional and organic agriculture [55]. Volatile oils from leaves of *Eucalyptus citriodora* caused severe damage to the noxious weed *Parthenium hysterophorus* [56]. The essential oil extracts from *Origanum syriacum*, *Micromeria fruticosa*, and *Cymbopogon citratus* had inhibitory effects on seed germination of wheat, *Amaranthus palmeri*, and *Brassica nigra* [57]. Other plants that have essential oils with allelopathic effects include aromatic plants such as *Rosmarinus officinalis*, *Laurus nobilis*, *Xanthoxylum rhesta*, *Cunila spicata*, and *Artemisia* spp. [58–62].

3. Factors Affecting the Efficacy of Bioherbicide

The efficacy of bioherbicides is the main limiting factor for their use, often due to environmental factors. The humidity requirements for establishment and spread of many foliar and stem fungal pathogens for weed control necessitate the development of special formulations to ensure the effectiveness of agents applied in the field [11]. A long dew period is required by some pathogens for infection on the aerial surfaces of target weeds [63]. Some organisms have limited shelf lives, and they are not suited for long-term storage [64]. *Xanthomonas campestris* pv. *Poannua*, a pathogen causing bacterial wilt of annual bluegrass, was not successfully commercialized due to low performance and variability in efficacy under different environmental conditions [65]. Soil moisture can be an important factor affecting pathogens attacking weeds. Application of a jute fabric to cover soil areas treated with a *Sclerotinia minor* granular bioherbicide to reduce water loss significantly enhanced control of dandelion (*Taraxacum* spp.), white clover (*Trifolium repens*), broadleaf plantain (*Plantago major*), buckhorn plantain (*Plantago major*), ground ivy (*Glechoma hederacea*), and prostrate knotweed (*Polygonum aviculare*) [66]. The influence of moisture was reduced by addition of an invert oil emulsion to conidial suspensions of *Colletotrichum truncatum*, which resulted in 100% control of hemp sesbania (*Sesbania exaltata*) in the absence of moisture in the greenhouse, and in 95% control of hemp sesbania in the field [67]. *Phoma macrostoma* has been registered as a bioherbicide to control broadleaved weed species, and its efficacy on dandelion was significantly increased by 10%–20% by amendment with nitrogen fertilizers [68].

The bioherbicide application method should be considered for enhancing efficacy of the biocontrol agent, including attention to spray droplet size, droplet retention and distribution, spray application volume, and the equipment used [69]. The application distribution pattern and pressure are important considerations for determining the quantity of bioherbicide applied [70]. Retention of spray droplets is affected by surface characteristics and morphology of the weed, its biotypes, the adjuvants used in the solutions, travel speed, and droplet size [71]. Smaller droplet sizes of *Colletotrichum truncatum* resulted in greater efficacy in controlling scentless chamomile (*Matricaria perforata*) [72]. Application of bioherbicides with different nozzles affected the disease incidence and development on waterhemp [73]. Innovations such as dual nozzle sprayers, and the use of compressed air rather than CO_2 to minimize the acidification of the spray solution, may have impacts on bioherbicide efficacy [69].

Other factors, such as the spectrum of the bioherbicide, whether broad or targeted to specific species, the type of formulation, and if it involves amino acid-excreting strains, can significantly affect efficacy. Broad-spectrum bioherbicides may show different efficacies in different regions. That can be altered, as the spectrum of *Alternaria crassa* was broadened by combining it with fruit pectin and plant filtrates [74]. Another method to broaden the spectrum of bioherbicide is to combine multiple pathogens. By combining *Alternaria cassiae*, *Phomospsis amaranthicola* and *Colletotrichum dematium*, weeds such as pigweed (*Amaranth* spp.), sicklepod (*Senna obtusifolia*), and showy crotolaria (*Crotalaria spectablis*) were effectively controlled [75]. Chandramohan and Charudattan (2003) [76] also found that a mixture of three pathogens, *Drechslera gigantia*, *Exserohilum longirostratum*, and *Exserohilum rostratum*, successfully suppressed the growth of seven weeds in citrus groves in Florida. Amendment of bacterial pathogen aqueous suspensions with surfactants has been studied for helping bacteria efficiently invade plant leaves and broaden host range [25]. Types of formulations using emulsions, organosilicone surfactants, and hydrophilic polymers have advantages and disadvantages in enhancing the efficacy of biotic agents and ease of application [69]. Emulsions may improve efficacy and consistency of weed control by predisposing weeds to a bioherbicide agent [69]. Organosilicone surfactants, such as Silwet L-77, facilitate direct entry of bacterial cells and small spores into weed tissues [69]. Hydrophilic polymers, including numerous types of natural and synthetic polymers, have different levels of water-holding qualities. However, formulations composed of expensive materials increase the cost of bioherbicide products. In addition, some materials used in these formulation are toxic to human health [69]. An abundant quantity of amino acids has the potential to terminate plant growth. Therefore, the selection of fungal strains that are able to produce significant quantities of amino acids is becoming a new technique to control weeds [77]. Valine excretion by mutants of *Fusarium oxysporum* controlled *Cannabis sativa* by 70%–90% compared to 25% by a wild type isolate [77].

In addition to bioherbicide efficacy, the high cost and the potential human health threats are some other limitations for use of bioherbicides. Although some pathogens are highly effective in controlling a number of weeds, they may also produce undesirable mammalian and avian toxins [11]. *Myrothecium verrucaria* was effective for weed control as a result of the production of herbicidal metabolites; however, the mammalian-toxic macrocyclic tricothecenes were also simultaneously produced, presenting a severe human health hazard [78]. A fungal pathogen, *Fusarium tumidum*, a potential bioherbicide for gorse (*Ulex europaeus*) and broom (*Cytisus scoparius*), also produced tricothecenes [79]. With the relatively small market at present, and the high cost of maintaining registration, bioherbicides may be dropped from production, like DeVine (*Phytophthora palmivora*) that provided 95%–100% control of strangler vine [80,81]. Although the demand for more environmentally-friendly strategies and bioherbicides for weed control is increasing, there have been few bioherbicides successfully registered and commercialized in North America due to these limitations.

4. Conclusions

Lacking few effective bioherbicides, the integration of biological controls into current weed management systems may be an effective alternative for organic horticultural production. Bioherbicide technology could be used as a component in integrated weed management strategies to help avoid herbicide resistance, reduce production costs, and increase crop yield in organic horticulture. While there have been significant efforts to develop bioherbicides, few have been registered for use. Future research should focus on the development of more cost-effective and efficient bioherbicides, as well as the optimization of their use in production systems.

Conflicts of Interest: The authors declare no conflict of interest.

References

1. Schonbeck, M. Principles of sustainable weed management in organic cropping systems. In *Workshop for Farmers and Agricultural Professionals on Sustainable Weed Management*, 3rd ed.; Clemson University: Clemson, SC, USA, 2011.
2. Altland, J.E.; Gilliam, C.H.; Wehtje, G. Weed control in field nurseries. *HortTechnology* **2003**, *13*, 9–17.
3. Webber, C.L.; Shrefler, J.W.; Brandenberger, L.P. Organic weed control. In *Herbicides—Environmental Impact Studies and Management Approaches*; Fernandez, R.A., Ed.; InTech: Rijeka, Croatia, 2012; pp. 186–198.
4. Evans, G.J.; Bellinder, R.R. The potential use of vinegar and a clove oil herbicide for weed control in sweet corn, potato, and onion. *Weed Technol.* **2009**, *23*, 120–128. [CrossRef]
5. Harpster, T.; Sellmer, J.; Kuhns, L.J. *Controlling Weeds in Nursery and Landscape Plantings*; PennState Cooperative Extension, College of Agricultural Sciences: State College, PA, USA, 2012.
6. Booth, A.L.; Skelton, N.W. The use of domestic goats and vinegar as municipal weed control alternatives. *Environ. Pract.* **2009**, *11*, 3–16. [CrossRef]
7. Granatstein, D.; Kirby, E.; Willer, H. Organic horticulture expands globally. *Chron. Horticult.* **2010**, *504*, 31–38.
8. Risku-Norja, H.; Maenpaa, I. MFA model to assess economic and environmental consequences of food production and consumption. *Ecol. Econ.* **2007**, *60*, 700–711. [CrossRef]
9. Worthington, V. Nutritional quality of organic *vs.* conventional fruits, vegetables, and grains. *J. Altern. Complement. Med.* **2001**, *7*, 161–173. [CrossRef] [PubMed]
10. Willer, H.; Lernoud, J. *The World of Organic Agriculture—Statistics and Emerging Trends 2015*; Research Institute of Organic Agriculture (FiBL), International Federation of Organic Agriculture Movements (IFOAM): Frick, Switzerland; Bonn, Germany, 2015.
11. Kremer, R.J. The role of Bioherbicides in weed management. *Biopestic. Int.* **2005**, *1*, 127–141.
12. Green, S. A review of the potential for the use of bioherbicides to control forest weeds in the UK. *Forestry* **2003**, *76*, 285–298. [CrossRef]
13. Frantzen, J.; Paul, N.D.; Müller-Schärer, H. The system management approach of biological weed control: Some theoretical considerations and aspects of application. *BioControl* **2001**, *46*, 139–155. [CrossRef]
14. Thomas, M.B.; Willis, A.J. Biocontrol—Risky but necessary? *TREE* **1998**, *13*, 325–329. [CrossRef]
15. Charudattan, R. Use of plant pathogens as bioherbicides to manage weeds in horticultural crops. *Proc. Fla. State Hort. Soc.* **2005**, *118*, 208–214.
16. Ayres, P.; Paul, N. Weeding with fungi. *New Sci.* **1990**, *732*, 36–39.
17. Walton, J.D. Host-selective toxins: Agents of compatibility. *Plant Cell* **1996**, *8*, 1723–1733. [CrossRef] [PubMed]
18. Phatak, S.C.; Summer, D.R.; Wells, H.D.; Bell, D.K.; Glaze, N.C. Biological control of yellow nutsedge with the indigenous rust fungus *Puccinia canaliculata*. *Science* **1983**, *219*, 1446–1447. [CrossRef] [PubMed]

19. Prasad, R. Development of bioherbicides for integrated weed management in forestry. In Proceedings of the 2nd International Weed Control Congress, Department of Weed Control and Pesticide Ecology, Slagelse, Denmark, 25–28 June 1996; Brown, H., Ed.; pp. 1197–1203.

20. Weidemann, G.J.; TeBeest, D.O.; Templeton, G.E. Fungal plant pathogens used for biological weed control. *Ark. Farming Res.* **1992**, *41*, 6–7.

21. Bailey, K.L.; Derby, J. Fungal Isolates and Biological Control Compositions for the Control of Weeds. U.S. Patent Application Serial No. 60/294,475, 20 May 2001.

22. Jones, R.W.; Hancock, J.G. Soilborne fungi for biological control of weeds. In *Microbes and Microbial Products as Microbial Herbicides*; Hoagland, R.E., Ed.; American Chemical Society: Washington, DC, USA, 1990; pp. 276–286.

23. Héraux, F.M.G.; Hallett, S.G.; Ragothama, K.G.; Weller, S.C. Composted chicken manure as a medium for the production and delivery of *Trichoderma virens* for weed control. *HortScience* **2005**, *40*, 1394–1397.

24. Hoagland, R.E.; Weaver, M.A.; Boyette, C.D. *Myrothecium verrucaria* fungus: A bioherbicide and strategies to reduce its non-target risks. *Allelopath. J.* **2007**, *19*, 179–192.

25. Johnson, D.R.; Wyse, D.L.; Jones, K.L. Controlling weeds with phytopathogenic bacteria. *Weed Technol.* **1996**, *10*, 621–624.

26. Norman, M.; Patten, K.; Gurusiddaiah, S. Evaluation of a phytotoxin from *Pseudomonas syringae* for weed control in cranberries. *HortScience* **1994**, *29*, 1475–1477.

27. Boyette, C.D.; Hoagland, R.E.; Abbas, H.K. Evaluation of the bioherbicide *Myrothecium verrucaria* for weed control in tomato (*Lycopersicon esculentum*). *Biocontrol Sci. Technol.* **2007**, *17*, 171–178. [CrossRef]

28. Ortiz-Ribbing, L.M.; Glassman, K.R.; Roskamp, G.K.; Hallett, S.G. Performance of two bioherbicide fungi for waterhemp and pigweed control in pumpkin and soybean. *Plant Dis.* **2011**, *95*, 469–477. [CrossRef]

29. Hopen, H.J.; Bewick, T.A.; Caruso, F.L. Control of dodder in cranberry *Vaccinium macrocarpon* with a pathogen-based bioherbicide. *Acta Hort.* **1997**, *446*, 427. [CrossRef]

30. Boydston, R.A.; Collins, H.P.; Vaughn, S.F. Response of weeds and ornamental plants to potting soil amended with dried distillers grains. *HortScience* **2008**, *43*, 191–195.

31. Boydston, R.A.; Anderson, T.; Vaughn, S.F. Mustard (*Sinapis alba*) seed meal suppresses weeds in container-grown ornamentals. *HortScience* **2008**, *43*, 800–803.

32. Liu, D.; Christians, N. Inhibitory activity of corn gluten hydrolysate on monocotyledonous and dicotyledonous species. *HortScience* **1997**, *32*, 243–245.

33. Bingaman, B.R.; Christians, N.E. Green-house screening of corn gluten meal as a natural control product for broadleaf and grass weeds. *HortScience* **1995**, *30*, 1256–1259.

34. Boydston, R.A.; Morra, M.J.; Borek, V.; Clayton, L.; Vaughn, S.F. Onion and weed response to mustard (*Sinapis alba*) seed meal. *Weed Sci.* **2011**, *59*, 546–552. [CrossRef]

35. Borek, V.; Morra, M.J. Ionic thiocyanate (SCN⁻) production from 4-hydroxybenzyl glucosinolate contained in *Sinapis alba* seed meal. *J. Agric. Food Chem.* **2005**, *53*, 8650–8654. [CrossRef] [PubMed]

36. Brown, P.D.; Morra, M.J. Control of soilborne plant pests using glucosinolate-containing plants. *Adv. Agron.* **1997**, *61*, 167–231.

37. Yu, J.; Morishita, D.W. Response of seven weed species to corn gluten meal and white mustard (*Sinapis alba*) seed meal rates. *Weed Technol.* **2014**, *28*, 259–265. [CrossRef]

38. Handiseni, M.; Brown, J.; Zemetra, R.; Mazzola, M. Use of Brassicaceous seed meals to improve seedling emergence of tomato and pepper in *Pythium ultimum* infested soils. *Arch. Phytopathol. Plant Protect.* **2012**, *45*, 1204–1209. [CrossRef]

39. Snyder, A.; Morra, M.J.; Johnson-Maynard, J.; Thill, D.C. Seed meals from brassicaceae oilseed crops as soil amendments: Influence on carrot growth, microbial biomass nitrogen, and nitrogen mineralization. *HortScience* **2009**, *44*, 354–361.

40. Banuelos, G.S.; Hanson, B.D. Use of selenium-enriched mustard and canola seed meals as potential bioherbicides and green fertilizer in strawberry production. *HortScience* **2010**, *45*, 1567–1572.

41. Fennimore, S.A.; Martin, F.N.; Miller, T.C.; Broome, J.C.; Dorn, N.; Greene, I. Evaluation of a mobile steam applicator for soil disinfestation in California strawberry. *HortScience* **2014**, *49*, 1542–1549.

42. Intanon, S.; Hulting, A.G.; Mallory-Smith, C.A. Field evaluation of meadowfoam (*Limnanthes alba*) seed meal for weed management. *Weed Sci.* **2015**, *63*, 302–311. [CrossRef]

43. Shrestha, A.; Rodriguez, A.; Pasakdee, S.; Banuelos, G. Comparative efficacy of white mustard (*Sinapis alba* L.) and soybean (*Glycine max* L. Merr.) seed meals as bioherbicides in organic broccoli (*Brassica oleracea* Var. Botrytis) and spinach (*Spinacea oleracea*) production. *Commun. Soil Sci. Plant Anal.* **2015**, *46*, 33–46. [CrossRef]

44. Rothlisberger, K.L.; Hons, F.M.; Gentry, T.J.; Senseman, S.A. Oilseed meal effects on the emergence and survival of crop and weed species. *Appl. Environ. Soil Sci.* **2012**, *2012*, 1–10. [CrossRef]

45. Johnson, W.C.; Boudreau, M.A.; Davis, J.W. Combinations of corn gluten meal, clove oil, and sweep cultivation are ineffective for weed control in organic peanut production. *Weed Technol.* **2013**, *27*, 417–421. [CrossRef]

46. Russo, V.M.; Webber, C.L. Peanut pod, seed, and oil yield for biofuel following conventional and organic production systems. *Ind. Crop Prod.* **2012**, *39*, 113–119. [CrossRef]

47. Liu, D.; Christians, N. Isolation and identification of root inhibiting compounds from corn gluten hydrolysate. *J. Plant Growth Regul.* **1994**, *13*, 227–230. [CrossRef]

48. Tsao, R.; Romanchuk, F.; Peterson, C.J.; Coats, J.R. Plant growth regulatory effect and insecticidal activity of the extracts of the tree of heaven (*Ailanthus altissima* L.). *BMC Ecol.* **2002**, *2*, 1. [CrossRef] [PubMed]

49. Ahn, J.K.; Chung, I.M. Allelopathic potential of rice hulls on germination and seedling growth of barnyardgrass. *Agron. J.* **2000**, *92*, 1162–1167. [CrossRef]

50. Nieves, J.A.; Acevedo, L.J.; Valencia-Islas, N.A.; Rojas, J.L.; Dávila, R. Fitotoxicidad de extractos metanólicos de los líquenes Everniastrum sorocheilum, Usnea roccellinay Cladonia confusa. *Glalia* **2011**, *4*, 96.

51. Tigre, R.C. Investigação dos Mecanismos de Ação Alelopática de Cladonia Verticillaris Sobre Lactuca Sativa e Solanum lycopersicum. Ph.D. Theses, Department of Geographical Sciences, Federal University of Pernambuco, Brazil, 2014.

52. Shrestha, A. Potential of a black walnut (*Juglans nigra*) extract product (NatureCur) as a pre- and post-emergence bioherbicide. *J. Sustain. Agric.* **2009**, *33*, 810–822. [CrossRef]

53. Ramezani, S.; Saharkhiz, M.J.; Ramezani, F.; Fotokian, M.H. Use of essential oils as bioherbicides. *Jeobp* **2008**, *11*, 319–327. [CrossRef]

54. Onen, H.; Ozer, Z.; Telci, I. Bioherbicidal effects of some plant essential oils on different weed species. *J. Plant Dis. Prot.* **2002**, *18*, 597–605.

55. Dayan, F.E.; Howell, J.L.; Marais, J.P.; Ferreira, D.; Koivunen, M. Manuka oil, a natural herbicide with preemergence activity. *Weed Sci.* **2011**, *59*, 464–469. [CrossRef]

56. Singh, H.P.; Batish, D.R.; Setia, N.; Kohli, R.K. Herbicidal activity of volatile oils from *Eucalyptus citriodora* against *Parthenium hysterophorus*. *Ann. Appl. Biol.* **2005**, *146*, 89–94. [CrossRef]

57. Dudai, N.; Poljakoff-Mayber, A.; Mayer, A.M.; Putievsky, E.; Lerner, H.R. Essential oils as allelochemicals and their potential use as bioherbicides. *J. Chem. Ecol.* **1999**, *25*, 1079–1089. [CrossRef]

58. Ahmad, A.; Misra, L.N. Terpenoids from *Artemisia annua* and constituents of its essential oil. *Phytochemistry* **1994**, *37*, 183–186. [CrossRef]

59. Hogg, J.W.; Terhune, S.J.; Lawrence, B.M. Dehydro-1,8-cineole: A new monoterpene oxide in *Laurus noblis* oil. *Phytochemistry* **1974**, *13*, 868–869. [CrossRef]

60. Manns, D. Linalool and cineole type glucosides from *Cunila spicata*. *Phytochemistry* **1995**, *39*, 1115–1118. [CrossRef]

61. Naves, Y.R.; Ardizio, P. Etudes sur les matieres vegetales volatiles CI. Sur la composition de l'essence de *Xanthoxylum rhetsa*, D.C. *Mem. Soc. Chim.* **1950**, *1950*, 673–678. (In French).

62. Zaouali, Y.; Messaoud, C.; Ben Salah, A.; Boussaïd, M. Oil composition variability among populations in relationship with their ecological areas in Tunisian Rosmarinus officinalis L. *Flav. Fragr. J.* **2005**, *20*, 512–520. [CrossRef]

63. Auld, B.A.; Hethering, S.D.; Smith, H.E. Advances in bioherbicide formulation. *Weed Biol. Man.* **2003**, *3*, 61–67. [CrossRef]

64. Ghosheh, H.Z. Constraints in implementing biological weed control: A review. *Weed Biol Manag.* **2005**, *5*, 83–92. [CrossRef]

65. Johnson, B.J. Biological control of annual bluegrass with *Xanthomonas campestris* pv. poannua in bermudagrass. *Hort. Sci.* **1994**, *29*, 659–662.

66. Abu-Dieyeh, M.H.; Watson, A.K. Increasing the efficacy and extending the effective application period of a granular turf bioherbicide by covering with jute fabric. *Weed Technol.* **2009**, *23*, 524–530. [CrossRef]

67. Boyette, C.D.; Quimby, P.C., Jr.; Bryson, C.T.; Egley, G.T.; Fulgham, F.E. Biological control of hemp sesbania (*Sesbania exaltata*) under field conditions with *Colletotrichun truncatum* formulated in emulsion. *Weed Sci.* **1993**, *41*, 497–500.

68. Bailey, K.L.; Falk, S.; Derby, J.; Melzer, M.; Boland, G.J. The effect of fertilizers on the efficacy of the bioherbicide, *Phoma macrostoma*, to control dandelions in turfgrass. *Biol. Control.* **2013**, *65*, 147–151. [CrossRef]

69. Charudattan, R. Biological control of weeds by means of plants pathogens: Significance for integrated weed management in modern agroecology. *Biocontrol* **2001**, *46*, 229–260. [CrossRef]

70. Klein, T. The application of mycoherbicides. *Plant Prot. Quart.* **1992**, *7*, 161–162.

71. Singh, M.; Tan, S.Y.; Sharma, S.D. Adjuvants enhance weed control efficacy of foliar-applied diuron. *Weed Technol.* **2002**, *16*, 74–78. [CrossRef]

72. Byer, K.N.; Peng, G.; Wolf, T.M.; Caldwell, B.C. Spray retention and its effect on weed control by mycoherbicides. *Biol. Control.* **2006**, *37*, 307–313. [CrossRef]

73. Doll, D.A.; Sojka, P.E.; Hallett, S.G. Effect of nozzle type and pressure on the efficacy of spray applications of the bioherbicidal fungus *Microsphaeropsis amaranthi*. *Weed Technol.* **2005**, *19*, 918–923. [CrossRef]

74. Boyette, C.D.; Abbas, H.K. Host range alteration of the bioherbicidal fungus *Alternaria crassa* with fruit pectin and plant filtrates. *Weed Sci.* **1994**, *42*, 487–491.

75. Chadramohan, S.; Charudattan, R.; Sonoda, R.M.; Singh, M. Field evaluation of a fungal mixture for the control of seven weedy grasses. *Weed Sci.* **2002**, *50*, 204–213. [CrossRef]

76. Chandramohan, S.; Charudattan, R. A multiple-pathogen system for bioherbicidal control of several weeds. *Biocontr. Sci. Technol.* **2003**, *13*, 199–205. [CrossRef]

77. Tiourebaev, K.S.; Nelson, S.; Zidak, N.K.; Kaleyva, G.T.; Pilgeram, A.L.; Anderson, T.W.; Sands, D.C. Amino acid excretion enhances virulence of bioherbicides. In Proceedings of the X International Symposium on Biological Control of Weeds, Montana State University, Bozeman, MT, USA, 4–14 July 1999; Spencer, N.R., Ed.; pp. 295–299.

78. Anderson, K.I.; Hallett, S.G. Herbicidal spectrum and activity of *Myrothecium verrucaria*. *Weed Sci.* **2004**, *52*, 623–627. [CrossRef]

79. Morin, L.; Gianotti, S.F.; Lauren, D.R. Trichothecene production and pathogenicity of *Fusarium tumidum*, a candidate bioherbicide for gorse and broom in New Zealand. *Mycol. Res.* **2000**, *104*, 993–999. [CrossRef]

80. Kenney, D.S. DeVine-the way it was developed—An industrialist's view. *Weed Sci.* **1986**, *34*, 15–16.

81. Karim Dagno, R.L.; Diourté, M.; Jijakli, M.H. Present status of the development of mycoherbicides against water hyacinth: Successes and challenges. A review. *Biotechnol. Agron. Soc. Environ.* **2012**, *16*, 360–368.

2

Effect of Bio-Organic and Inorganic Nutrient Sources on Growth and Flower Production of African Marigold

Gaurav Sharma *, Naresh Prasad Sahu and Neeraj Shukla

Department of Floriculture and Landscape Architecture, Indira Gandhi Agricultural University, Krishak Nagar, Raipur, Chhattisgarh 492012, India; nareshsahu008@gmail.com (N.P.S.); shuklaniraj@rediffmail.com (N.S.)
* Correspondence: gauravhort@gmail.com

Abstract: African marigold (*Tagetes erecta* L.) is one of the most important flower crops grown commercially throughout India as a loose flower for worshipping, garland making, and garden display. The productivity and quality of flowers is greatly influenced by the quantity and source of nutrients. At present, these nutrients are primarily supplied through chemical fertilizers. The indiscriminate use and complete reliance on the use of chemical fertilizers has also led to deterioration of soil health, thereby affecting sustainable flower production. Keeping these points in view, a field experiment was conducted on African marigold cv. "Orange Culcuttia" at the Horticultural Research Farm, Indira Gandhi Agricultural University, Raipur, Chhattisgarh, India. The experiment was laid out in a randomized block design with three replications and twelve treatment combinations comprised of bio-organics (Cow Urine and Vermicompost), bio-fertilizers (*Azospirillum* and Phosphate-Solubilizing Bacteria) and NPK fertilizers. Application of *Azospirillum* + Phosphate-Solubilizing Bacteria + 5% Cow Urine + 50% recommended dose of "N" through Vermicompost + 50% recommended dose of NPK fertilizer was most effective in increasing vegetative growth parameters, such as plant height, number of branches, plant spread, as well as flower yield parameters like number of flowers, flower diameter, fresh and dry weight of flowers, flower yield, flowering duration, shelf life, and it also had the maximum benefit:cost ratio. Thus, use of inorganic fertilizers conjointly with bio-fertilizers and organic manures resulted in excellent vegetative growth and flower yield attributes in African marigold.

Keywords: *Azospirillum*; bio-fertilizer; cow urine; vermicompost

1. Introduction

Among flower crops, African marigold (*Tagetes erecta* L.) is one of the most important commercially-exploited flowers throughout the world. It is in demand for loose flower production, for garland making, garden display and decorative purposes. Nutrient management plays an important role in determining flowering in African marigold. Among the various reasons behind low productivity, poor soil and nutrient management is a major cause. Therefore, nutrient management has prime importance for successful cultivation. The use of organic manures and biofertilizers along with balanced use of chemical fertilizers is known to improve the physico-chemical and biological properties of soil, besides improving the efficiency of applied fertilizers [1]. Biofertilizers improve crop growth and quality by fixation of atmospheric nitrogen and also by dissolving insoluble forms of phosphorus. *Azospirillum* (Azo) fixes atmospheric nitrogen to some extent and makes available the fixed soil nitrogen to the crop, whereas phosphorus-solubilising bacteria (PSB) possesses the ability to convert insoluble phosphorus into soluble forms in the soil by secreting organic acids. Arancan and Edwards [2] demonstrated the effects of vermicomposts as an important organic manure for petunias, marigolds, asters, and chrysanthemums. Recently, the use of cow urine (CU) has been given importance as it may act as a growth promoter of plants and is a vital component in improving soil fertility. It not only possesses an inherent property of acting as a fertilizer but also is a mild biocide [3]. Thus, the present experiment was carried out to determine the effect of bio-organic and inorganic nutrient sources on the flower yield of marigold in view of maintaining soil health and the environment.

2. Experimental Section

The present investigation was conducted at the Department of Horticulture, Indira Gandhi Agricultural University, Raipur, Chhattisgarh, India, during the winter season of 2013–2014. The experiment was laid out in a randomized block design with three replications and twelve treatments. Individual plot size was 1.2 m $\times$ 1.2 m with a spacing of 30 cm $\times$ 30 cm. The recommended management practices of raising a healthy crop were followed. The treatments included an inorganic form of N in the form of urea (200 kg·ha^{-1}), P_2O_5 as superphosphate (200 kg·ha^{-1}) and K_2O as muriate of potash (200 kg·ha^{-1}) as 100% of the recommended dose of fertilizer (RDF). The rates of N and P were reduced in some treatments to 50% or 75% of the RDF. The NPK was applied in two parts, half the N and the full amount of P and K at the time of transplanting, and the remaining half of N was applied 40 days after transplanting. Vermicompost (VC) was applied in quantities equivalent to 50% of the recommended dose of N before planting. A slurry of 200 g of a culture of Azo and PSB were prepared in 1000 mL of water individually, and also combinations of both 100 g Azo and 100 g PSB were prepared in 1000 mL of water. Azo (200 g·ha^{-1}) and PSB (200 g·ha^{-1}) were applied by seedling root treatment for 30 min before transplanting. CU (5%) spray was applied at 30 days after transplanting. The treatment combinations were: 100% RDF (RDF), 50% RD"N" through VC + 50% RDF (VC), *Azo* + 75% RD"N" + 100% RD"P" and "K"(AZO), PSB + 75% RD"P" + 100% RD"N" and "K" (PSB), Cow Urine (5%) + 75% RD"N" + 100% RD"P" and "K" (CU), *Azo* + 50% RD"N" through VC + 50% RDF (AZVC), PSB + 50% RD"N" through VC + 50% RDF (PSVC), Cow Urine (5%) + 50% RD"N" through VC + 50% RDF (CUVC), *Azo* + PSB + 50% RD"N" and "P" + 100% RD"K" (AZPS), *Azo* + PSB + Cow Urine (5%) + 50% RD"N" and "P" + 100% RD"K" (APC), *Azo* + PSB + 50% RD"N" through VC + 50% RDF (APV), *Azo* + PSB + Cow Urine (5%) + 50% RD"N" through VC + 50% RDF (APCV). Data were recorded from five randomly-selected plants from each treatment, and included flowering behaviour, yield attributes and flower yield. Data were subjected to statistical analysis using SPSS 10.0 statistical software (SPSS Inc., Chicago, IL, USA).

3. Results and Discussion

3.1. Vegetative Growth Parameters

The different nutrient sources affected various vegetative parameters of African marigold (Table 1). The treatment APCV had the maximum plant height (53.31 cm), plant spread (37.78 cm), number of primary branches/plant (19.47) and number of secondary branches/plant (39.53) APV which, however, had at similar values. Combined application of bio-organic nutrient sources along with 50% inorganic nutrient sources proved to be beneficial for robust growth of plants as compared to other treatments. Bioinoculants like Azo and PSB may have been beneficial by fixing atmospheric nitrogen and solubilizing fixed phosphorous in the soil, making it available to plants, and also by secretion of growth substances like auxin which might have stimulated plant metabolic activity and photosynthetic efficacy leading to better growth and development. These results are in conformity with the findings of Mittal et al. [4] and Mohanty et al. [5] in marigold.

Table 1. Effect of bio-organic and inorganic nutrient sources on growth parameters of African marigold after 90 DAT.

Treatments	Plant Height (cm)	Plant Spread (cm)	No. of Primary Branches Plant^{-1}	No. of Secondary Branches Plant^{-1}
RDF	49.22	33.96	17.60	36.07
VC	4.49	29.00	15.47	28.73
AZO	46.27	30.05	16.50	31.43
PSB	46.94	31.58	16.80	31.50
CU	45.03	29.24	15.80	30.83
AZVC	47.37	32.63	17.33	32.30
PSVC	50.16	35.25	18.13	37.97
CUVC	48.59	33.74	17.40	35.40
AZPS	40.02	27.20	13.33	28.43
APC	44.23	28.94	15.53	29.27
APV	52.29	35.80	19.13	38.33
APCV	53.31	37.78	19.47	39.53
CD ($p = 0.05$)	6.31	4.83	2.60	5.09

3.2. Flowering and Yield Attributes

Data presented in Table 2, on flowering and yield attributes, show significant responses to different treatments of bio-organic and inorganic nutrients sources. With respect to days required for 50% flowering, the application of APCV and APV recorded the minimum number of days for 50% flowering (41.85 and 46.13, respectively). The present findings are similar to the finding of Gupta et al. [6] with marigold. The maximum flower diameter (6.18 cm) was with the application of APCV followed by APV. The same trend was observed for the highest number of flowers/plant (32.80), fresh flower weight/plant (198.83 g), and dry flower weight/plant (44.96 g), with the maximum byAPCV which also had the highest flower yield/plot (3.18 kg) and flower yield/ha (22.09 t), with APV similar.

The higher values recorded for flowering attributes and yield may be due to active and rapid multiplication of bacteria, especially in the rhizosphere, creating favourable conditions for nitrogen fixation and phosphorus solubilisation at higher rates making it available to the plants leading to more uptakes of nutrients and water. This in turn increases photosynthesis and enhances food accumulation and also diversion of photosynthates towards sinks resulting in better growth and subsequently higher number of flowers/plant and flower yield/ha [1]. The present findings are support those of Mohanty et al. [5] and Owayez Idan et al. [7] in marigold. The maximum duration of flowering (76.16 days) and longer shelf life of flowers (7.89 days) was recorded with APCV and APV. The results are in agreement with Patanwar and Sharma [8] in chrysanthemum and [6] Gupta et al. in marigold. However, the maximum B:C ratio (3.56) was found with APCV. Hence, the treatment of APCV proved to be most profitable. The optimum fertilizer use was by reducing the RDF to 50% and supplementing the deficit by using 50% vermicompost equivalent to the RDN along with Azo, PSB, and 5% cow urine , resulted in higher profit without depleting the soil macronutrients, results that agree with those of Thumar et al. [9] with African marigold.

Table 2. Effect of bio-organic and inorganic nutrient sources on yield and yield attributes of marigold.

Treatments	Days to 50% Flowering	Flower Diameter (cm)	No. of Flowers/Plant	Fresh Flower Weight/Plant (g)	Dry Flower Weight/Plant (g)	Flower Yield (kg·plot^{-1})	Flower Yield (t·ha^{-1})	Duration of Flowering (Days)	Shelf Life (Days)	Benefit Cost Ratio
RDF	48.39	5.51	30.33	175.07	37.94	2.80	19.45	72.80	7.08	3.23
VC	52.94	4.73	25.87	131.47	29.01	2.10	14.61	67.88	4.84	2.12
AZOAZO	50.91	5.15	26.60	141.74	31.64	2.27	15.75	70.15	6.25	2.40
PSBPSB	50.89	5.15	27.47	150.38	31.90	2.41	16.71	71.81	6.28	2.68
CUCU	51.08	5.01	26.53	140.17	31.58	2.24	15.57	69.08	6.24	2.35
AZVCVC	49.46	5.31	28.20	155.99	34.76	2.50	17.33	72.46	6.66	2.65
PSVCVC	46.35	6.05	31.07	180.81	39.34	2.89	20.09	73.05	7.79	3.23
CUVCVC	48.78	5.40	28.67	160.93	37.27	2.57	17.88	72.71	6.83	2.75
AZPS	53.29	4.27	22.87	106.68	28.03	1.71	11.85	62.13	4.73	1.75
APCAPC	52.48	4.89	26.33	136.79	29.94	2.19	15.20	68.25	5.23	2.45
APVVC	46.13	6.07	32.20	189.40	39.78	3.03	21.04	74.48	7.51	3.40
APCVVC	41.85	6.18	32.80	198.83	44.96	3.18	22.09	76.16	7.89	3.56
CD (p = 0.05)	6.39	0.90	5.15	32.09	7.09	0.51	3.57	7.24	1.55	-

4. Conclusions

African marigold growth and yield was higher when inorganic fertilizers were supplemented with biofertilizers and organic manures like vermicompost and cow urine as compared to only inorganic fertilizers. Application of Azo + PSB + Cow Urine (5%) + VC was found to be the most effective in increasing vegetative growth, yield attributes, and yield, and also gave the highest B:C ratio.

Author Contributions: Gaurav Sharma, Major advisor of the thesis research work carried out; Naresh Prasad Sahu, Student who carried out this masters' thesis research work; Neerj Shukla, Member, advisory committee of the thesis research work carried out.

Conflicts of Interest: The authors declare no conflict of interest.

References

1. Verma, S.K.; Angadi, S.G.; Patil, V.S.; Mokashi, A.N.; Mathad, J.C.; Mummigatti, U.V. Growth, yield and quality of chrysanthemum (*Chrysanthemum morifolium* Ramat.) *cv.* Raja as influenced by integrated nutrient management. *Karnataka J. Agric. Sci.* **2011**, *24*, 681–683.

2. Arancon, N.Q.; Edwards, C.A. The utilization of vermicomposts in horticulture and agriculture. In Proceedings of Indo-U.S. Workshop on Vermitechnology in Human Welfare, Coimbatore, India, 4–7 June 2007; pp. 12–13.

3. Singh, V.J.; Sharma, S.D.; Kumar, P.; Bhardwaj, S.K. Effect of bio-organic and inorganic nutrient sources to improve leaf nutrient status in apricot. *Indian J. Hortic.* **2012**, *69*, 45–49.

4. Mittal, R.; Patel, H.C.; Nayee, D.D.; Sitapara, H.H. Effect of integrated nutrient management on growth and yield of African marigold (*Tagetes erecta* L.) *cv.* "Local" under middle Gujarat agro-climatic conditions. *Asian J. Hortic.* **2010**, *5*, 347–349.

5. Mohanty, A.; Mohanty, C.R.; Mohapatra, P.K. Studies on the response of integrated nutrient management on growth and yield of marigold (*Tagetes erecta* L.). *Res. J. Agric. Sci.* **2013**, *4*, 383–385.

6. Gupta, P.; Kumari, S.; Dikshit, S.N. Effect of integrated nutrient management in marigold on its seed vigour and soil health. *Ann. Agric. Biol. Res.* **2012**, *17*, 56–58.

7. Owayez Idan, R.; Prasad, V.M.; Saravanan, S. Effect of organic manures on flower yield of African marigold (*Tagetes erecta* L.) *cv.* Pusa Narangi Gainda. *Int. J. Agric. Sci. Res.* **2014**, *4*, 39–50.

8. Patanwar, M.; Sharma, G. Flowering attributes and yield of chrysanthemum (*Dendranthema grandiflora* tzvelev) as influenced by integrated nutrient management. *Ecol. Environ. Conserv.* **2015**, *21*, 353–356.

9. Thumar, B.V.; Barad, A.V.; Neelima, P.; Bhosale, N. Effect of integrated system of plant nutrition management on growth, yield and flower quality of African marigold (*Tagetes erecta* L.) *cv.* Pusa Narangi. *Asian J. Hortic.* **2013**, *8*, 466–469.

3

A Review on Organic Food Production in Malaysia

Chandran Somasundram *, Zuliana Razali and Vicknesha Santhirasegaram

Institute of Biological Sciences & Centre for Research in Biotechnology for Agriculture (CEBAR),
Faculty of Science, University of Malaya, Kuala Lumpur 50603, Malaysia; zuliana@um.edu.my (Z.R.);
vicknesha19@gmail.com (V.S.)
* Correspondence: chandran@um.edu.my

Abstract: The consumption of organic food has grown remarkably, both in developed and developing countries. Although organic food comprises only a small fraction of the food market, its rapid growth has generated much interest among consumers and businesses, as well as researchers. For products to be called organic, the production must conform to a certain established organic standard and be certified by a recognized certifying body. In Malaysia, the local organic food industry is still small, as more than 60% of organic food products are imported. Most of the organic products are sold domestically, while some are exported to Singapore. The perception and understanding of organic food production is based mainly on not using synthetic fertilizers and pesticides. In general, there is a lack of awareness among producers, retailers, and consumers of the wider extent of organic production and processing standards in local markets. The organic food industry is facing several challenges in Malaysia. Although the demand for organic food in Malaysia is growing, the supply of local organic products is not able to keep up with the increased demand. In addition to the inconsistent supply, the variety of local organic food is also limited. Another problem faced by local organic food consumers is the price difference between organic and conventional food. Hence, to match the recent increases in demand, the Malaysian Agricultural Research and Development Institute is actively developing the organic farming sector through various programs and activities.

Keywords: organic; food; production; consumers and Malaysia

1. Introduction

Consumers worldwide are increasingly concerned about nutrition, health, and the quality of their food. Major concerns have appeared because of recurrent food crises involving pesticide residues on fresh produce, food contamination by chemicals in dairy and seafood products, and the unregulated use of additives in processed foods. With health-related problems such as obesity, type 2 diabetes, and coronary heart disease on the rise, consumers are becoming more aware of the effects of their eating habits on their health. In addition, the increase in environmental awareness and threats posed by pesticide use are also related to the growing interest in organic food production. As a result, organic food is perceived as safer by consumers because they believe it is chemical-free when compared to products from conventional farming. Therefore, health can be considered as an important factor and has a positive relationship to consumers' decision to buy organic food [1].

Although organic food comprises only a small fraction of the food market, its rapid growth has generated much interest among consumers and businesses, as well as researchers. The demand for organic food has increased tremendously, especially in developed countries. This trend has also moved to developing countries, including Malaysia. The movement towards organic food is reflected by the increasing number of countries producing organic food and the increase in total sales [2].

Due to growing organic food product development, it is important to identify the issues and challenges associated with organic food production in Malaysia. This review will provide an overview of issues related to the organic food industry, organic certification, and the challenges faced by the emerging organic food industry. This information is needed to enhance the level of awareness and marketing strategies that could be adopted to improve organic food products in the country.

2. Organic Food Industry

2.1. Organic Agriculture and Food

Organic agriculture improves agro-ecosystem health, including biodiversity, biological cycles, and soil biological activity. It emphasizes the use of cultural, biological, and mechanical practices as opposed to using synthetic materials to fulfil any specific functions within the system. It promotes the health of soil, water, and air by minimizing all forms of pollution that may result from agricultural practices. Hence, organic agriculture is an integrated farming approach which gives importance to both technical and economic aspects, as well as human health [3].

The National Organic Standards Board of the U.S. Department of Agriculture established a national standard for the term "organic" in December 2000. Organic food is produced by farmers who emphasize the use of renewable resources and the conservation of soil and water to enhance environmental quality for future generations. Organic meat, poultry, eggs, and dairy products come from animals that are given no antibiotics or growth hormones. Organic food is produced without the use of most conventional pesticides, synthetic fertilizers, bioengineering, or ionizing radiation. For products to be called organic, the production must conform to a certain established organic standard and be certified by a recognized certifying body [3,4].

2.2. Status of the Global Market

According to the latest survey by The Research Institute of Organic Agriculture (FiBL) and the International Federation of Organic Agriculture Movements (IFOAM) in 2015, there were 43.1 million hectares (ha) of organic agricultural land in 2013. The regions with the largest areas of organic agricultural land are Oceania (40%), Europe (27%), and Latin America (15%). Asia contributed 8% to the world's agricultural land, followed by North America (7%) and Africa (3%).

There were almost 2 million producers in 2013. Thirty-six percent of the world's organic producers are in Asia, followed by Africa (29%) and Europe (17%). About a quarter of the world's agricultural land (11.7 million ha) and more than 80% (1.7 million ha) of the producers are in developing countries and emerging markets. Global sales of organic food reached 72 billion U.S. dollars in 2013. As shown in Figure 1, revenues have increased almost five-fold since 1999. The growth of organic product sales has consistently increased over the last decade [5].

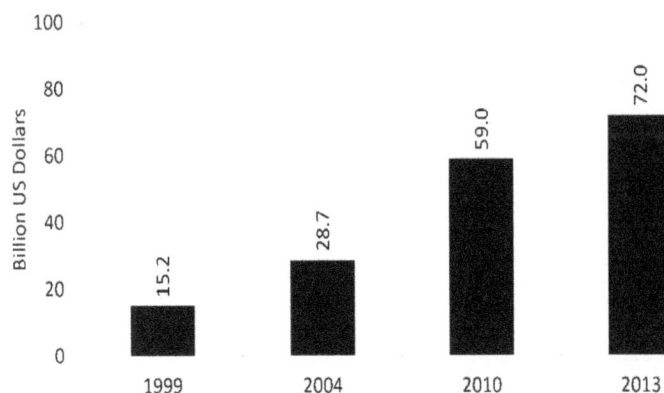

Figure 1. Growth of the global market for organic food, 1999–2013 (Source: [5]).

2.3. The Malaysian Scenario

2.3.1. Organic Food Production and Market

Organic agriculture in Malaysia has developed on two concurrent paths, led by non-governmental organizations (NGOs) and the private sector. The Centre for Environment, Technology and Development Malaysia (CETDEM) is an example of an NGO that has played a pioneering and prominent role in identifying problems with conventional agricultural practices. CETDEM has focused on environment degradation, the health of plantation workers from pesticide use, food safety, and sustainable agriculture. However, it was only in the 1990s that many pioneering organic farms were established, such as those in Penang and Kuantan, Sustainable Living Centre in Perak, Lifestyle farmhouse in Melaka, Ecofarm in Negeri Sembilan, and Nakim Farm in Negeri Sembilan.

Currently, organic food remains a niche market, but one that is slowly growing. In 2001, only 131 ha in Malaysia were organic farms. In a span of just five years, the land area for organic farms grew by an incredible 18-fold to 2367 ha, of which 962 ha are certified organic, as surveyed by the FiBL and the German Foundation Ecology & Farming (SOEL) in 2007. In 2001, the Malaysian Department of Agriculture (DOA) reported that there were 27 organic producers in the country. In East Malaysia, the production of organic products is limited to vegetables and fruit. In 2013, the DOA reported that there was a total of 89 farms occupying 1634 ha of land under organic farming and 49 farmers had valid certification. In Malaysia, the local organic food industry is still small, as more than 60% of organic food products are imported. Most of the organic products are sold domestically, while some are exported to Singapore [1,2,6].

The Third National Agriculture Policy (NAP3) realized a key benefit of organic farming—export opportunities in the niche organic market that could bring high revenue to the country. One of the strategies undertaken by the government to realize this plan was to encourage small-scale producers to participate in organic farming. It has also been the government's strategy to increase producers' incomes. In the ninth Malaysia Plan (2006–2010), the government targeted organic farming, which was reported to have a potential value of more than U.S. $200 million over 5 years. The Ministry of Agriculture planned to have 20,000 ha of organic farms by 2010 and to increase local production by 4000 ha per year [6,7].

The hub of vegetable farming in Malaysia can be found in Cameron Highlands, where Grace Cup Pte Ltd. (Pahang, Malaysia) and Cameron Organic Produce Pte Ltd. (Pahang, Malaysia) have established organic vegetable farms. Large local organic retailers, such as Country Farm Organics and Zenxin have established a foothold in selling and distributing organic food. The distribution of organic food in Malaysia has been fragmented due to the existence of specialty shops which operate all over the country. About 70% of the organic food products have been distributed through specialized organic food stores and supermarket chains. The balance has been distributed through traditional retail shops, wet markets, or home deliveries [7].

2.3.2. Organic Certification and Standards

The Malaysian government has also introduced national standards for organic farming and organic foods. In 2002, the Malaysian DOA outlined national standards and a government certification program. The scheme was revised and renamed the Malaysia Organic Scheme (MOS) in 2003. Organic products produced according to the MS 1529 national standard display the Organic Malaysia logo. The MOS is limited in that it only covers plant-based products, whereas livestock products and processed foods are not yet included [2,3].

The Scheme is open for participation by all farmers who are engaged in the primary production of fresh organic food products. A group of trained agricultural officers has been assigned to carry out field inspection to verify that the farm operations or practices are in compliance with the organic standards. In 2002, the Ministry of Agriculture noted that support services such as extension, research, and development would be devoted to developing organic agriculture in Malaysia. Export of certified organic food also depends on the laws of the receiving country. Presently, more than 70 farmers have MOS accreditation to produce organic products for the local as well as the export market [2,6].

2.3.3. Challenges and Alternatives

The organic food industry is facing several challenges in Malaysia. Although the demand for organic food is growing, the supply of local organic product is not keeping up with the increased demand. Aside from an inconsistent supply, the variety of local organic food is also limited. Consequently, Malaysia still needs to heavily import organic food products from other countries, especially from the United States, Japan, Australia, New Zealand, and China [2,7].

Another problem faced by local organic food consumers is the price difference between organic and conventional food. Although it is well known that organic food is more expensive than conventional food, their price difference in Malaysia is particularly substantial—by as much as 100% to 300%—compared to only a 25% to 30% price gap in the United States and European Union (EU). The higher price results from the higher cost of production, especially labour costs and the loss of income or opportunity costs when farmers convert their conventional farms to organic farms. The concern a customer has for the environment may drive them to purchase an organic product; however, the high prices may limit their ability to buy the product. Conversely, there is still an increased demand for organic product globally, despite the price. This is mainly due to the potential of increased nutritional value and/or reduced food safety risks [2,8].

Finally, there is still a distinct lack of trust among consumers towards produce that is labelled "organic", even with the Government's certification efforts. More needs to be done to overcome the prevailing issue of credibility and trust in the marketplace. In general, there is a lack of awareness among producers, retailers, and consumers of the wider extent of organic production and processing standards in local markets [8].

3. Conclusions

The recent progress of the organic food industry has compelled the Malaysian Agricultural Research and Development Institute to actively develop the organic farming sector through various programs and activities. Organic food has been considered a new industry that contributes to economic growth. The production of organic food has become an innovative strategy for the Malaysian agricultural structure to sustain its competitive advantages. It was timely for the Malaysian government to review its organic farming policy and change its paradigm to ensure that the organic food industry becomes the engine of growth for the agricultural sector. The success of establishing this industry demands comprehensive strategies and collaboration from the government, producers, and marketers.

Acknowledgments: The authors would like to thank University of Malaya for supporting this research.

Author Contributions: All authors have contributed equally towards the research and the writing of the paper.

Conflicts of Interest: The authors declare no conflict of interest.

References

1. Mohamad, S.S.; Rusdi, S.D.; Hashim, N.H. Organic food consumption among urban consumers: Preliminary results. *Procedia Soc. Behav. Sci.* **2014**, *130*, 509–514. [CrossRef]
2. Dardak, R.A.; Abidin, A.Z.Z.; Ali, A.K. Consumers' perception, comsumption, and preference on organic product: Malaysian perspective. *Econ. Technol. Manag. Rev.* **2009**, *4*, 95–107.
3. A New Beginning for Organic Food Industry. Available online: http://www.asean.org/communities (accessed on 7 August 2015).
4. Organic Food Standards and Labels: The Facts. Available online: http://www.ams.usda.gov (accessed on 6 August 2015).
5. Willer, H.; Lernoud, J. *The World of Organic Agriculture: Statistics and Emerging Trends 2015*; Research Institute of Organic Agriculture (FiBL): Frick, Switzerland; IFOAM—Organics International: Bonn, Germany, 2015.

6.   Tiraieyari, N.; Hamzah, A.; Samah, B.A. Organic Farming and Sustainable Agriculture in Malaysia: Organic Farmers' Challenges towards Adoption Asian Social Science. *Asian Soc. Sci.* **2014**. [CrossRef]

7.   Stanton, E.S. Malaysia's Markets for Functional Foods, Nutraceuticals and Organic Foods: An Introduction for Canadian Producers and Exporters. *The Counsellor and Regional Agri-Food Trade Commissioner, Southeast Asia, and the High Commission of Canada in Malaysia*; Agriculture and Agri-Food Canada: Ottawa, ON, Canada, 2011.

8.   Saleki, Z.S.; Saleki, S.M.S. The main factors influencing purchase behaviour of organic products in Malaysia. *Interdiscip. J. Contemp. Res. Bus.* **2012**, *4*, 98–116.

4

Adoption and Income Effects of Public GAP Standards: Evidence from the Horticultural Sector in Thailand

Henning Krause [1],*, Rattiya Suddeephong Lippe [2] and Ulrike Grote [1]

[1] Institute for Environmental Economics and World Trade, Leibniz Universität Hannover, Königsworther Platz 1, 30167 Hannover, Germany; grote@iuw.uni-hannover.de

[2] Institute of Development and Agricultural Economics, Leibniz Universität Hannover, Königsworther Platz 1, 30167 Hannover, Germany; lippe@ifgb.uni-hannover.de

* Correspondence: krause@iuw.uni-hannover.de

Abstract: To reduce potential food hazards and increase the image of Thai horticultural products abroad, the Thai government introduced public standards of Good Agricultural Practices (Q-GAP). What makes orchid and mango producers in Thailand adopt Q-GAP standards and how do these affect their income and export shares? Primary data from 400 certified and non-certified orchid and mango producers was collected from main exporting provinces in Thailand. The binary probit model estimations show that it is the orchid and mango producers with higher education, and more physical and social capital who tend to comply with Q-GAP standards. Results from the Propensity Score Matching approach reveal that adoption of public GAP standards results in positive income effects for mango producers, but not for orchid producers. This can be explained by the fact that certified mango producers can sell their products to high-value retail chains which offer higher prices for their products, while certified and non-certified orchid producers cooperate with traders from the same value chain.

Keywords: Good Agricultural Practice; mango; orchid; Thailand; income impacts; Heckman model; Propensity Score Matching

1. Introduction

The demand for fresh horticultural products is continually rising, both in domestic and international markets [1]. Consumers from industrialized countries, such as some countries in the European Union, desire the year-round availability of a variety of horticultural products including tropical fruits and flowers [2]. In response to these trends, the interest in accessing these markets has significantly increased among producers and traders of horticultural products in developing countries. However, a more frequent occurrence of food safety problems and the rising health consciousness of consumers have led to an increasing demand for safe horticultural products [3]. At the same time, consumers have increasingly paid attention to environmental concerns and the way food is being produced including its farming practices. In addition, there was a gradual shift from concerns about only edible goods to other agricultural goods, such as flowers or timber products.

The scope of national legislation in many developing countries is considered insufficient for transnational trade [4]. Thus, trade liberalization and globalization in the last decades have resulted in institutional adaptation, driving the development of internationally recognized standards and regulatory schemes [5]. Standards and certification systems which define whether products have been produced in an environmentally-friendly way or whether they are free of pesticides are considered as viable solutions to sort out the problem of information asymmetry and reduce transaction costs among different actors in the supply chain [6]. Certification systems also help to increase the communication between actors in the supply chain on production processes. Similarly, labeling is a form of directly conveying information of quality aspects of the product to end consumers [7].

In global horticultural value chains, private actors, such as big supermarkets and/or retail chains, play an increasingly large role. They have become rule-setters and often take the initiative to ensure certain safety and quality levels and create market demand for exotic produce [8]. Good Agricultural Practices (GAP) are food safety and quality standards of on-farm and post-farm activities, encompassing a set of management regulations of how to produce environmentally friendly and socially-acceptable products [9]. GAPs are the most important standards in global horticultural value chains and they have been especially promoted by the private sector initiative GLOBALG.A.P. However, the trend towards private GAP standards still poses many challenges as compliance can be a rather complex task; it also involves high investment costs and additional skilled laborers. As a result, producers without appropriate financial resources are not likely to adopt those private standards [10]. In contrast to private GAP standards, public GAP certification schemes are introduced by governments in export-oriented emerging nations. They are based on their own quality and safety standards to position themselves in competitive global markets [11]. They can also be an alternative for export-oriented producers due to lower cost of compliance and lower complexity of its schemes. Besides the easier access for the individual producers, public GAP standards can be a way of adding value to the country's agricultural goods for export as a whole [12]. The empowerment of public standards is often the only necessary step to unleash the potential of exporting horticultural goods [13,14]. To enhance export growth, some developing countries have already successfully upgraded their export sectors through better compliance with the increasing public food standards in global markets [15].

The Thai government follows the long-term strategy to expand Thai exports of horticultural products [16]. Thai authorities have actively promoted voluntary public Q-GAP standards (Q denotes quality) as part of their strategy to maintain and expand export markets. The aim of this scheme is to assure product quality and compliance with certain environmental and social regulations [3,16]. A full implementation of the national GAP program (Q-GAP) started in 2004, as an important component in the road map of food safety policy [16,17]. The national Q-GAP standards contain eight principles (water source, cultivation site, use of agro-chemicals, product storage and on-site transportation, data records, pest-free products, quality management, harvesting and post-harvest handling) and to date emphasize pesticide control and Maximum Residue Limits (MRLs) monitoring. Producers are allowed to label their produce with the golden Q-mark if they comply with all required principles. The entire certification process is carried out by government agencies [16]. In this regard, costs of compliance, specifically for training, external auditing and the annual certification fee, are mostly covered by government agencies and are thus free for producers. However, producers still have to cover other costs of upgrading their facilities (i.e., farm infrastructure and equipment such as storage room for pesticides) and buying protective clothing for farm workers [10].

While the number of certified producers in the horticultural food sector is growing, recently the non-food sector (i.e., flowers) has shown a reduction in Q-GAP adoption rates [18]. It is therefore questionable whether producers do benefit from complying with this public scheme. However, only a small number of in-depth analyses exist in the context of public GAP standards and for the horticulture sector in general. Up to the point of this study, no impact analysis from a public GAP certification scheme on income could be found. This study therefore aims at assessing the determinants of adopting a public GAP standard among producers of horticulture products and the standard's impact on producers' income and export shares by focusing on mango and orchids as representative for food and non-food horticultural products.

Orchid and mango were chosen due to their economic importance and high export potential to overseas markets [19]. The advantage of comparing orchids and mangoes is that they are rather similar in terms of their presence in Thai domestic and international markets. While they are both rather common products in domestic markets bought by everyone, they are also exported in large quantities [19]. This allows us to draw conclusions from the results focusing only on the aspect of edible vs. inedible horticultural crops. The problems that producers of edible horticultural products face when they want to enter the export market are quite different to the ones that producers of inedible goods like flowers face. Food safety, for example, is a major concern for producers of fresh edible horticultural products, while for inedible crops quality aspects, such as the appearance, are more important. It is thus questionable if producers of these two types of products have similar adoption patterns for the same certification program. We expect that the standards in the certification program, which are identical for mango and orchid producers, will have different effects because of the different needs among producers and the demand of consumers of edible and inedible horticultural crops.

This study will help Thai government officials and other actors in the Q-GAP administration to evaluate the success of the program and seek strategies to enhance it. Next to the Thai government, all governmental and non-governmental agencies which design or plan to design certification schemes are expected to benefit from this study. They will be more aware about the factors that drive producers to adopt certain schemes but they will be also informed about the impact of the schemes in terms of income and export facilitation.

The paper is structured as follows. The next section gives an overview of the current adoption and impact literature for standards and certification in horticultural products, followed by the conceptual framework. Data collection and methodology are presented in the third section. The last section concludes with a summary of the main results of the study and highlights the need for further empirical research.

2. Literature Review

2.1. The Impact of Certification on Horticultural Producers' Welfare

In spite of the rapid integration of developing countries into the global horticultural markets, not all producers and exporters could fulfill demands from markets associated with standards and certification compliance. Previous studies mainly focused on the impact of private standards and certification on individual producers [20–22]. For instance, certified Brazilian producers receive a higher net income due to the price premium for their certified products in contrast to those who are not certified. Besides increasing income levels, other benefits of compliance relate to environmental, health, food safety aspects and higher productivity [22]. EurepGAP-certified Turkish producers have been found to sell their tomatoes at a higher price, having a net income which is around 2.8 times higher per unit area compared to non-certified ones. The adoption rate of EurepGAP certification even accelerates if producers already exhibit experience with the certification system [21]. Another example is Kenya, where EurepGAP producers could significantly improve their financial performance by exporting certified vegetables. Further benefits from compliance are long-term relations with their buyers, maintaining the share in lucrative export markets, or increased awareness of agrochemical handling practices [20]. However, complying with standards and certifications entails some costs, such as an increase in investment costs due to production inputs, facilities and additional labor needs. In that sense, it is likely that small-scale producers are unable to comply with the emerging standards, increasing the probability of their exclusion from lucrative export markets [1,20,23–25].

To the knowledge of the authors, no study exists that analyses the direct impact of public GAP schemes on producers' income. However, several studies investigated some benefits of these schemes for the farmers. Thai Mangosteen producers following the Q-GAP scheme stated that they did not receive a direct price premium for the certified product, but they could increase their share of high-quality fruits and this way achieved higher prices [26]. Furthermore, exporters preferred business links with Q-GAP certified farmers and this way Q-GAP facilitated the export of Mangosteen, even though the production costs were higher due to the increased need for labor [26]. In contrast to this, Q-GAP was not demanded in the Thai domestic coffee market and thus participation of coffee farmers was low [27]. Other hampering factors were that requirements were considered too complicated and the frequency of inspections was low, which shed doubt on the reliability of the scheme among the farmers [27]. VietGAP is a public GAP scheme rather similar to the Thai Q-GAP scheme, which has been implemented by the Vietnamese government. After receiving training for VietGAP implementation, Vietnamese citrus producers found that they could reduce input costs and increase yield and quality thanks to the implemented standards [28]. The VietGAP program was also found to reduce health hazards from the spraying of harmful pesticides [29]. This way, the farmers were not only healthier thanks to the scheme, but they also significantly reduced their loss of income [29].

2.2. Conceptual Framework

International trading companies often demand quality assurance systems such as Q-GAP from producers of agricultural products [26]. Currently, Q-GAP is the most important and recognized public scheme for export-oriented Thai horticultural products. Implementing the Q-GAP scheme could thus facilitate the access to high-value export markets. Especially if implementation and recurring costs of the certification are low, as in the case of Q-GAP, selling their products at a higher price than the domestic market will lead to a higher household income for orchid and mango producers. The impact of Q-GAP certification is thus evaluated through the direct change in income from mango and orchid production and the share of mangoes and orchids sold to the high-value/export market.

The present study focuses on the impact of these standards and the decision process of compliance, based on the five-stage innovation decision model from Rogers [29]. The adoption of a certification program can be considered an innovation aiming to differentiate products [30]. Rogers [31] states that certain household characteristics and the expected benefits of the certification scheme influence the adoption decision (Figure 1). Based on the livelihood framework of Scoones [32], the farm decision maker will decide to adopt certification depending on the livelihood assets available [33]. According to this framework, demographic factors, human resource factors, physical capital, social capital and public infrastructure have an influence on the adoption decision [32]. These five dimensions of livelihood assets and the results of empirical adoption literature up to this point thus guide the variable choices for the model describing the adoption process of Q-GAP.

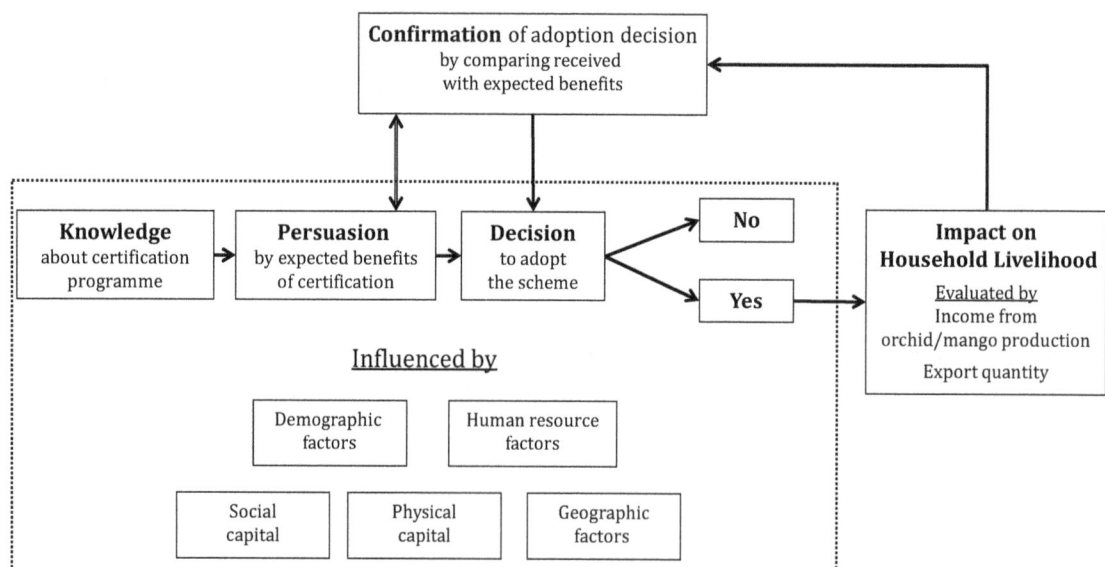

Figure 1. Relation between impact of certification and decision process towards certification. Source: Own picture based on [29,32].

2.3. Choice of Variables for Adoption Model

Age and gender of the household head are included as demographic factors (Table 1). Research found a significant influence of age on the decision to become Fairtrade certified in the Latin American banana and coffee sector, although the direction of influence was positive in some cases and negative in others [34–36]. For Thai small-scale fruit and vegetable producers, age had a negative influence on the decision to adopt GLOBALG.A.P. [37]. A male household head had a positive influence on adopting a geographical identification for Thai Jasmine rice [33]. They see reasons for these findings in the more dominant behavior and faster decision making of male household heads. Another variable included as a demographic factor is the household size, which had a positive influence in the case of geographical identification adoption [33] and for Kenyan vegetable producers adopting EurepGAP certification [38].

Table 1. Variables used for the choice model with expected signs.

Group	Variable	Description	Expected Signs	Sources
Demographic factors	Age	Age of HH head	+/−	[34–38]
	Gender	Gender of HH head (binary: 1 = female)	−	[33]
	Household size	Number of nucleus household members	+	[33,38]
Human resources	Education	years of formal education of HH head	+	[20,39,40]
	Farming experience	years of experience in orchid/mango cultivation	+	[27,29,39]
Physical capital	Agricultural assets	Total value of agricultural assets (THB)	+	[20,25]
	Land under main crop	Area under orchid/mango cultivation (rai)	+	[27,38,41]
Social capital	Producer group	producer group participation (binary: 1 = HH head participates in at least one producer group)	+	[20,33]
	Know Q-GAP from DoAE	Respondent knows about Q-GAP certification from DoAE (binary: 1 = yes)	+	[20,39,40]
Public infrastructure	Distance to next town	Distance to next town (km)	−	[32]

1 THB = 1 Thai Baht = 0.026 Euro (24 October 2016); 1 rai = 0.16 hectare, DoAE = Department of Agricultural Extension; HH = household.

As human resource indicators, years of formal education of the household head and experience in orchid or mango cultivation are included in the model. Higher levels of education of household heads had a positive influence on GLOBALG.A.P. adoption in Kenyan vegetable [20] and Peruvian mango production [39]. Years of experience of the household head also facilitated adoption in the Peruvian case. The formal education facilitated initial Q-GAP adoption of Thai rice producers in Ayutthaya Province [40]. Among coffee farmers in Chumphon Province in Thailand, years of formal education and cultivation experience positively influenced Q-GAP adoption [27]. Younger age, higher education and longer cultivation experience influenced adoption of VietGAP, a public GAP standard like Q-GAP, by Vietnamese vegetable producers [29]. This is why a positive influence of the human resource variables is expected.

High levels of agricultural assets [20,25] or area under main crop [38] show a positive influence on standard adoption in Kenya. Cultivation area was also positively correlated with adoption of Q-GAP among Thai coffee producers [27] and adoption of ChinaGAP among Chinese agricultural cooperatives [41]. Thus, agricultural assets and area under main crop are included in the choice model to cover the physical capital dimension of livelihood choices and are expected to have a positive influence on the adoption decision.

Group membership can be considered as a form of social capital, because it gives farmers access to a network of other farmers, thus facilitating the process of information spreading. This has been confirmed in the process of geographical identification adoption among Thai farmers [33], as well as for adoption of GLOBALG.A.P. certification by Kenyan vegetable producers [20]. In both cases, information about the certification scheme itself also had a positive influence on the adoption decision. This relation was also found in the case of Q-GAP adoption for Thai coffee producers [27] and

VietGAP adoption for Vietnamese vegetable producers [29]. Thus, knowledge about the scheme is included as a binary variable, one of which being if the household received knowledge about the scheme from the Agricultural extension service. Access to extension services was positively related to GLOBALG.A.P. adoption among Peruvian mango producers [39] and to Q-GAP adoption among Thai rice producers [40].

3. Data and Methodology

3.1. Data Collection

Mango and orchid were chosen as representative of edible and inedible produce due to the important role in the Thai floricultural and horticultural sectors, respectively. Thailand is one of the leading export countries for tropical flowers, and the most commercial one is orchid. In 2015, the total harvested areas of orchids amounted to 22,169 rai (1 rai = 0.16 ha) with a total production quantity at 50,873 tons. Production quantity declined from 2010 to 2011 due to severe flooding events in the second part of 2011, having detrimental effects on local infrastructure and production facilities. Nevertheless, orchid is still the most important economic floricultural product of Thailand with an export value of 2716 million THB (1 THB = 0.026 Euro (24 October 2016)) where 77% are cut fresh orchids, where 80% of export cut orchid flowers are the Dendrobium variety [42]. The price achieved for exporting cut orchids strongly depends on the destination. The average export price per kg in 2015 for cut orchid flower was around 84 THB per kg. Thai cut orchids are appreciated in many parts of the world with the significant trade partners being Japan, USA, China, Italy, Russian Federation, India and the Netherlands [43]. Whereas orchid flower is considered as a common flower in the domestic market. They are mainly sold on local markets and often used for religious purposes in Thailand [44]. Due to the Free Trade Agreement under the Asian Economic Community (AEC), it is remarkable that the major import markets for Thai cut orchid flowers would be concentrated more in Asian countries, whereas the export rate to the USA and the European market would decline as a result of the global economic recession.

Mango is also one important economic produce in the Thai fruit sector. More than 100 mango varieties have been cultivated all over the country with a total harvested area amounting to 2.1 million rai and of the total production quantity at 3.1 million tons in 2015 [42]. At the same period, Thailand exported fresh mango of around 33,900 tons, valued at 1211 million THB [42]. Among these, Nam Dok Mai is one of the significant commercial varieties and is renowned in overseas markets due to its taste. The major importing markets of Thai fresh mango are concentrated in Asian countries, especially South Korean, Vietnam and Japan with an average export price at 35,723 THB per ton (Office of Agricultural Extension, 2015).

The cross-sectional data was collected at the household level, obtained from a stratified randomly sampled survey of 256 orchid producers. Initially, five provinces (Nakornpatom, Samutsakorn, Bangkok, Ratchaburi and Chonburi) were selected based on the intensity of production for both domestic and export markets as well as the number of certified national Q-GAP standard producers in the region in the major Thai orchid production areas. The main orchid genus cultivated as cut flower for export is *dendrobium*. Subsequently, all orchid producers in these provinces were stratified into three strata: certified Q-GAP, formerly certified Q-GAP and non-certified Q-GAP. The final sampling units were randomly selected from these groups (Table 2). Interestingly, almost a third of all interviewed orchid producers had not been recertified in the year of the study. Some of them did not want to continue with the certification; some of them could not be recertified because the Department of Agricultural Extension (DoAE) did not have enough capacities available for the auditing. To not lose the information but to nevertheless distinguish them from the producers that were still certified or never obtained the Q-GAP certification, the stratum "Formerly Q-GAP" has been introduced. For the analyses in this paper, only producers from the certified and non-certified strata have been used.

Table 2. Number of interviewed producers in 2012.

Certifying Status	Orchid		Mango	
	Population	Sampling	Population	Sampling
Q-GAP producers	75	68	228	80
Formerly Q-GAP producers *	207	76	-	-
Non Q-GAP producers	868	112	4180	64
Total	1150	256	4408	144

* Sample producers who belong to the formerly certified orchid producer group are the ones who used to be certified with the national Q-GAP standard but left certification during the time period of our field survey. Source of orchid/mango population numbers: Thai Department of Agricultural Extension (DoAE).

For the case of mangoes, Chachengsao province was chosen as the study area. It is one of the major mango producing provinces focusing on *Nam Dok Mai* (*Mangifera indica* L.), a variety especially produced for export. This variety is predominantly grown in the eastern part of Thailand due to suitable climate and soil conditions and low transportation costs due to the proximity to Bangkok [45]. Similar to orchids, the total population from all mango producing districts was stratified into certified Q-GAP and non-certified Q-GAP. In Chachengsao province there were no formerly certified mango producers who had left the scheme in the year of the study. Consequently, 144 mango producers were randomly selected from these strata (Table 2).

The data collection was conducted in 2012 by employing a structured questionnaire with the following six sections: land information, production, marketing, attitude towards Q-GAP standard, household socio-economic and additional section for training record, environmental concern, infrastructure, and borrowing. The questionnaire was translated into Thai language and pre-tested three times. In addition, local experts were consulted before the finalization. All enumerators were trained by the responsible researchers and practiced some interviews before the actual survey started. Overall, the producer survey for orchid and mango lasted around three months.

3.2. Methodology

3.2.1. Adoption Model

Two probit models were used to estimate the factors, which influence the adoption of Q-GAP among orchid and mango producers, respectively.

Let the binary variable y_t be the dependent variable, being 1 if the household is certified and 0 otherwise. The conceptual framework showed that the farm decision maker's demographic factors, human resource factors, physical capital, social capital and public infrastructure have an influence on the adoption decision. According to decision theory [33,46] the household will try to maximize its utility U_t, which is based on the socioeconomic characteristics discussed above and the characteristics of being certified or not. Thus, the farmer will choose to get certified with Q-GAP if the latent variable model $y_t^* = U_1 - U_0 > 0$ and will reject certification if $y_t^* = U_1 - U_0 \leq 0$, ceteris paribus. If the socioeconomic characteristics are translated into covariates x_t, the latent variable y_t^* can be written as the linear model

$$y_t^* = x_t'\beta + \varepsilon_t^*$$

(1)

Thus, the probability of being certified can be written as

$$P(y_t = 1|x_t) = P(y_t^* > 0|x_t).$$

(2)

and

$$P(y_t = 0|x_t) = P(y_t^* \leq 0|x_t)$$

(3)

The covariates x_t have the same direction of influence on y_t^* as on y_t [47].

3.2.2. Propensity Score Matching (PSM)

This study investigates the impact of Q-GAP certification on income from mango and orchid production and the share of mangoes and orchids sold to the export market. Since farmers choose by themselves whether they want to be certified or not, treated and untreated households are not randomly assigned and thus may differ systematically in certain household characteristics or resource endowments, which determined the decision to adopt the scheme. This problem is called selection bias [48,49] and may overlay the treatment effect [50].

PSM is a good tool to estimate treatment effects by reducing the selection bias. It is especially feasible in the case of certification, because there is already a well-established body of literature and empirical evidence available on the adoption of certifications in general. This prior knowledge allows building a selection equation, for which the Conditional Independence Assumption (CIA) holds [51]. Furthermore, PSM does not assume a certain functional form of the outcome equation [51] and treatment effects estimated by PSM are robust if the sample size is as small as in our case [52]

The basic idea of PSM in our case is to compare every certified orchid or mango producer with one or several uncertified producers with relatively similar characteristics. To evaluate the similarity of characteristics, the Propensity Score (PS) $p(X)$ is used as a balancing score, describing the probability of being certified given the observed characteristics X [53]. $p(X)$ is the result of a binary choice model describing the probability of being certified or non-certified given the covariates X. The average effect of certification on the certified individuals would then be written as

$$ATT_{PSM} = E_{p(X)|C=1}(Imp|C=1) = E[Y(1)|C=1, p(X)] - E_1[Y(0)|C=1, p(X)] \qquad (4)$$

The formula adopted from Caliendo and Kopeinig [54] shows that the Average Treatment Effect on the Treated (ATT) describes the expected impact (Imp) of certification only for certified growers. It is derived by subtracting the expected outcome they would have if they were not certified ($C = 0$) from the expected outcome if they were certified ($C = 1$), weighted by the probability $p(X)$ of being certified or not. To assure robustness of the results to the matching algorithm, three different algorithms were used and evaluated parallel to each other: nearest neighbor (NN) matching with replacement, radius matching and kernel matching. Optimal bias reduction in caliper matching is reached with a caliper equaling to a fifth of the propensity score's standard deviation [55] and is thus set at 0.062 for orchid and 0.042 for mango. The bandwidth for Kernel matching was set to 0.05 for orchid and 0.2 for mango to ensure optimal bias reduction without too much loss of information. For a detailed discussion on the different algorithms, see [54].

In PSM, variance and thus standard errors of the ATT are overestimated, because it also includes the variance of the PS and common support estimation [56]. To reduce this problem, bootstrapped standard errors were used, where 500 repetitions are considered a good compromise between theory and practicability [57]. The ATT is only defined in the region of common support, in which all observations with the same covariates X have a positive probability of being certified or not [53]. To ensure optimal bias reduction, the sample size was reduced by the observations within the areas of lowest PS distribution density (Table A3 in Appendix A) [58]. For the matching quality's evaluation, standardized bias and t-test of all covariates X are investigated. Table A4 (Appendix A) shows that mean bias across all covariates could not be reduced to recommended levels of less than 5% after matching [54]. Due to a relatively small sample size, too rigid imposition of common support would have resulted in too high loss of information. However, Pseudo-R^2's approximating zero and insignificant F-test results indicate that there is no longer any significant difference between the covariates [53,59]. Thus, matching on the covariates can be considered successful with all three matching algorithms.

One major problem in PSM is the so-called hidden bias resulting from unobserved variables affecting both certification status and outcome. In this paper, this unobserved heterogeneity was investigated with Rosenbaum bounds [60]. Only statistically significant results were tested for sensitivity to hidden bias [61,62]. Table A5 (Appendix A) shows that the significant effects of Q-GAP certification among mango producers are very robust against hidden bias, since even a more than two times increase in hidden bias influence does not affect this significance.

To calculate the propensity score, the initial adoption model has been used in a slightly adapted form, because the covariates have to hold the Conditional Independence Assumption (CIA) to dedicate the differences in outcome clearly to the fact of being certified or not. They must not be influenced by the certification status, although they need to influence the participation decision and/or the outcome variable. Therefore, covariates need to be either fixed over time or measured before participation in the certification scheme [54]. The agricultural assets bought five years prior to the survey have been taken as an asset indicator, because Q-GAP certification was not fully implemented at that time and thus these assets are not influenced by Q-GAP certification. Furthermore, household size was taken out to prevent endogeneity with income per capita as outcome variable. Results of the adoption model for PSM can be found in Table A2 in Appendix A.

Additionally, the Heckman's two stage model has been estimated to confirm the PSM estimates. The choice models discussed above were used as the first stage (Appendix B). The reason for using Heckman as a robustness check is that we can include control variables in the second stage equation to rule out the influence of effects such as the floods that affected some orchid producing areas in the year of the survey.

3.2.3. Outcome Variables

The income from orchids and mango is directly calculated from the survey data. First, orchid/mango sales are calculated, multiplying the average price per unit of orchids or mango achieved by the farmer within the year of survey with the quantity sold. Farmers could state orchid sales in kg or stems and mango sales in kg, but had to state the average price in the same unit (THB/stem or THB/kg). From this sales figure, variable costs of labor, machinery and pesticide and fertilizer expenditures are deducted. To increase accuracy, these costs have been stated by the farmers for one representative plot from which they had best knowledge or most consistent records and subsequently calculated up to the overall area under orchid cultivation.

The export share is calculated as the share of the total sales of orchids and mangoes in THB, which was sold to buyers in the export market.

4. Results and Discussion

4.1. Descriptive Results

The demographic parameters do not differ substantially between orchid and mango producers (Table A1 in Appendix A). About a third of all household heads are female. The number of household members ranges in both crops around 4, whereas certified orchid producers have significantly larger households than uncertified ones. More than two-thirds of both orchid and mango producers participate in producer groups. This share is significantly higher for certified producers of both crops. Almost a hundred percent of certified mango producers are in producer groups. Orchid producers seem to have more contact to the Department of Agricultural Extension, because a higher share of them heard about Q-GAP certification from this institution as compared to mango producers.

Orchid producers have a much higher annual income than the average mango producer, both in terms of total annual income and per capita income (Table 3). Because orchids are produced in greenhouses with very intensive production methods, it does not surprise that orchid producers' income per rai is much higher than the one from mango producers, who produce in orchards. Orchid producers also have higher levels of agricultural assets and sell substantially more of their products to the high-value and export market. Interestingly, these three indicators are significantly lower for uncertified mango producers compared to certified ones, while there is no significant difference between certified and uncertified orchid producers. Both certified mango and orchid producers employ more workers next to family workers, though the difference is not significant. The average daily wages paid are slightly higher among non-certified producers in both crops.

Table 3. Selected economic indicators of the sample.

Variable	Orchid			Mango		
	Non-cert	Cert	Total	Non-cert	Cert	Total
Annual household income (THB)	1,979,539 (4,004,664)	1,950,701 (2,779,954)	1,968,537 (3,578,294)	63,760 *** (369,905)	1,332,287 (3,134,011)	773,425 (2,433,825)
Income per rai (THB)	185,682 (335,227)	156,986 (381,120)	174,671 (352,721)	4521 *** (19,391)	17,949 (35,267)	12,033 (30,014)
Income per capita (THB)	555,708 (1,086,166)	456,055 (682,796)	517,690 (951,630)	24,812 *** (88,600)	325,741 (688,997)	193,164 (538,523)
Share of products sold to export market	0.649 (0.378)	0.682 (0.342)	0.662 (0.364)	0.012 *** (0.062)	0.340 (0.363)	0.195 (0.319)
Total value of assets (THB)	493,675 (491,718)	588,735 (324,542)	529,587 (437,601)	159,472 *** (219,255)	479,215 (397,984)	338,349 (366,720)
Land under orchid/mango cultivation (rai)	16.76 (29.05)	17.41 (15.37)	17.01 (24.73)	15.88 *** (20.84)	64.25 (54.52)	42.94 (49.24)
Average number of workers employed	5.3 (19.7)	10.3 (48.8)	7.2 (33.8)	5.0 (9.3)	7.7 (8.4)	6.7 (8.8)
Average daily wage paid to workers (THB)	516.56 (2754.86)	455.15 (1876.30)	489.69 (2402.72)	372.58 * (283.29)	294.17 (170.19)	321.25 (217.97)

Standard error in brackets, *** = significant at the 1%-level, 1 rai and 0.16 ha.

When certified orchid and mango producers are asked about the advantages of Q-GAP certification, both orchid and mango producers mention better farm management as the major advantage of being certified (Figure 2). Some farmers said that, thanks to the required book keeping, they can handle inputs and calculate their costs better. These findings go along with an earlier study about Mangosteen producers under Q-GAP, who said that Q-GAP enhanced their farm management and reduced their input costs [26]. Vegetable producers following VietGAP certification came to the same statement [29].

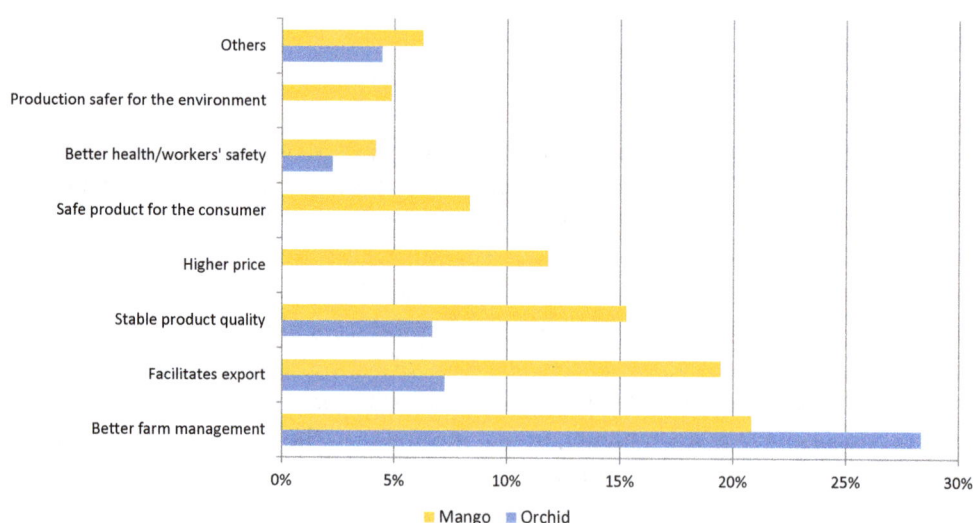

Figure 2. What do you consider as an advantage of Q-GAP? (open-ended question to all certified producers, answers grouped according to topics; multiple answers possible; N_{orchid} = 68, N_{mango} = 80).

The other perceived advantages differ significantly between orchid and mango producers. Overall, mango producers find much more positive points to say about Q-GAP certification than orchid producers. Only around 7% of orchid producers think that Q-GAP helps them to stable product quality and facilitate the export. Among mango producers, 19% and 15% consider this as an advantage. They also mention higher prices and that their products are safer for consumers and the environment—points that have not been mentioned by the orchid producers at all. Thus, while management related factors seem to be positively influenced by Q-GAP standards for both crops, only mango producers perceive a positive influence on product marketing and product safety.

One factor for this could be different customer demand for the different products. Since food safety is an ongoing issue in the Thai fresh horticultural market [3], producers of food crops would place more value on a certification scheme that enables them to fulfill higher food safety requirements. Furthermore, earlier findings on public GAP schemes suggest that producers of edible crops appreciate the increase in product safety and quality [26,28,29]. Orchid producers, on the other hand, may not have to pay attention to product safety at all. This is supported by the fact that the main reason for non-certified orchid producers to not get certified with Q-GAP was that exporters did not require the certification.

What are the real constraints that orchid and mango producers face when they want to produce for the high-value/export market? Figure 3 shows that challenges are quite different for orchid and mango producers. While the chemical residues on the product are the second most important constraint for mango producers, it is the least important for orchid producers. Since most orchids are used for decorative purposes, low levels of chemical residues play only a minor role as a sign of quality. Furthermore, orchid producers are less worried about their capability to comply with international standards as compared to mango producers.

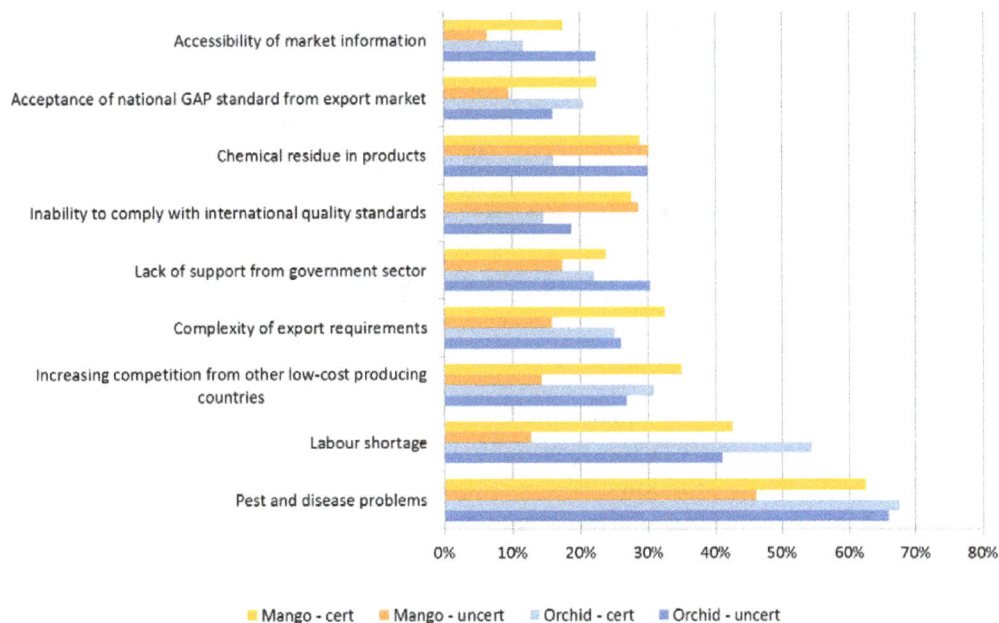

Figure 3. Main constraints to produce for/access the high-value export market (multiple answers possible; all producers, N_{orchid} = 180, N_{mango} = 144).

About 30% of both certified and uncertified orchid producers consider the increasing competition from low-cost producing countries as a threat. Among mango producers, this share is much higher for certified producers than for uncertified ones. This could indicate that mango producers certified with Q-GAP hope to use the certification to differentiate their product in the market.

The most important aspects for both orchid and mango producers are pest and disease problems and labor shortage, while in both cases a higher percentage of orchid producers compared to mango producers see this as a major constraint. Certified mango producers consider this more often as an obstacle than uncertified ones, indicating that the certification makes pest and disease control more difficult as well as requiring more work. This confirms earlier research on the public VietGAP certification, where increased labor requirements due to the standard were a major hampering factor for adoption among litchi farmers [63].

4.2. Factors Influencing the Adoption Decision of Q-GAP

The estimated results of the binary probit are presented in Table 4. The age of the household head, family size and gender have been included as demographic indicators, while the influence of the latter two are statistically insignificant for both mango and orchid producers. For the case of orchid, age of household head has an effect on the decision of Q-GAP adoption with the parabolic relationship indicated by the positive and negative significance of age and age square variables, respectively. This means that the older a farmer is, the more likely he or she is to certify, but with increasing age, this likelihood decreases again. The influence from age is significant at the 10% level. The same parabolic relation between age and household income was found in the case of coffee certification [47].

Table 4. Factors influencing the adoption of Q-GAP for mango and orchid producers.

Group	Variable	Mango		Orchid	
		Coef.	(SE)	Coef.	(SE)
Demographic factors	Age HH head	0.187	(0.140)	0.173 **	(0.067)
	Age to the square	-0.002	(0.001)	-0.001 **	(0.001)
	Gender HH head	0.239	(0.314)	-0.209	(0.242)
	Household size	-0.050	(0.107)	0.107	(0.071)
Human resources	Education HH head	0.016	(0.049)	0.054 **	(0.025)
	Farming experience	-0.001	(0.014)	-0.011	(0.012)
Physical capital	Agricultural assets	1.7×10^{-6} **	(6.7×10^{-7})	3.8×10^{-7}	(2.7×10^{-7})
	Land under main crop	0.024 ***	(0.007)	-0.005	(0.005)
Social capital	Producer group	1.311 ***	(0.363)	0.809 ***	(0.220)
	Know QGAP from DoAE	0.601	(0.659)	0.653 **	(0.259)
Public infrastructure	Distance to next town	0.018	(0.012)	0.011	(0.012)
	Constant	-7.502 *	(3.925)	-6.675 ***	(1.786)
	Goodness of fit	Pseudo-R^2: 0.4737 Corr. Classified: 86.01%		Pseudo-R^2: 0.1710 Corr. Classified: 71.67%	

se = standard error, HH = household, * = significant at the 10%-level, ** = significant at the 5%-level, *** = significant at the 1%-level.

Years of formal education and years of experience in growing orchids or mangoes, respectively, are used to indicate human resource factors determining adoption. Both variables are not significant in the adoption decision of mango producers. For orchid producers, the years of formal education have a significant positive influence. This confirms earlier findings on public GAP standards that are easier adopted by farmers with higher education [27,29] and goes along with results from GLOBALG.A.P. adoption among vegetable and mango producers in Kenya and Peru [20,43]. The distance to the next town as a proxy for public infrastructure did not have any significant influence.

While productive asset endowments, such as land and agricultural assets, have no significant effect on Q-GAP adoption for orchid producers, they have a significant positive influence in the case of mangoes. Higher land size has also been found as adoption influence for ChinaGAP [41] and among coffee producers for Q-GAP [27]. Furthermore, mango producers that were part of a producer group were significantly more likely to adopt certification. This positive influence of physical and social capital was also seen in earlier studies about private standards [20,35,42]. It shows that Q-GAP actually does not differ in the adoption patterns as compared to private standards. One advantage of Q-GAP is that it is free of charge and requirements are easier to accomplish than in private standards such as GLOBALG.A.P. and ThaiGAP, so that small farmers can also implement it [3]. At least for the case of mango and orchid production for export, it is still the better-educated, richer households that are getting certified.

4.3. Impact on Producers' Income and Export Shares

The ATT in all three matching algorithms show a significant positive influence of Q-GAP certification on the overall household income and per capita income from mango producers (Table 5). The production efficiency in terms of income per rai also increases for certified mango producers, though this effect is only significant in one out of three matching algorithms. For orchid producers, however, Q-GAP certification does not have a significant influence on the income.

Table 5. Impact of Q-GAP certification on income and export share.

Variable	Method	Mango		Orchid	
		ATT	(SE)	ATT	(SE)
Annual household income (THB)	N	995,855 ***	(331,337)	−350,411	(1,065,111)
	R	1,014,486 ***	(348,456)	512	(879,661)
	K	1,059,040 ***	(322,164)	−66,752	(703,702)
Income per rai (THB)	N	7512	(6923)	−154,077	(114,441)
	R	9056	(5813)	−33,090	(87,826)
	K	11,510 **	(5661)	−35,251	(76,418)
Income per capita (THB)	N	248,989 ***	(82,953)	−439,151	(779,674)
	R	250,226 ***	(77,727)	−299,563	(647,944)
	K	258,532 ***	(78,993)	−266,960	(629,682)
Share export/high-value market (%)	N	29.75 ***	(5.17)	4.65	(9.88)
	R	30.30 ***	(5.31)	0.03	(7.03)
	K	30.69 ***	(4.95)	1.01	(7.38)

ATT = Average treatment on the treated; SE = bootstrapped standard errors; *** = significant at the 1%-level; N = Nearest Neighbor, Matching with replacement; R = Radius, Matching with caliper of 0.1; K = Kernel, Matching with bandwidth of 0.05.

The insignificant influence of Q-GAP certification on orchid producers could be explained by the fact that Q-GAP certification does not facilitate orchid producers' access to the export market. All three matching algorithms in PSM and the Heckman model show no effect of Q-GAP certification on the share of sold value going to export. Descriptive results showed that orchid producers already sell about two-thirds of their harvest to the export market, showing that Q-GAP is not required in the orchid export sector. Earlier research also found that non-certified and certified orchid producers are using the same value chains [10]. In contrast, certified mango producers sell a significantly higher share of their product to high-value markets than uncertified ones. Another study on Thai Mangosteen producers leads to similar results: Q-GAP certification itself does not allow for a direct price premium, but it increased the share of high-quality fruits and is needed to enter the high-value export market, thus indirectly increasing farmers' income [26]. Thus, our study confirms earlier results that much of the income effect from the implementation of GAP standards is based on access to the high-value and export market [64].

The results of the PSM are supported by those of the second step Ordinary Least Squares (OLS) regression according to Heckman (Tables B1 and B2 in Appendix B). While the certification status has a positive influence on income and share of product sold to the export market for mango producers, it is insignificant on the outcomes for orchid producers. The influence of certification is even negative, though weakly significant, on orchid producers' income (Tables B1 and B2 in Appendix B).

Customers in Thai high-value retail and export are concerned about food safety, mainly due to end-customer concerns in Thailand and abroad. The reduction of food hazards in the domestic market and improvement of the image of Thai fresh produce in the export markets were among the major motivations of the Thai government to initiate Q-GAP [3]. While food safety plays an important role for fresh mangoes, it does not do so for orchids, because they are not an edible crop. In terms of food safety, the program indeed established certain credibility among Thai high-value retailers and middle-class end-customers [3]. Thus, the difference between the mango and the orchid case can be interpreted as a difference in how well the certification is already established as a brand in these crops in the Thai high-value and export market.

Another reason for the insignificant influence of Q-GAP on orchid producers' income could be the effect of floods in the main orchid production area in 2011. To control for the effect of the flooding on individual households, it was included as a binary variable in the second stage OLS in the Heckman correction. Results indicate that being directly affected by the floods did not have a significant influence on income and export shares.

5. Conclusions

The present study evaluates the impact of voluntary public GAP standards on income and export shares of mango and orchid producers. Based on primary data from 400 certified and non-certified orchid and mango producers from the major exporting areas in Thailand, the binary probit model estimations show that the better-off orchid and mango producers in terms of education, physical and social capital tend to comply with Q-GAP standards. Compared to previous literature, there is no difference between the characteristics of horticultural adopters of private GAP schemes and this public GAP scheme. This study focuses on producers with export-oriented crops. Further study is necessary to evaluate the situation for the domestic market in order to get a full picture of the adoption patterns of Q-GAP.

The results of the impact models indicate that mango producers who comply with Q-GAP standards obtain a significantly higher net income than non-certified ones. For orchid producers, however, certification status did not have any influence on export quantities or income. It seems like the program is not designed to fit the needs of orchid producers. The Q-GAP certification focuses on pesticide reduction and compliance to international standards [3]. Our data shows that while these are indeed obstacles to access the export market for mango producers, these points are not an issue for orchid producers. They perceive the inability to comply with international standards as the least of their problems and none of the orchid producers perceived product safety or environmental protection as an advantage of the Q-GAP program. Instead, major obstacles to enter the export market, such as shortage of labor and pests and diseases on the product, are further worsened by the standard's requirements of record keeping and pesticide control. Nevertheless, a demand for excluding non-edible crops from Q-GAP certification might be premature. The case of VietGAP shows that producers actually gain from pesticide reduction because they significantly reduce the loss of income due to health hazards from pesticide handling [29]. This and the environmental benefits to reduce and ban certain pesticides are good reasons to continue the promotion of GAP standards in non-edible crops as well.

Conclusively, if Q-GAP was to be successful for non-edible horticultural crops, the certification needs to provide an additional value beyond product safety, which by itself is currently not remunerated by the buyers in international markets. The promotion of health benefits of following this standard for the farmer themselves could be one possibility to increase acceptance. Furthermore, since in the long run, Q-GAP seeks to be benchmarked to other international standards such as ASEAN-GAP or GLOBALG.A.P., producers already implementing this standard now may have an advantage in the future.

The major limitation of this study is that, so far, only cross-sectional data is available to gain insights into the certification's impact on producers' income. Even though two different established methods for impact analysis have been applied to check for robustness of the results, variation over time is not visible. An analysis based on panel data could reveal the real magnitude of impact and control better for hidden bias. While the analysis at hand showed clear differences between edible and non-edible horticultural crops regarding the adoption decision and impact of Q-GAP, further research is needed to understand the impacts of certification on other crops such as fresh vegetables for export. Furthermore, the mechanism as to why Q-GAP certified mango producers achieve higher income could be investigated with a follow-up study. Does Q-GAP enable the producers to access the high-value export market through better farm management and product safety, or did they adopt the certification simply to continue exporting, because exporters started to demand Q-GAP? Further research should also extend to other stakeholders in the value chain to understand their perception and expectations of Q-GAP, which could help the Thai government to ameliorate the scheme's performance further.

Acknowledgments: The authors would like to thank the Federal Ministry of Education and Research (BMBF) and the Federal State Ministries: Lower Saxony Ministry of Science and Culture, Brandenburg Ministry of Infrastructure and Agriculture and Bavarian Ministry of State of Science, Research and Art for financial support through the research program "Network of Excellence in Horticulture (WEGA)".

Author Contributions: Henning Krause, Rattiya Suddeephong Lippe and Ulrike Grote conceived and designed the experiments; Henning Krause and Rattiya Suddeephong Lippe performed the experiments; Henning Krause and Rattiya Suddeephong Lippe analyzed the data; Henning Krause, Rattiya Suddeephong Lippe and Ulrike Grote wrote the paper.

Conflicts of Interest: The authors declare no conflict of interest. The founding sponsors had no role in the design of the study; in the collection, analyses, or interpretation of data; in the writing of the manuscript, and in the decision to publish the results.

Appendix A. Additional results and robustness checks for Propensity Score Matching

Table A1. Selected household characteristics of the sample.

Variable	Orchid			Mango		
	Non-Cert	Cert	Total	Non-Cert	Cert	Total
Age HH head	49.04 (12.32)	51.15 (9.58)	49.83 (11.38)	54.48 (11.38)	54.85 (9.48)	54.69 (10.33)
Gender HH head (1 = female)	0.321 (0.469)	0.221 (0.418)	0.283 (0.452)	0.413 ** (0.496)	0.250 (0.436)	0.322 (0.469)
Number of HH members	4.18* (1.43)	4.62 (1.56)	4.34 (1.49)	4.08 (1.56)	4.20 (1.56)	4.15 (1.56)
Education HH head (year)	9.21 (4.74)	10.24 (4.93)	9.59 (4.83)	6.60 (3.67)	6.88 (4.03)	6.76 (3.86)
Experience in orchid/mango production	19.91 (9.50)	20.78 (10.33)	20.24 (9.80)	19.57 (10.86)	22.10 (11.12)	20.99 (11.04)
HH participates in producers group (1 = yes)	0.518 *** (0.502)	0.765 (0.427)	0.611 (0.489)	0.460 *** (0.502)	0.938 (0.244)	0.727 (0.447)
HH head knows QGAP from DoAE	0.134 *** (0.342)	0.338 (0.477)	0.211 (0.409)	0.032 * (0.177)	0.113 (0.318)	0.077 (0.267)
Distance to next town	10.03 (9.32)	11.40 (8.17)	10.55 (8.90)	11.61 * (10.43)	15.52 (15.25)	13.80 (13.44)

Standard error in brackets, * = significant at the 10%-level, ** = significant at the 5%-level, *** = significant at the 1%-level, HH = household.

Table A2. Covariates for choice model to calculate the likelihood of being certified as propensity score.

Variable	Mango		Orchid	
	Coef.	(SE)	Coef.	(SE)
Age HH head	0.231 *	(0.124)	0.198 ***	(0.068)
Age to the square	−0.002 *	(0.001)	−0.002 **	(0.001)
Gender HH head	−0.264	(0.270)	−0.163	(0.242)
Education HH head	0.006	(0.044)	0.066 **	(0.026)
Experience in farming	0.012	(0.012)	−0.012	(0.013)
Assets bought 5 years ago	0.000 ***	(0.000)	0.000	(0.000)
Producer group	1.486 ***	(0.315)	0.764 ***	(0.222)
Know QGAP from DoAE	1.045 *	(0.557)	0.657 **	(0.259)
Distance to next town	0.011	(0.010)	0.013	(0.012)
Constant	−7.552 **	(3.524)	−6.996 ***	(1.826)
Goodness of fit	Pseudo R^2: 0.3724		Pseudo R^2: 0.2610	

HH = household, DoAE = Department of Agricultural Extension, SE = standard errors, * = significant at the 10%-level, ** = significant at the 5%-level, *** = significant at the 1%-level.

Table A3. Imposition of common support in the different matching procedures.

Matching Algorithm		Mango		Orchid	
		Off Support	On Support	Off Support	On Support
NN	Untreated	9	54	1	106
	Treated	12	68	10	56
	Total	21	122	11	162
Radius	Untreated	9	54	4	103
	Treated	12	68	10	56
	Total	21	122	14	159
Kernel	Untreated	9	54	1	106
	Treated	12	68	10	56
	Total	21	122	11	162

NN = Nearest Neighbor, Matching with replacement; Radius, Matching with caliper of 0.1; Kernel, Matching with bandwidth of 0.05.

Table A4. Summary of bias reduction for the different matching algorithms.

Sample	Pseudo-R^2	F-Test	Mean Bias	Median Bias
Mango (unmatched)	0.333	0.000	37.0	29.9
NN	0.064	0.207	14.3	7.6
Radius	0.047	0.455	15.1	17.0
Kernel	0.031	0.755	9.8	11.1
Orchid (unmatched)	0.155	0.000	23.1	19.1
NN	0.076	0.246	10.3	7.6
Radius	0.017	0.980	11.9	12.6
Kernel	0.015	0.986	9.5	7.1

NN = Nearest Neighbor, Matching with replacement; Radius, Matching with caliper of 0.1; Kernel, Matching with bandwidth of 0.05.

Table A5. Hidden bias increase (%) until which results for mango producers remain significant at the 10%-level.

Outcome Variable	Matching Algorithm	%
Income per capita	NN	250
	Radius	270
	Kernel	300
Share of export market	NN	380
	Radius	380
	Kernel	370

NN = Nearest Neighbor, Matching with replacement; Radius, Matching with caliper of 0.1; Kernel, Matching with bandwidth of 0.05.

Appendix B. Robustness Check with Heckman's Two-Stage Model

The Heckman model addresses the selection bias by using an instrumental variable technique to remove the dependence of adoption on the welfare related variables [20]. The procedure of the Heckman model comprises (i) a selection equation and (ii) an outcome equation [20]. The selection equation is depicted as:

$$c_i^* = \alpha X_i + u_i \qquad u \sim N(0,1) \tag{B1}$$

c_i^* is the latent variable for Q-GAP adoption. c denotes the observable part of a dummy variable for Q-GAP adoption which $c = 1\ if\ c_i^* > 0$ otherwise $c = 0$. X_i denotes a vector of observed farm and non-farm characteristics determining Q-GAP adoption. The parameters in the selection equation are estimated using a probit regression for all observations. Based on the binary probit model, the correction-factor, the so-called "Inverse Mills Ratio (IMR)" is derived from the "univariate standard normal probability function" and the "cumulative distribution function" for each observation. The calculated IMRs are inserted in the second stage equations to test for potential selection bias. This is done through the coefficient of IMR testing the null hypothesis that there is no correlation between the error terms in Equations (B1) and (B2) [47].

$$y_{O,M} = \beta_0 + \beta_1 c_i + \beta_2 X_i + IMR + e_i \tag{B2}$$

y_O denotes net income/share of product sold to export markets from orchid and y_M from mango production, respectively. For the second stage, it reruns the regression with the primary interest to see whether Q-GAP adoption positively impacts on producers' income and share of product sold to export markets. Participation in a producer group is used as an exclusion restriction, because it strongly influences Q-GAP adoption, but does not have a significant influence on the outcome variables.

Table B1. Results Heckman two stage for mango.

	Treatment (Probit)	Outcome (OLS)			
Dependent Variable	Q-GAP	Income (log)	Income p. rai (log)	Income p. cap (log)	Export Share
Independent Variables					
Age HH head (year)	0.187 (0.140)	0.075 (0.109)	0.113 (0.106)	0.052 (0.108)	0.013 (0.021)
Age to the square	−0.002 (0.001)	−0.001 (0.001)	−0.001 (0.001)	−0.0007 (0.001)	−0.0002 (0.0002)
Gender HH head (1 = female)	0.239 (0.314)	−0.071 (0.272)	0.220 (0.264)	−0.067 (0.271)	−0.067 (0.052)
Household size	−0.050 (0.107)	−0.051 (0.086)	−0.082 (0.084)	−0.319 *** (0.086)	−0.006 (0.016)
Formal education HH head (year)	0.016 (0.049)	0.022 (0.037)	0.005 (0.036)	0.025 (0.037)	−0.004 (0.007)
Experience in farming (year)	−0.001 (0.014)	0.018 (0.011)	0.011 (0.011)	0.019 (0.011)	0.000 (0.002)
Agricultural assets	1.7×10^{-6} ** (6.7×10^{-7})	2.2×10^{-7} (5×10^{-7})	4.7×10^{-7} (4×10^{-7})	2.4×10^{-7} (4.6×10^{-7})	9×10^{-8} (8×10^{-8})
Land under main crop (rai)	0.024 *** (0.007)	0.014 *** (0.003)	0.001 (0.003)	0.014 *** (0.003)	0.001 (0.001)
Producer group participation	1.311 *** (0.363)				
Household knows QGAP from DoAE	0.601 (0.659)	0.168 (0.468)	0.300 (0.454)	0.109 (0.466)	0.115 (0.087)
Distance to next town	0.018 (0.012)	0.004 (0.009)	0.011 (0.009)	0.005 (0.009)	0.001 (0.002)
Q-GAP certification		1.265 *** (0.325)	0.783 ** (0.316)	1.245 *** (0.324)	0.221 *** (0.063)
IMR		−0.373 (0.280)	0.030 (0.272)	−0.388 (0.279)	−0.066 (0.050)
Constant	−7.502 * (3.925)	9.694 *** (3.188)	6.001 * (3.097)	9.918 *** (3.174)	0.039 (0.618)
R^2/Pseudo-R^2	0.4737	0.6305	0.2216	0.6277	0.3593

HH = household, DoAE = Department of Agricultural Extension, * = significant at the 10%-level, ** = significant at the 5%-level, *** = significant at the 1%-level.

Table B2. Results Heckman two stage for orchid.

	Treatment (Probit)	Outcome (OLS)		Outcome (OLS)	
Dependent Variable	**Q-GAP**	**Income p. cap (log)**	**Export Share**	**Income p. cap (log)**	**Export Share**
Independent Variables					
Age HH head (year)	0.187 (0.140)	−0.377 (0.278)	−0.323 (0.238)	−0.373 (0.253)	0.017 (0.019)
Age to the square	−0.002 (0.001)	0.003 (0.003)	0.003 (0.002)	0.003 (0.002)	0.000 (0.000)
Gender HH head (1 = female)	0.239 (0.314)	0.196 (0.907)	0.133 (0.785)	0.192 (0.821)	−0.097 (0.063)
Household size	−0.050 (0.107)	−0.281 (0.290)	−0.272 (0.250)	−0.441 * (0.262)	−0.032 (0.020)
Formal education HH head (year)	0.016 (0.049)	−0.005 (0.106)	0.002 (0.091)	−0.002 (0.095)	−0.007 (0.007)
Experience in farming (year)	−0.001 (0.014)	0.145 *** (0.047)	0.121 *** (0.040)	0.131 *** (0.042)	−0.008 ** (0.003)
Agricultural assets	1.7×10^{-6} ** (6.7×10^{-7})	-6×10^{-8} 1.1×10^{-6}	-3×10^{-7} 9.5×10^{-7}	-1.4×10^{-7} (9×10^{-7})	-5.3×10^{-9} (-7.2×10^{-8})
Land under main crop (rai)	0.024 *** (0.007)	0.037 (0.028)	−0.002 (0.024)	0.039 * (0.024)	0.002 * (0.001)
Producer group participation	1.311 *** (0.363)	−1.134 (1.143)	−1.084 (0.984)	−1.144 (1.031)	0.012 (0.079)
Household knows QGAP from DoAE	0.601 (0.659)	0.042 (0.047)	0.030 (0.040)	0.042 (0.042)	0.001 (0.003)
Distance to next town	0.018 (0.012)	−0.377 (0.278)	−0.323 (0.238)	1.487 (0.805)	0.018 (0.061)
Q-GAP certification		1.770 (0.897)	1.181 (0.771)	−0.417 * (0.715)	0.018 (0.061)
Household affected by floods (1 = yes)		−0.401 (0.792)	−0.493 (0.680)	−0.647 (1.281)	−0.009 (0.055)
IMR		−0.427 (1.415)	−0.408 (1.216)	19.207 (8.244)	−0.046 (0.098)
Constant	−7.502 * (3.925)	19.108 (9.084)	17.121 (7.776)	−0.373 ** (0.253)	0.680 (0.631)
R^2/Pseudo-R^2	0.4737	−0.401	−0.493	0.1189	0.1087

HH = household, DoAE = Department of Agricultural Extension, * = significant at the 10%-level, ** = significant at the 5%-level, *** = significant at the 1%-level.

References

1. Malorgio, G.; Felice, A. Trade and logistics: The fruit and vegetables industry. In *MediTerra 2014: Logistics and Agro-Food Trade, A Challenge for the Mediterranean*; Les Presses de Sciences Po: Paris, France, 2014; pp. 149–171.
2. European Commission Directorate General for Agriculture and Rural Development. Agricultural Commodity Markets Past Developments: Fruits and Vegetables. Available online: http://ec.europa.eu/ agriculture/analysis/tradepol/worldmarkets/fruitveg/072007_en.pdf (accessed on 23 June 2016).
3. Wongprawmas, R.; Canavari, M.; Waisarayutt, C. A multi-stakeholder perspective on the adoption of Good Agricultural Practices in the Thai fresh produce industry. *Br. Food J.* **2015**, *117*, 2234–2249. [CrossRef]
4. Hatanaka, M.; Bain, C.; Busch, L. Third-party certification in the global agrifood system. *Food Policy* **2005**, *30*, 354–369. [CrossRef]
5. Gebreeyesus, M. Firm adoption of international standards: Evidence from the Ethiopian floriculture sector. *Agric. Econ.* **2015**, *46*, 139–155. [CrossRef]
6. Josling, T.; Roberts, D.; Orden, D. Food regulation and trade: Towards a safe and open global system. In Proceedings of the American Agricultural Economics Association Annual Meeting, Denver, CO, USA, 1–4 August 2004; Institute for International Economics: Washington, DC, USA, 2004.
7. Dankers, C. *Environmental and Social Standards, Certification and Labelling for Cash Crops*; Food and Agriculture Organization of the United Nations: Rome, Italy, 2003.

8. Fuchs, D.; Kalfagianni, A.; Havinga, T. Actors in private food governance: The legitimacy of retail standards and multistakeholder initiatives with civil society participation. *Agric. Hum. Values* **2011**, *28*, 353–367. [CrossRef]

9. Food and Agriculture Organization of the United Nations (FAO). Incentive for the adoption of good agricultural practices. In *FAO GAP Working Paper Series 3*; Background Paper for the FAO Expert Consultation on a Good Agricultural Practice Approach; FAO: Rome, Italy, 2003.

10. Lippe, R.S.; Grote, U. Costs and Benefits of GAP Standards Adoption in Thai Horticulture. In *Aktuelle Forschung in der Gartenbauökonomie, Proceeding of the 1st Symposium für Ökonomie im Gartenbau, Göttingen, Germany, 27 November 2013*; Dirksmeyer, W., Theuvsen, L., Kayser, M., Eds.; Johann Heinrich von Thünen-Institut: Braunschweig, Germany, 2015; pp. 205–214.

11. Henson, S.; Jaffee, S. Understanding developing country strategic responses to the enhancement of food safety standards. *World Econ.* **2008**, *31*, 548–568. [CrossRef]

12. Henson, S.; Humphrey, J. Understanding the complexities of private standards in Global Agri-Food Chains as they impact developing countries. *J. Dev. Stud.* **2010**, *46*, 1628–1646. [CrossRef] [PubMed]

13. Giovannucci, D.P.; Sterns, P.A.; Eustrom, M.; Haantuba, H. *The Impact of Improved Grades and Standards for Agricultural Products in Zambia*; PFID-F&V Report No. 3; Michigan State University and United States Agency for International Development: East Lansing, MI, USA, 2001.

14. Ouma, S. Global Standards, local realities: Private agrifood governance and the restructuring of the Kenyan Horticulture Industry. *Econ. Geogr.* **2010**, *86*, 197–222. [CrossRef]

15. Maertens, M.; Swinnen, J.F.M. Food standards, trade and development. *Rev. Bus. Econ.* **2009**, *54*, 313–326.

16. Schreinemachers, P.; Schad, I.; Tipraqsa, P.; Williams, P.; Neef, A.; Riwthong, S.; Sangchan, W.; Grovermann, C. Can public GAP standards reduce agricultural pesticide use? The case of fruit and vegetable farming in northern Thailand. *Agric. Hum. Values* **2012**, *29*, 519–529. [CrossRef]

17. Wannamolee, W. *Development of Good Agricultural Practices (GAP) for Fruit and Vegetables in Thailand*; Office of Commodity and System Standards Accreditation; National Bureau of Agricultural Commodity and Food Standards: Bangkok, Thailand, 2008.

18. Department of Agricultural Extension; Bangkok, Thailand. Personal Communication, 2011.

19. Centre for Agricultural Information (CAI); Office of Agricultural Economics (OAE). *Thailand Foreign Agricultural Trade Statistics 2014*; Ministry of Agriculture and Co-operatives: Bangkok, Thailand, 2015.

20. Asfaw, S.; Mithöfer, D.; Waibel, H. What impact are EU supermarket standards having on developing Countries' export of high-value horticultural products? Evidence from Kenya. *J. Int. Food Agribus. Mark.* **2010**, *22*, 252–276. [CrossRef]

21. Bayramoglu, Z.; Gundogmus, E.; Tatlidil, F.F. The impact of EurepGAP requirements on farm income from greenhouse tomatoes in Turkey. *Afr. J. Agric. Res.* **2010**, *5*, 348–355.

22. Dörr, A.C.; Grote, U. Impact of certification on fruit producers in the Sao Francisco Valley in Brazil. *Ann. Dunărea de Jos Univ.* **2009**, *15*, 5–16.

23. Jenson, M.F. Developing New Exports from Developing Countries: New Opportunities and New Constraints. Ph.D. Thesis, The Royal Veterinary and Agricultural University, Copenhagen, Denmark, 2004.

24. Kersting, S. Adaptation of Smallholder Farmers to the Increasing Demand for Standards and Their Impacts-Case Study of the Horticultural Sector in Thailand. Diploma Thesis, University of Passau, Passau, Germany, 2009.

25. Okello, J.J. Compliance with International Food Safety Standards: The Case of Green Bean Production in Kenyan Family Farms. Ph.D. Thesis, Michigan State University, East Lansing, MI, USA, 2005.

26. Pongvinyoo, P.; Yamao, M.; Hosono, K. Cost efficiency of Thai National GAP (QGAP) and mangosteen farmers' understanding in Chanthaburi Province. *Am. J. Rural Dev.* **2015**, *3*, 15–23. [CrossRef]

27. Pongvinyoo, P.; Yamao, M.; Hosono, K. Factors affecting the implementation of Good Agricultural Practices (GAP) among Coffee Farmers in Chumphon province, Thailand. *Am. J. Rural Dev.* **2014**, *2*, 34–39. [CrossRef]

28. Nicetica, O.; van de Flierta, E.; Chienb, H.V.; Maic, V.; Cuongb, L. Good Agricultural Practice (GAP) as a vehicle for transformation to sustainable citrus production in the Mekong Delta of Vietnam. In Proceedings of the 9th European IFSA Symposium, Vienna, Austria, 4–7 July 2010.

29. Tran, H.B.C.; Le, T.Q.A. *Impact of VietGAP Vegetable Production on the Health of Farmers Thua Thien Hue Province, Vietnam*; EEPSEA Research Report No. 2015-RR19; Economy and Environment Program in Southeast Asia: Laguna, Philippines, 2015.

30. Kleinwechter, U.; Grethe, H. The adoption of the Eurepgap standard by mango exporters in Piura. In Proceeding of the International Association of Agricultural Economists Conference, Gold Coast, Australia, 12–18 August 2006.

31. Rogers, E.M. *Diffusion of Innovations*, 5th ed.; The Free Press: New York, NY, USA, 2003.

32. Scoones, I. *Sustainable Rural Livelihoods: A Framework for Analysis*; IDS Working Paper 72; University of Sussex: Brighton, UK, 1998.

33. Ngokkuen, C.; Grote, U. Geographical indication for jasmine rice: Applying a logit model to predict adoption behavior of Thai farm households. *Q. J. Int. Agric.* **2012**, *51*, 157–185.

34. Fort, R.; Ruben, R. Impact of fair trade certification on coffee producers in Peru. In *The Impact of Fair Trade*; Ruben, R., Ed.; Wageningen Academic Publishers: Wageningen, The Netherlands, 2008; pp. 75–98.

35. Saenz-Segura, F.; Zuniga-Arias, G. Assessment of the effect of fair trade on smallholder producers in Costa Rica: A comparative study in the coffee sector. In *The Impact of Fair Trade*; Ruben, R., Ed.; Wageningen Academic Publishers: Wageningen, The Netherlands, 2008; pp. 117–135.

36. Zuniga-Arias, G.; Saenz-Segura, F. The impact of fair trade in banana production of Costa Rica. In *The Impact of Fair Trade*; Ruben, R., Ed.; Wageningen Academic Publishers: Wageningen, The Netherlands, 2008; pp. 99–116.

37. Kersting, S.; Wollni, M. New institutional arrangements and standard adoption: Evidence from small-scale fruit and vegetable farmers in Thailand. *Food Policy* **2012**, *37*, 452–462. [CrossRef]

38. Muriithi, B.W.; Mburu, J.; Ngigi, M. Constraints and determinants of compliance with EurepGap standards: A case of smallholder french bean exporters in Kirinyaga District, Kenya. *Agribusiness* **2011**, *27*, 193–204. [CrossRef]

39. Lemeilleur, S. Smallholder compliance with private standard certification: The case of GlobalGAP adoption by mango producers in Peru. *Int. Food Agribus. Manag. Rev.* **2013**, *16*, 159–180.

40. Srisopaporn, S.; Jourdain, D.; Perret, S.R.; Shivakoti, G. Adoption and continued participation in a public good agricultural practices program: The case of rice farmers in the central plains of Thailand. *Technol. Forecast. Soc. Chang.* **2015**, *96*, 242–253. [CrossRef]

41. Jin, S.; Zhou, J. Adoption of food safety and quality standards by China's agricultural cooperatives. *Food Control* **2011**, *22*, 204–208. [CrossRef]

42. Office of Agricultural Extension. *Statistical Information on Agricultural Commodities*; Office of Agricultural Extension: Bangkok, Thailand, 2015.

43. Thammasiri, K. Current status of orchid production in Thailand. In *Acta Horticulturae 1078*; Uthairatanakij, A., Wannakrairoj, S., Eds.; International Society for Horticultural Science (ISHS): Leuven, Belgium, 2015.

44. Lekawatana, S. Thai Orchid: Current situation. In *Proceeding of the Taiwan International Orchid Symposium, Tainan, Taiwan, 5 March 2010*; Flower and Ornamental Plant Production Promotion Group, Department of Agricultural Extension: Bangkok, Thailand, 2010.

45. Postharvest Technology Research Institute (PHTRI). *Thai Mango's Value Chain Analysis and Supply Chain Management System to EU Markets*; Final Report to The European Commission's Asia Invest Program "Integrated Supply Chain Management of Exotic Fruits from the ASEAN Region"; European Commission: Brussels, Belgium, 2009.

46. Dreier, J. Rational preference: Decision theory as a theory of practical rationality. *Theory Decis.* **1996**, *40*, 249–276. [CrossRef]

47. Wooldridge, J.M. *Econometric Analysis of Cross Selection and Panel Data*, 2nd ed.; The MIT Press: Cambridge, MA, USA, 2010.

48. Jena, P.R.; Chichaibelu, B.B.; Stellmacher, T.; Grote, U. The impact of coffee certification on small-scale producers' livelihoods: A case study from the Jimma Zone, Ethiopia. *Agric. Econ.* **2012**, *43*, 429–440. [CrossRef]

49. Blackman, A.; Rivera, J. Producer-Level Benefits of Sustainability Certification. *Conserv. Biol.* **2011**, *25*, 1176–1185. [CrossRef] [PubMed]

50. Heras-Saizarbitoria, I.; Molina-Azorín, J.F.; Dick, G.P.M. ISO 14001 certification and financial performance: Selection-effect versus treatment-effect. *J. Clean. Prod.* **2011**, *19*, 1–12. [CrossRef]

51. Bryson, A.; Dorsett, R.; Purdon, S. *The Use of Propensity Score Matching in the Evaluation of Active Labour Market Policies*; Her Majesty's Stationary Office: London, UK, 2002.

52. Pirracchio, R.; Resche-Rigon, M.; Chevret, S. Evaluation of the Propensity score methods for estimating marginal odds ratios in case of small sample size. *BMC Med. Res. Methodol.* **2012**. [CrossRef] [PubMed]

53. Rosenbaum, P.R.; Rubin, D.B. The central role of the propensity score in observational studies for causal effects. *Biometrilca* **1983**, *70*, 41–55. [CrossRef]

54. Caliendo, M.; Kopeinig, S. Some practical guidance for the implementation of propensity score matching. *J. Econ. Surv.* **2008**, *22*, 31–72. [CrossRef]

55. Austin, P.C. Optimal caliper widths for propensity-score matching when estimating differences in means and differences in proportions in observational studies. *Pharm. Stat.* **2010**, *10*, 150–161. [CrossRef] [PubMed]

56. Heckman, J.; Ichimura, H.; Todd, P. Matching as an econometric evaluation estimator. *Rev. Econ. Stud.* **1998**, *65*, 261–294. [CrossRef]

57. Lechner, M. Some practical issues in the evaluation of heterogeneous labour market programmes by matching methods. *J. R. Stat. Soc.* **2002**, *165*, 59–82. [CrossRef]

58. Smith, J.; Todd, P. Does matching overcome LaLonde's critique of nonexperimental estimators? *J. Econom.* **2005**, *125*, 305–353. [CrossRef]

59. Sianesi, B. An evaluation of the Swedish system of active labour market programmes in the 1990s. *Rev. Econ. Stat.* **2004**, *86*, 133–155. [CrossRef]

60. Rosenbaum, P.R. Observational studies. In *Springer Series in Statistics*, 2nd ed.; Springer Science & Business Media: New York, NY, USA, 2002.

61. Hujer, R.; Caliendo, M.; Thomsen, S.L. New evidence on the effects of job creation schemes in Germany—A matching approach with threefold heterogeneity. *Res. Econ.* **2004**, *58*, 257–302. [CrossRef]

62. Kassie, M.; Shiferaw, B.; Muricho, G. Adoption and impact of improved groundnut varieties on rural poverty. In *Discussion Paper Series 10–11*; The Environment for Development (EfD) Initiative: Gothenburg, Sweden, 2010.

63. Han, V.T.; Dung, N.M.; Santi, S. Litchi farmers' preference for the adoption of Vietnamese Good agricultural practices in Luc Ngan district, Vietnam. *J. ISSAAS* **2016**, *22*, 64–76.

64. Maertens, M.; Dries, L.; Dedehouanou, F.A.; Swinnen, J.F.M. High-value supply chains, food standards and rural households in Senegal. In *Global Supply Chains, Standards and the Poor*; Swinnen, J.F.M., Ed.; Catholic University of Leuven: Leuven, Belgium, 2007; pp. 159–172.

5

Hydraulic Performance of Horticultural Substrates—1. Method for Measuring the Hydraulic Quality Indicators

Uwe Schindler *, Lothar Müller and Frank Eulenstein

Leibniz Centre for Agricultural Landscape Research (ZALF), Institute of Landscape Hydrology, Eberswalder St. 84, Müncheberg D15374, Germany; mueller@zalf.de (L.M.); feulenstein@zalf.de (F.E.)
* Correspondence: uschindler@zalf.de

Abstract: Besides nutrient composition, the hydraulic performance of horticultural substrates is a main issue for evaluating their quality for horticultural purposes. Their water and air capacity and their suitability for transporting water are important hydraulic quality indicators. Shrinkage and water repellency could have a negative impact on storing and transporting water and solutes. The commonly used methods and devices for quantifying the water retention properties of horticultural substrates (sand box, pressure plate extractor) are outdated. The measurements are time-consuming, the devices are expensive, and the results are affected by uncertainties. Here, the suitability of the extended evaporation method (EEM) and an associated HYPROP (HYdraulic PROPerty analyser, device was successfully tested for very loosely-bedded horticultural substrates. EEM and HYPROP enabled the simultaneous and effective measurement of the water retention curve and the unsaturated hydraulic functions. The measurement time of horticultural substrates ranges between 7 and 10 days. Furthermore, the shrinkage properties and the water rewetting time can be measured with the HYPROP system. Results using 18 horticultural substrates are presented. These results are discussed and compared with natural organic and mineral soils showing the specific hydraulic performance of substrates for horticultural applications.

Keywords: horticultural substrates; growing media; hydraulic properties; water retention curve; unsaturated hydraulic conductivity; water repellency; water drop penetration time; shrinkage; extended evaporation method (EEM); HYPROP

1. Introduction

Horticultural substrates are specially designed media for horticultural applications. Bog peat is the main basis for creating horticultural substrates [1,2]. Special organic and mineral ingredients are added to improve the physical and technological properties of the substrates. Besides the nutrient composition, the hydraulic performance of horticultural substrates is a main issue for evaluating their quality for horticultural purposes. The water and air capacity and suitability for transporting water are important hydraulic quality indicators [1–6]. The basic properties are the water retention curve and the hydraulic conductivity function. Shrinkage and water repellency could have a negative impact on storing and transporting water and solutes [1,7].

The measurement of the water retention curve of horticultural substrates is generally executed with the sand box and the pressure plate extractor [1,8,9]. These methods and devices are outdated, the measurement is time-consuming, the equipment is expensive, and the results are affected by uncertainties [10]. Only a few unsaturated hydraulic conductivity measurements in substrates have been presented, but they are required for an overall evaluation of horticultural substrate quality [1,11]. In some cases, the one-step outflow method has been used [2,12]. Raviv and Lieth [1] concluded that there is a lack of technologies and methods for the effective physical characterization of substrates in horticulture.

The extended evaporation method (EEM) and an associated HYPROP (HYdraulic PROPerty analyzer [13]) system enables the simultaneous measurement of the hydraulic functions, the water retention curve, and the hydraulic conductivity function. Here, the suitability of the EEM and the HYPROP was tested using very loosely-bedded horticultural substrates. Furthermore, the HYPROP was used to measure the shrinkage behaviour and the water rewetting properties of the horticultural substrates. These results are discussed and compared with natural organic and mineral soils, showing the specific performance of horticultural substrates for horticultural applications.

2. Experimental Section

2.1. Samples and Preparation

The hydraulic measurements were conducted on 18 commercial horticultural substrates and compared with 10 mineral and organic soils (Table 1). The horticultural substrates mainly consisted of 30% to 100% of bog peat (degree of decomposition between H3 and H7 [14], and different proportions of organic residuals (garden (G) and forest (F) compost), coir (Co) and mineral additives such as perlite (P), lime (K), clay (C) and sand (S). One of the horticultural substrates (no. 6) was totally free of bog peat. The natural soils are not used for horticultural purposes, and were only collected and analysed to show their hydraulic differences to the special commercial horticultural substrates. The fen soil material was collected from Muencheberg, Rotes Luch, Brandenburg, Germany. The degree of decomposition was quantified according to Von Post [14]. The mineral soils were formed by glacial processes and were collected at different arable sites in the state of Brandenburg, Germany.

Table 1. Collection of horticultural substrates (HS) and fen samples.

HS No.	Ingredients, Texture Class	Ash Content (%)	C_{org} (%)
	Horticultural substrates		
1	Hh (H3–H8), R, G	68.1	
2	Hh (H2–H4) H7–H9, G, R	16.8	
3	Hh (H2–H5), P, C, K	35.1	
4	Hh (H3–H8), R, G, P, C, K	21.4	
5	95% Hh (H3–H7), P, Co	24.4	
6	R, C, Co, Guano	25.3	
7	90% Hh (H4–H8), 10% C, K	41.0	
8	Hh (H3–H8), C,P	35.9	
9	75% Hh (H3–H5 and H6–H7), Co, C, K	25.1	
10	80% Hh (H3–H5 and H6–H7), Co, C	39.9	
11	Hh (H2–H5), G, R, P, C, K	48.3	
12	Hh (H3–H8), G, R, P, C, C	42.7	
13	Hh (H3–H5 and H7–H9)	15.1	
14	Hh (H2–H5), G, R, K	35.8	
15	Hh (H3–H8), G, P, K	10.8	
16	60% Hh (H3–H5 and H6–H7), R, G, Co, K	25.5	
17	60% Hh (H3–H5 and H6–H7), Co, C, P	42.8	
18	50% Hh (H3–H5), G, R, C	36.2	
	Natural organic and mineral soils		
19	Fen peat (Hn, H7)	55.0	
20	Half-fen (Aa)		11.6
21	Sand (Ss, strong humic		2.9
22	Weak silty sand (Su2)		0.9
23	Weak loamy sand Sl2		1.0
24	Medium loamy sand (Sl3)		1.1
25	Strong loamy sand (Sl4)		1.2
26	Medium clayey silt (Ut3)		1.3
27	Medium sandy loam (Ls3)		1.5
28	Sandy clayey loam (Lts)		1.6

Hh—bog peat, H degree of decomposition; R—compost of forest residuals; G—compost of garden residuals; Co—coir (raw coconut fibre); P—perlite; K—lime; C—clay; S—sand; Hn—fen peat; C_{org}—organic carbon; Texture class acc. to Boden, A.G. [15].

Sample preparation: a plastic pipe (diameter: 15 cm, height: 65 cm) was loosely filled with the substrate up to 5 cm below the upper edge. Water was added at the surface until water left at the bottom of the pipe. The pipe was placed for 2 days in a pan with a 3-cm water level. The substrate compacted itself hydraulically and after 2 days the capillary equilibrium was reached. At this time, the tension at the surface layer was about 50 hPa. The substrate material of the upper 5 cm layer of the pipe was taken, mixed and used to loosely fill 250 cm^3 HYPROP steel cylinders. During the filling procedure, the cylinder was stamped 10 times. The prepared sample was saturated and ready for the hydraulic measurements with the HYPROP system. This procedure was derived from DIN EN 13041 [9] and Verdonck and Gabriels [16] and guarantees a high reproducibility. It enables the hydraulic comparability of growing media though the substrates in the cylinder are of different basic moistures.

2.2. Hydraulic Criteria

The hydrological evaluation of the suitability of the tested substrates for horticultural applications depends on their hydraulic properties and the kind of cultivation (containers with different heights or bed cultivation). The most important aspects are (i) the amount of easily plant-available water (EAW) and (ii) the air capacity depending on the kind of cultivation. The capillary rise is an additional indicator for characterizing transport properties. The (iv) rewetting time and (v) shrinkage dynamics could negatively influence hydraulic substrate quality. In this study, the evaluation was carried out as an example of growing in 20-cm-high containers (P20).

2.3. Measurement of the Water Retention Curve and the Unsaturated Hydraulic Conductivity Function

The extended evaporation method [10] was tested for the very loosely-bedded horticultural substrates. The EEM enables the simultaneous measurement of the water retention curve and the hydraulic conductivity function. Using new cavitation tensiometers and applying the air entry value of the tensiometer's ceramic cup, it allowed the range to be extended almost up to the wilting point. The measurements were carried out using the HYPROP system. HYPROP [13] is the commercial device used to implement the EEM. The total measurement time depends on the soil or substrate and the evaporation conditions and ranges between 3 and 10 days. Multiple samples can be measured simultaneously.

2.3.1. Short Description of the Procedure

The substrate samples were slowly saturated with water. Two tensiometers were inserted from the bottom and the core was sealed at the bottom by clamping the cylinder with the assembly. The core was placed on weighing scales, and the mass and pressure conditions in the soil core were controlled online. The soil surface was exposed to free evaporation and the measurement cycle started. There was no uncontrolled water loss, not at the bottom and not at the surface. Figure 1 shows the principle of the experimental setup. Tensions and the sample mass were recorded at selected time intervals. The hydraulic gradient was calculated on the basis of the tensions recorded during the time interval. The water flux was derived from the associated soil water volume difference. Individual points on the water retention curve were calculated on the basis of the water loss per volume of the sample at a time t_i and were related to the mean tension in the sample at this time. The unsaturated hydraulic conductivity (K) was calculated according to the Darcy-Buckingham law (Equation (1)).

$$K(\Psi_{mean}) = \frac{\Delta m}{a A \, \Delta t \, i_m} \tag{1}$$

where Ψ_{mean} is the mean tension over the upper tensiometer at position z_1 (3.75 cm above the bottom of the sample) and the lower tensiometer at position z_2 (1.25 cm above the bottom), geometrically averaged over a time interval of $\Delta t_J = t_{i+1} - t_i$, with $i = 1 \ldots n$, $j = 1 \ldots n - 1$; Δm is the sample mass difference in the time interval (assumed to be equal to the total evaporated water volume ΔV_{H2O} of the whole sample in the interval); a is the flux factor (in the case of rigid soils $a = 2$); A is the cross-sectional area of the sample; i_m is the hydraulic gradient averaged over the time interval.

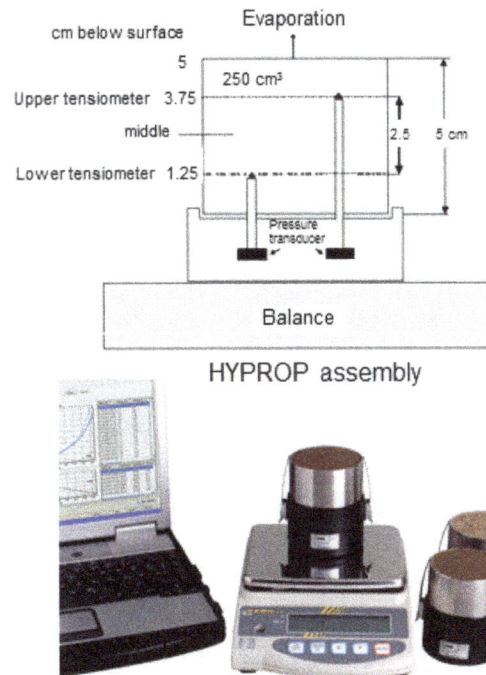

Figure 1. Schematic illustration of the evaporation method (Photo: UMS GmbH, Munich, Germany).

Data points of the water retention curve were pairs of mean tension at times t_i and t_{i+1} for $i = 1 \ldots n$ and the corresponding volumetric water content. The soil was assumed to be rigid. An EEM data set of a single sample consisted of multiple user-defined water retention and hydraulic conductivity data pairs. At the end of the measurement cycle, the residual amount of storage water was derived from the water loss upon oven drying (105 °C), and the initial water content was calculated. The dry bulk density was derived from the dry soil mass. Furthermore, soil hydraulic functions could be measured under consideration of shrinkage [17].

2.4. Rewetting Properties

The Water Drop Penetration Time Method (WDPT) [18] was used in this study to quantify the rewetting properties. The method is based on the time taken for a drop of water to infiltrate into the substrate. Using a pipette, one drop of water was added to the sample and the water penetration time was measured. The measurement was executed at different times during the evaporation experiment to gain WDPT values at tensions of approximately 100 hPa. The measurement was repeated 3 times to calculate the average value. This procedure is easy to handle and does not need a great deal of technical effort.

2.5. Shrinkage Measurement

Shrinkage was estimated during the evaporation experiment. The diameter of the sample's surface was measured at a tension of about 100 hPa using a calliper. The shrinkage from the bottom to the top of the sample is linear in this tension range [17]. In conclusion, the shrinkage of the sample was calculated (Equation (2)). Isotropic conditions were assumed.

$$V_s = \frac{\pi}{4} * \left[\frac{d_i + d_s}{2} \right]^2 * h_s \tag{2}$$

where V_s is the volume of the shrunken sample, d_i is the initial sample diameter, d_s is the diameter at the sample's surface at 100 hPa and h_s is the height of the shrunken sample.

A more accurate but also more complicated method is described by Schindler et al. [17]. Here the shrinkage was measured online during the evaporation process.

3. Results and Discussion

The water retention curves of all tested natural mineral and organic soils and three example horticultural substrates are illustrated in Figure 2. The easily plant-available water (EAW) in the tension range between 20 and 100 hPa and the air capacity (Air) are marked in the figure. All other required water and air capacities (for different kinds of cultivation) could be calculated based on these functions. As expected, due to the low dry bulk density, the saturated water content of the horticultural substrates was very high (Table 2). It varied between 71.8% (HS1) and 87.1% (HS6) by vol., averaged 82.4% by vol., and was comparable to the fen peat (Hn, 84.0% by vol.). However, the EAW and the air capacity demonstrated the special performance of horticultural substrates for horticultural applications. The tested horticultural substrates provided between 26.5 and 44.2 mm of water per 10 cm substrate depth (an average 30.8 mm). The natural mineral and organic soils could only store between 5.7 and 20.7 mm. However, what was more dramatic was the air capacity. The horticultural substrates provided an average 10.4 vol. % air with a major variation between the single samples (HS13: 2.7 vol. % and the peat-free HS6: 31.7% by vol.). Only about half of the tested substrates provided sufficient air (threshold value 10 vol. %, according to Raviv and Lieth [1]). The natural soils were even worse (0.6%–4.6% by vol., an average 2.0% by vol.) and were far from achieving the threshold value of 10%. The suitability for storing and transporting water and solutes is strongly influenced by shrinkage and rewetting properties [1,2]. The rewetting properties of most substrates were sufficient, with the exception of HS4, HS12 and HS14 which exceeded the threshold value for the WDPT of 5 s [19]. The shrinkage also showed great variability, ranging between 0.4 and 9.1 vol. % within the substrates. Only about half of the samples achieved or exceeded the threshold value for the capillary rise of 30 cm height for a 5 mm·day^{-1} rate. The same situation was observed for the mineral sandy (Ls3) and clayey loam (Lts) substrates.

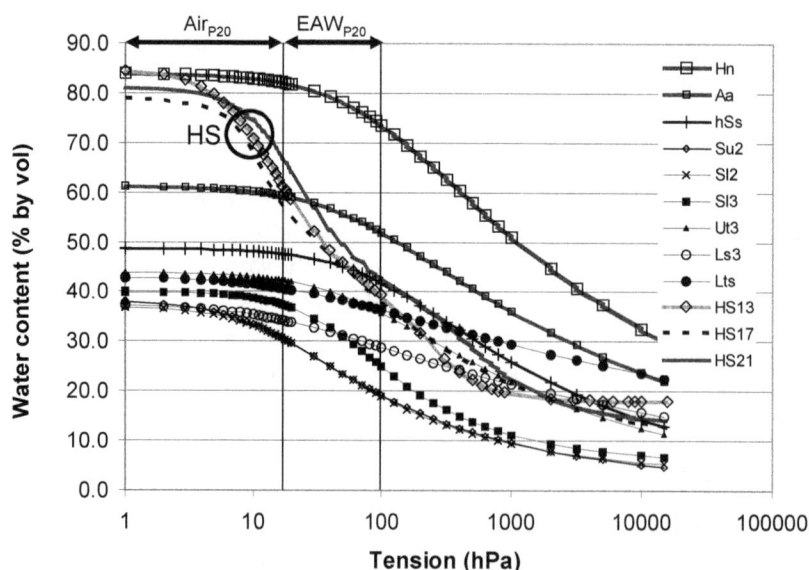

Figure 2. Water retention curves of the natural mineral and organic soils and examples of three horticultural substrates (HS), Air$_{P20}$—average air volume in 20-cm-high containers, EAW$_{P20}$—easily plant-available water in 20-cm-high containers.

Table 2. Hydraulic properties of the horticultural substrates.

No.	Θ_s	FC	Air_{P20}	EAW_{P20}	S	DBD	CR_5	WDPT
	Vol. %					$g \cdot cm^{-3}$	cm	Sec.
Horticultural substrates								
1	71.8	38.7	9.3	28.2	2.1	0.43	10.1	0.1
2	86.0	48.0	6.5	36.6	6.2	0.17	45.7	1
3	79.6	46.4	9.4	27.9	5.4	0.30	54.7	0.1
4	79.0	46.2	11.8	26.7	3.3	0.26	26.7	15
5	86.2	45.1	11.7	34.4	2.1	0.20	42.9	2
6	87.1	31.6	31.7	26.7	0.4	0.18	17.9	0.1
7	84.2	53.8	4.5	31.1	9.1	0.21	24.4	1
8	81.2	50.6	5.7	29.1	6.6	0.25	45.7	0.1
9	80.7	38.8	13.9	33.5	0.8	0.22	29.3	0.1
10	84.4	44.1	13.6	31.4	6.2	0.18	29.3	0.1
11	83.1	54.7	7.2	26.5	6.2	0.31	13.1	2
12	75.8	40.8	6.8	32.8	1.0	0.30	26.7	6
13	84.5	55.1	2.7	32.5	0.6	0.21	47.7	1
14	78.8	43.2	11.6	29.2	0.8	0.28	15.9	6
15	83.4	52.0	6.9	29.6	7.0	0.19	36.4	2
16	81.1	39.8	13.7	31.5	3.3	0.19	12.7	1
17	80.8	48.8	13.8	32.1	6.2	0.23	29.3	0.1
18	81.0	47.3	7.6	30.6	7.0	0.26	79.9	2
Natural organic and mineral soils								
19	84.0	77.2	1.1	9.4	7.2	0.43	80	6
20	61.3	54.9	1.1	8.2	4.1	1.05	56	9
21	48.7	44.3	0.6	20.7	2.1	1.35	51	3
22	38.2	22.2	4.6	14.5	nm	1.63	32	nm
23	36.9	22.2	3.6	14.4	nm	1.66	46	nm
24	40.0	29.4	1.4	13.8	nm	1.58	123	nm
25	34.5	25.6	3.0	7.6	nm	1.72	30	nm
26	43.9	38.4	0.9	7.3	nm	1.48	61	nm
27	37.5	30.5	2.0	6.7	nm	1.64	4	nm
28	43.0	37.8	1.4	5.4	1.9	1.50	17	nm

Θ_s—Water content at saturation; FC—Field capacity at pF 1.8; Air_{P20}—air capacity in a 20-cm-high container; EAW_{P20}—easily plant-available water in 20-cm-high containers; S—shrinkage at a tension of 100 hPa; DBD—Dry bulk density; CR_5—steady-state capillary height for a 5 mm·day^{-1} rate; WDPT—water drop penetration time in seconds at 100 hPa; nm—not measured.

The results for the specially composed horticultural substrates in this study showed their superiority for horticultural applications. However, there were differences in their hydraulic suitability. Generally, the water demand in shallow containers was sufficiently covered by most samples. The most sensitive element, however, was the air supply, especially for cultivation in shallow containers. High water penetration times and substrate shrinkage is of key relevance for sustainable, resource-saving water and nutrient management.

4. Conclusions

The hydraulic evaluation of horticultural substrates is an important part of an overall assessment of their suitability for horticultural applications. The applied hydraulic measurement techniques and methods (EEM, HYPROP, WDPT) proved to be suitable for characterizing the hydraulic properties of horticultural substrates. Lack of air was the main critical factor. The results with the specially composed horticultural substrates in this study showed their superiority for horticultural applications. The development of a rating system to evaluate the hydraulic quality of horticultural substrates should be the topic of further studies.

Author Contributions: Uwe Schindler made the HYPROP measurements and evaluated the results. Lothar Müller made result evaluation. Frank Eulenstein was responsible for selecting the samples and was involved in the discussion and evaluation of the results.

Conflicts of Interest: The authors declare no conflict of interest.

References

1.   Raviv, M.; Lieth, J.H. *Soilless Culture*; Elsevier Publications: London, UK, 2008; p. 608.

2.   Caron, J.; Pepin, S.; Periard, Y. Physics of growing media in green future. International Symposium on Growing Media and Soilless Cultivation. *Acta Hortic.* **2014**, *1034*, 309–317. [CrossRef]

3.   Karlovich, P.T.; Fonteno, W.C. Effect of soil moisture tension and soil water content on the growth of chrysanthemum in 3 container media. *J. Am. Soc. Hortic. Sci.* **1986**, *111*, 191–195.

4.   Milks, R.R.; Fonteno, W.C.; Larson, R.A. Hydrology of horticultural substrates: I. Mathematical models for moisture characteristic curves of horticultural container media. *J. Am. Soc. Hortic. Sci.* **1989**, *114*, 48–52.

5.   Milks, R.R.; Fonteno, W.C.; Larson, R.A. Hydrology of horticultural substrates: II. Predicting physical properties of media in containers. *J. Am. Soc. Hortic. Sci.* **1989**, *114*, 53–56.

6.   Milks, R.R.; Fonteno, W.C.; Larson, R.A. Hydrology of horticultural substrates: III. Predicting air and water content in limited-volume plug cells. *J. Am. Soc. Hortic. Sci.* **1989**, *114*, 57–61.

7.   Ritsema, C.J.; Dekker, L.W. Water repellency and its role in forming preferred flow paths in soils. *Aust. J. Soil Res.* **1996**, *34*, 475–487. [CrossRef]

8.   Al Naddafa, O.; Livieratos, I.; Stamatakisa, A.; Tsirogiannisb, I.; Gizasb, G.; Savvasc, D. Hydraulic characteristics of composted pig manure, perlite, and mixtures of them, and their impact on cucumber grown on bags. *Sci. Hortic.* **2011**, *129*, 135–141. [CrossRef]

9.   DIN EN 13041. *Bodenverbesserungsmittel und Kultursubstrate—Bestimmung der Physikalischen Eigenschaften— Rohdichte (Trocken), Luftkapazität*; Wasserkapazität, Schrumpfungswert, und Gesamtporenvolumen, Beuth Verlag GmbH: Berlin, Germany, 2012.

10. Schindler, U.; Durner, W.; von Unold, G.; Mueller, L.; Wieland, R. The evaporation method—Extending the measurement range of soil hydraulic properties using the air-entry pressure of the ceramic cup. *J. Plant Nutr. Soil Sci.* **2010**, *173*, 563–572. [CrossRef]

11. Heiskanen, J. Physical properties of two-component growth media based on Sphagnum peat and their implications for plant-available water and aeration. *Plant Soil* **1995**, *172*, 45–54. [CrossRef]

12. Bibbiani, C.; Campiotti, C.A.; Incrocci, L. Estimation of hydraulic properties of growing media with a one-step outflow technique. International Symposium on Growing Media and Soilless Cultivation. *Acta Hortic.* **2014**, *1034*, 319–325. [CrossRef]

13. UMS GmbH Munich, HYPROP©—Laboratory Evaporation Method for the Determination of pF-Curves and Unsaturated Conductivity. Available online: http://www.ums-muc.de/en/products/soil_laboratory.html (accessed on 29 August 2011).

14. Von Post, L. Sveriges Geologiska Undersöknings torvinventering och nogra av des hittils vunna resultat (SGU peat inventory and some preliminary results). *Svenska Mosskulturforeningens Tidskrift Jonköping* **1922**, *36*, 1–37.

15. Boden, A.G. *Bodenkundliche Kartieranleitung. 5. Aufl. (KA 5); 438 S.*; Bundesanstalt für Geowissenschaften und Rohstoffe: Hannover, Germany, 2005.

16. Verdonck, O.; Gabriels, R.I. Reference method for the determination of physical properties of plant substrates. II. Reference method for the determination of chemical properties of plant substrates. *Acta Hortic.* **1992**, *302*, 169–179. [CrossRef]

17. Schindler, U.; Doerner, J.; Müller, L. Simplified method for quantifying the hydraulic properties of shrinking soils. *J. Plant Nutr. Soil Sci.* **2015**, *178*, 136–145. [CrossRef]

18. Letey, J. Measurement of contact angle, water drop penetration time, and critical surface tension. In *Water-Repellent Soils*; DeBano, L.F., Letey, J., Eds.; University of California: Berkeley, CA, USA, 1968; pp. 43–47.

19. Blanco-Canqui, H.; Lal, R. Extent of soil water repellency under long-term no-till soils. *Geoderma* **2009**, *149*, 171–180. [CrossRef]

6

The Impact of Production Technology on Plant Phenolics

Robert Veberic

Department of Agronomy, Biotechnical Faculty, University of Ljubljana, Jamnikarjeva 101, Ljubljana 1000, Slovenia; robert.veberic@bf.uni-lj.si

Abstract: Due to rising public pressure in recent decades, alternatives for large-scale and industrial farming are being sought. Environmental and sustainability issues and the rising awareness of the link between the overuse of pesticides/fertilizers and negative health effects have been key factors for creating the integrated production approach, which encompasses environmentally friendly technologies. Moreover, the demand for organically grown products is constantly growing. The organic production model is a step towards further restriction of synthetic chemical use in plant production. Limited use of pesticides may boost the plant's investment into its own defense systems, which may result in a higher content of secondary compounds. Synthesis of secondary metabolites is a common plant response to any form of stress (biotic or abiotic), and their function is to help the plant overcome unfavorable conditions. Many compounds, especially phenolics, are also considered beneficial for human health; therefore, numerous studies comparing different production systems have been conducted in the past 20 years. Generally, organically produced food may contain greater amounts of health beneficial compounds and diminished levels of pesticide residues and nitrates. However, the results are not always clear, as other factors may influence the composition of natural products (e.g., environmental and varietal factors, sampling, and the design of experiments). Therefore, controlled field trials, in which most of the factors can be either controlled or at least recorded, should be encouraged. The present paper synthesizes the function of phenolics as a response to different forms of stress, which can occur during plant growth, with a special emphasis on different production systems. Examples of diverse horticultural crops are presented.

Keywords: organic; biodynamic; integrated; conventional; phenolics

1. Introduction

In 2013, the UN forecast a world population of 9.6 billion people by 2050, presenting a major challenge for agriculture. Food production needs to increase in order to feed and sustain approximately two billion more people than there are today. At the same time, global awareness is rising, and consumers demand safe, high-quality food, produced according to environmental and animal friendly standards that take into account social and economic aspects [1]. This is a difficult task for any production technology whether based on conventional, integrated, or organic approaches and principles. High yields of excellent quality and the minimization of negative impacts on the environment seldom go hand in hand. Conventional agriculture strongly relies on the use of synthetic pesticides and mineral fertilizers as well as other off-farm inputs to control plant growth. Meanwhile, organic farming emphasizes sustainable technologies, replacing the synthetic pesticides and fertilizers with those of natural origin such as plant extracts, organic manures, and natural minerals. It also uses crop rotation, advanced orchard designs, and resistant cultivars [2]. Integrated production is a compromise between both production systems. It still relies on the use of synthetic pesticides and fertilizers; however, the application is restricted and the self-defense of plants is encouraged.

Food produced according to conventional or integrated practices can contain residuals of chemicals used for plant protection. However, their levels should not exceed maximum permitted limits. In a review by Smith-Spangler et al. [3], it was reported that organic fruit, vegetables, and grains have about a 30% lower risk of containing pesticide residues compared with conventional produce. Fourfold higher pesticide residues in conventional crops have also been reported by others [4]. Organic production is generally more energy-efficient, as energy use is much more effective compared with conventional farming [5]. However, the impact is not as significant in fruit production.

Increased organic production of various horticultural crops (especially fruits and vegetables) has been recorded over the last 20 years. Moreover, these production technologies have gained a great deal of scientific attention. The driving force behind such rapid increase in importance is the growing consumer demand for organic horticultural produce [2]. Consumers often believe that organic food is healthier and more nutritious compared with the conventional one and, for that reason, they are willing to pay more for the former [6]. However, are the differences really that evident? In the past several years, numerous studies have been undertaken to answer this question. Some were able to prove increased levels of health-beneficial compounds in organically grown natural products. In contrast, others have not been able to detect any differences between the two production systems, or have even attained less favorable results for organic crops. Different reviews consisting of diverse analytical approaches have been prepared. Even so, the results have not always been consistent. For example, Dangour et al. [6] based their review on 55 studies and summarized that organic versus conventional food does not conclusively show any differences in nutrient content except for nitrogen, which was higher in conventional food, while phosphorus and titratable acidity were higher in organic food. Smith-Spangler et al. [3] also reported higher phosphorus content in organic produce. Additionally (and beneficially), organic food is also characterized by higher phenolic levels. An increased amount of phytochemicals (particularly phenolics) in organically grown produce has similarly been reported by others [2,4].

2. Plant Secondary Metabolites with Special Emphasis on Phenolics

Plants synthesize an enormous variety of organic compounds that serve the plant in chemically mediated interactions in their biotic and abiotic environment [7]. Those compounds are often characterized as secondary metabolites to be distinguished from primary ones, which are involved in basic cell metabolic processes of the plant. Secondary metabolites may have beside other functions also an important protective role against different forms of stress. They are synthesized as a normal part of plant metabolism, often produced in specialized cells or tissues, and tend to be more complex than primary compounds [8]. Plants produce secondary metabolites as a response to non-favorable environmental conditions or in particular developmental stages. Their biosynthesis is initiated from a particular primary metabolite or from intermediates of primary metabolism. Many accumulate in surprisingly high levels in some species. So far, many thousands of molecules have been identified underlining the diversity and complexity of plant secondary metabolism. These compounds can be characteristic for a certain plant family, genus, or even species and therefore may be used as taxonomic tools in classifying plants [9].

Although ignored until recently, their function in plants is now attracting attention as some appear to have a key role in mutualistic interactions of plants and other organisms such as animals (pollinators, seed dispersals), fungi (mycorrhiza), and bacteria (formation of nitrogen-fixing root nodules in legumes). They are important in the protection of the plant against various abiotic stress forms (e.g., excessive UV radiation, drought, wounding) as well as in chemical plant defense against herbivores and pathogens [7]. They may also play an important role in plant–plant interactions, in which they function as allelochemicals with either positive or negative effects on the neighboring plants.

Secondary metabolites also represent an economic interest as they can be used for dyes, fibers, glues, oils, waxes, flavoring agents, drugs, and perfumes, and they are viewed as potential sources of new natural drugs, antibiotics, insecticides, and herbicides [10]. The role of selected secondary metabolites as protective dietary constituents has become an increasingly important area of human nutrition research in recent years. These compounds are not essential for short-term well-being, but there is increasing evidence that modest long-term intake can have favorable impacts on the development (via suppression) of many chronic diseases [11].

Based on their biosynthetic origins, plant secondary metabolites can be divided into three major groups: (a) phenolics; (b) terpenoids; and (c) nitrogen-containing compounds.

Plant phenolics are one of the most important groups of secondary metabolites found throughout the plant kingdom with more than 10,000 molecules reported. Phenolics are synthesized from aromatic amino acids, predominantly from phenylalanine, and therefore are characterized as having at least one aromatic ring with one or more hydroxyl groups attached [12]. Phenolics range from simple, low molecular-weight compounds with a single aromatic ring, such as phenolic acids, to large and complex molecules such as tannins. The most prominent group of phenolics is the flavonoids with two aromatic rings. Sugars, organic acids, and other compounds are often bound to the rings, and the compounds are water-soluble. However, some flavonoids form large water-insoluble polymers. They are mostly stored in vacuoles of the epidermal cell of plant organs or specialized cells fulfilling their function as secondary metabolites. Another important group of phenolics, chemically related to flavonoids, are stilbenes with resveratrol as the most compound in this group [13].

Similar to other secondary metabolites, phenolics have been linked to numerous ecological functions. They often act as a part of plant responses to biotic and abiotic stimuli. They are not only indicators of plant stress responses to various abiotic factors such as intensive light or mineral deficiencies, but are also one of the most important compounds in plant resistance and tolerance against pests and diseases. They represent a biochemical basis for successful reproduction of the plants, which synthesize them to attract pollinators or seed-dispersing animals. Phenolic-based polymers, such as lignin, suberin, or tannins, contribute substantially to the stability and robustness of plant tissues, towards mechanical or environmental damage, such as drought or wounding [12].

Phenolics can act as allelopathic compounds. A good example is juglone, a chemical released from the decomposing plant material of *Juglans nigra* (black walnut). Plants which synthesize and release allelochemicals mostly influence the performance of neighboring plants in a negative way, gaining advantage in nutrient, water, or light acquisition. Targeted positioning of allelopathic plants could be utilized in crop rotations with the aim of reducing pesticide use and enhancing crop yields. However, additional research on allelopathic compounds and their dynamics is needed to efficiently implement these measures [1].

3. Factors Influencing the Level of Phenolics

In principal, phenolic content differs between species of the same family, even between cultivars. In the study of Veberic et al. [14], 22 apple cultivars were analyzed and it was clearly demonstrated that phenolic content was far more genotype-dependent than influenced either by organic or integrated production. Moreover, diverse plant organs or even tissues can be characterized by different phenolic composition. Normally, phenolics are concentrated in the epidermal tissue where they fulfill their biological function.

Ripening is an important factor affecting the phenolic content of horticultural crops, especially fruit. In red-colored fruit, such as red apple cultivars, their anthocyanin content increases with maturity. At the same time, the level of other phenolic groups, especially those responsible for the astringent taste, decreases [15]. In some cases, phenolic content is higher in unripe produce, but these crops are seldom edible. Still, they can be considered as potential sources for phenolic extraction [11].

Any form of important deviation from the environment is considered stress for the plant and provokes responses in its phenolic metabolism. One of the major triggers upregulating phenolic content is the exposure of plant tissue to intensive light, especially UV-light. In nature, this is often accompanied by high temperatures and results in photo-oxidative damage to the photosystem complex. The recognition of the stress event at the cellular level triggers repair processes in the cells as well as de novo synthesis of protective substances. Many of these protective compounds belong to different groups of phenolics. The effect of excessive sun irradiation on phenolic accumulation has been confirmed in a study by Zupan et al. [16], in which light-exposed apples contained increased levels of total and several individual phenolics compared with the shaded part of the same fruit or fruit growing in full shade. The highest levels of phenolics have been measured in the part of the fruit damaged by sunburn. Jakopic et al. [17] demonstrated that anthocyanin content in apple skin was strongly light-dependent and that the use of reflective cover increased their synthesis. On the other hand, the use of hail nets can reduce their phenolic content as shown for the French bean (*Phaseolus vulgaris*) [18]. Both practices, reflective cover and various colors of anti-hail nets, are widely applied in different fields of horticulture production. Greenhouse covering can additionally affect the content of phenolics in vegetables, especially if it modifies UV radiation.

Stress factors can be prevented or, on the other hand, caused by different cultivation techniques. In this aspect, fertilization is especially important. It is well-known that high nitrogen doses improve plant growth but to some extent reduce the amount of phenolics, often making the plants more susceptible to pests and diseases. On the other hand, nutrient deficiency can cause accumulation of certain phenolics. One of the reasons could be that growth is inhibited with a deficiency and the carbon is allocated to secondary metabolites [10]. This was especially shown for phosphorus [12] where deficiency often results in increased accumulation of anthocyanins. However, foliar fertilization with a phosphorus–calcium mix significantly improved red color formation in apples [19]. Salt stress due to a high concentration of certain ions can increase the phenolic content as shown in a number of plant species [10]. An optimized water supply in cultivation technology is needed to guarantee normal growth and yield as well as phenolic composition of crops, where a mild water stress could be beneficial in this aspect [12].

4. Phenolics and Plant Disease

Plants are constantly in contact with other organisms including those that are pathogenic which try to get nutrients from plants. Plants have developed defense mechanisms to protect themselves, and pathogens have established countermeasures to overcome these mechanisms. Plants invest in physical (e.g., cell wall, cuticle, trichomes) and chemical barriers (different secondary compounds), which should protect them from invading organisms. These barriers can be either pre-formed (stored phytochemicals are called phytoanticipins) or induced with pathogen attack (de novo synthesized phytoalexins) [9].

Each species, or closely related species, can form characteristic molecules that act as defense compounds against specific or nonspecific pathogens. Most antimicrobial natural products have relatively broad spectrum activity. Specificity is often determined by whether or not a pathogen has the tools to enzymatically detoxify a particular host product.

The first line of the defense response to a pathogen attack is the recognition of the pathogen molecules. These elicitors (peptides, oligosaccharides, or glycoconjugates) originate from the pathogen or are plant-derived molecules released at wounding. The host recognition of pathogen elicitors initiates early cellular responses, such as alterations in plant cell walls, changes in ion fluxes, the accumulation of reactive oxygen species and reactive nitrogen species, and gene transcription [20]. The second line of defense responses leads to the induction of plant defense genes and biosynthesis of endogenous secondary metabolites, cell wall fortification, and, very often, hypersensitive responses and the activation of systemic acquired resistance [20,21].

As shown for apple scab fungus (*Venturia inaequalis*), susceptible and resistant cultivars contain similar amounts of pre-formed phenolic compounds [15]. After infection, susceptible apple cultivars start to accumulate phenolic compounds and the corresponding enzymes are upregulated, especially around the symptomatic spot. However, they are unable to prevent the spread of the fungus [22,23]. The authors conclude that phenolics are not the only resistance mechanism present in apple plants. Moreover, not just the quantity but also the speed of the plant response is crucial for its increased resistance. When the induced responses are triggered rapidly during a plant–pathogen interaction, the plant is resistant to the disease. Contrarily, post-infection defense mechanisms are established at a slower rate in susceptible plants.

5. Phenolics and Type of Production

In the organic cultivation of fruit crops, no synthetic chemicals such as fertilizers, pesticides, and fruit thinning agents are used. Only natural compounds are allowed, and their use is closely monitored. In theory, differences between organic and conventional cultivation may lead to enhanced content levels of phenolics in organically produced fruits. This was clearly demonstrated in apple leaves and fruits of four cultivars in a two-year trial, as the research reported a somewhat higher content of individual and total phenolics in leaves and fruit of organically grown apples compared with apples from integrated cultivation [24]. The authors concluded that this difference lay in the fact that trees exposed to various stress factors synthesized more protective compounds (i.e., phenolics). The lack of pesticides in organic production allowed the incidence of injuries and pathogen infections due to higher disease pressure and potentially elevated the phenolic content in plant tissue. Similarly, fully ripe biodynamic and organic mangoes, compared with conventional grown ones, were characterized by a significantly higher amount of flavonoids [25]. Lima et al. [26] compared different plant parts from organic and conventional production. They confirmed the hypothesis of higher total phenolic and flavonoid content in organically produced plants, but not in all cases.

Many studies have demonstrated the impact of fertilization on the phenolic content. In organic farming, nutrients are often supplied through compost, manure, and plant-derived byproducts. Organic nitrogen is transformed into the inorganic form by soil microflora, so the level of nutrient availability to plants may be difficult to control. Toor et al. [27] reported that tomatoes fertilized with chicken manure or grass-clover had lower shoot biomass compared with those fertilized by mineral nutrient solution, despite the fact that NPK macronutrient input per plant was higher in the organic fertilization treatments. However, the organic fertilization treatments led to a 17.6% higher soluble phenolic content. Toor et al. [27] summarized that the carbon surplus, which was not utilized for growth due to the slow release of nutrients from the organic manure, was allocated to the production of secondary metabolites, in particular phenolics. Bavec et al. [28] similarly demonstrated that biodynamic red beet (*Beta vulgaris* L. ssp. *vulgaris*) accumulated higher levels of total phenolics than did red beets from a conventional farming system. The authors also list, among other factors, lower availability of nutrients in the biodynamic farming system.

Nitrogen use efficiency and availability to plants is better in conventional production as inorganic nitrogen is applied. Roussos and Gasparatos [29] observed higher nitrogen content in conventionally grown apple fruit compared to organic apples. Increased width and higher fresh weight of the conventional fruit have been recorded, but there were no differences in major metabolites compared with organic apples. Additionally, organic apples contained more calcium, which influences the storability of the fruit. However, a higher copper content has been measured in conventionally grown apples due to the use of fungicides. Similarly, Jakopic et al. [30] reported that organically produced apples had 14% less weight than integrated ones. Integrated apples had higher amounts of sugars and higher organic acid levels, but decreased amounts of phenolics. In a trial on French beans, a yield reduction was noted in organic production compared with the integrated and conventional systems [31], although higher nitrogen input in the form of cattle manure was characteristic of the organic production. The sum of phenolics in the pods was not significantly different among the production systems, with a tendency for a higher amount in the integrated system. Meanwhile, ascorbic acid content was the lowest in the conventional system, and sugar levels were the highest in the organic system.

Orchard soil management is an important issue because of the reduction of weed occurrence, reduction of soil erosion, and optimal water and nutrient management. In conventional fruit production, herbicides are used under the trees to avoid any competition for nutrients. Living mulch (green cover plants) is particularly important due to its benefits for orchard biodiversity and soil quality, which is crucial in organic production. In addition to generally reported benefits of their utilization in the orchard, it has been established that different types of living mulch positively influence the content of phenolics in the fruit [32].

It would be interesting to determine whether the storage life of fruits can be influenced by different production types. Production technology had little effect on the storage of cashew nuts, as it did not influence the content of phenolic for up to 180 days of storage. The soluble solids were higher in conventional farming [33]. Various quality parameters have been monitored over six months of apple storage, and it was demonstrated that changes mainly occurred due to different storage conditions rather than possible effects of organic versus integrated production systems [34].

6. Conclusions

Despite the fact that the literature lacks unequivocal evidence of a higher content of primary and secondary metabolites in organically produced food, there are some indications that higher biotic and abiotic stresses can influence plant secondary metabolism, especially the synthesis and accumulation of phenolics. There are still many areas, such as the use of allelopathic plants, natural pesticides, and stimulation of plant immune system, that can improve the performance and yield of organically grown plants. Conventional production could include more organic approaches without the fear of yield reduction or loss of quality as has been done in integrated production systems.

Acknowledgments: This work is part of the program Horticulture No. P4-0013-0481, funded by the Slovenian Research Agency.

Conflicts of Interest: The author declares no conflict of interest.

References

1.   Wezel, A.; Casagrande, M.; Celette, F.; Vian, J.F.; Ferrer, A.; Peigne, J. Agroecological practices for sustainable agriculture. A review. *Agron. Sustain. Dev.* **2014**, *34*, 1–20. [CrossRef]

2.   Zhao, X.; Carey, E.E.; Wang, W.Q.; Rajashekar, C.B. Does organic production enhance phytochemical content of fruit and vegetables? Current knowledge and prospects for research. *Horttechnology* **2006**, *16*, 449–456.

3.   Smith-Spangler, C.; Brandeau, M.L.; Hunter, G.E.; Bavinger, C.; Pearson, M.; Eschbach, P.J.; Sundaram, V.; Liu, H.; Schirmer, P.; Stave, C. Are organic foods safer or healthier than conventional alternatives? *Ann. Intern. Med.* **2012**, *157*, 348–366. [CrossRef] [PubMed]

4.   Baranski, M.; Srednicka-Tober, D.; Volakakis, N.; Seal, C.; Sanderson, R.; Stewart, G.B.; Benbrook, C.; Biavati, B.; Markellou, E.; Giotis, C. Higher antioxidant and lower cadmium concentrations and lower incidence of pesticide residues in organically grown crops: A systematic literature review and meta-analyses. *Br. J. Nutr.* **2014**, *112*, 794–811. [CrossRef] [PubMed]

5.   Lynch, D.H.; MacRae, R.; Martin, R.C. The carbon and global warming potential impacts of organic farming: Does it have a significant role in an energy constrained world? *Sustainability* **2011**, *3*, 322–362. [CrossRef]

6.   Dangour, A.D.; Dodhia, S.K.; Hayter, A.; Allen, E.; Lock, K.; Uauy, R. Nutritional quality of organic foods: A systematic review. *Am. J. Clin. Nutr.* **2009**, *90*, 680–685. [CrossRef] [PubMed]

7.   Hartmann, T. The lost origin of chemical ecology in the late 19th century. *Proc. Natl. Acad. Sci. USA* **2008**, *105*, 4541–4546. [CrossRef] [PubMed]

8.   Bednarek, P.; Osbourn, A. Plant-microbe interactions: Chemical diversity in plant defense. *Science* **2009**, *324*, 746–748. [CrossRef] [PubMed]

9.   Piasecka, A.; Jedrzejczak-Rey, N.; Bednarek, P. Secondary metabolites in plant innate immunity: Conserved function of divergent chemicals. *New Phytol.* **2015**, *206*, 948–964. [CrossRef] [PubMed]

10.   Akula, R.; Ravishankar, G.A. Influence of abiotic stress signals on secondary metabolites in plants. *Plant Signal. Behav.* **2011**, *6*, 1720–1731. [CrossRef] [PubMed]

11.   Kalt, W. Effects of production and processing factors on major fruit and vegetable antioxidants. *J. Food Sci.* **2005**, *70*, R11–R19. [CrossRef]

12. Treutter, D. Managing phenol contents in crop plants by phytochemical farming and breeding-visions and constraints. *Int. J. Mol. Sci.* **2010**, *11*, 807–857. [CrossRef] [PubMed]

13. Bavaresco, L.; Mattivi, F.; De Rosso, M.; Flamini, R. Effects of elicitors, viticultural factors, and enological practices on resveratrol and stilbenes in grapevine and wine. *Mini Rev. Med. Chem.* **2012**, *12*, 1366–1381. [CrossRef] [PubMed]

14. Veberic, R.; Trobec, M.; Herbinger, K.; Hofer, M.; Grill, D.; Stampar, F. Phenolic compounds in some apple (*Malus domestica* Borkh.) cultivars of organic and integrated production. *J. Sci. Food Agric.* **2005**, *85*, 1687–1694. [CrossRef]

15. Slatnar, A.; Mikulic-Petkovsek, M.; Halbwirth, H.; Stampar, F.; Stich, K.; Veberic, R. Polyphenol metabolism of developing apple skin of a scab resistant and a susceptible apple cultivar. *Trees Struct. Funct.* **2012**, *26*, 109–119. [CrossRef]

16. Zupan, A.; Mikulic-Petkovsek, M.; Slatnar, A.; Stampar, F.; Veberic, R. Individual phenolic response and peroxidase activity in peel of differently sun-exposed apples in the period favorable for sunburn occurrence. *J. Plant Physiol.* **2014**, *171*, 1706–1712. [CrossRef] [PubMed]

17. Jakopic, J.; Stampar, F.; Veberic, R. Influence of hail net and reflective foil on cyanidin glycosides and quercetin glycosides in "Fuji" apple skin. *Hortscience* **2010**, *45*, 1447–1452.

18. Selan, M.; Kastelec, D.; Jakopič, J.; Veberič, R.; Mikulič-Petkovšek, M.; Kacjan-Maršić, N. Hail net cover, cultivar and pod size influence the chemical composition of dwarf French bean. *Sci. Hortic.* **2014**, *175*, 95–104. [CrossRef]

19. Bizjak, J.; Weber, N.; Mikulic-Petkovsek, M.; Slatnar, A.; Stampar, F.; Alam, Z.; Stich, K.; Halbwirth, H.; Veberic, R. Influence of phostrade Ca on color development and anthocyanin content of "Braeburn" apple (*Malus domestica* Borkh.). *Hortscience* **2013**, *48*, 193–199.

20. Newman, M.-A.; Sundelin, T.; Nielsen, J.T.; Erbs, G. MAMP (microbe-associated molecular pattern) triggered immunity in plants. *Front. Plant Sci.* **2013**, *4*, 139. [CrossRef] [PubMed]

21. Amil-Ruiz, F.; Blanco-Portales, R.; Muñoz-Blanco, J.; Caballero, J.L. The strawberry plant defense mechanism: A molecular review. *Plant Cell Physiol.* **2011**, *52*, 1873–1903. [CrossRef] [PubMed]

22. Mikulic-Petkovsek, M.; Stampar, F.; Veberic, R. Accumulation of phenolic compounds in apple in response to infection by the scab pathogen, Venturia inaequalis. *Physiol. Mol. Plant Pathol.* **2009**, *74*, 60–67. [CrossRef]

23. Slatnar, A.; Mikulic-Petkovsek, M.; Halbwirth, H.; Stampar, F.; Stich, K.; Veberic, R. Enzyme activity of the phenylpropanoid pathway as a response to apple scab infection. *Ann. Appl. Biol.* **2010**, *156*, 449–456. [CrossRef]

24. Mikulic Petkovsek, M.; Slatnar, A.; Stampar, F.; Veberic, R. The influence of organic/integrated production on the content of phenolic compounds in apple leaves and fruits in four different varieties over a 2-year period. *J. Sci. Food Agric.* **2010**, *90*, 2366–2378. [CrossRef] [PubMed]

25. Maciel, L.F.; Oliveira, C.D.S.; Bispo, E.d.S.; Spinola Miranda, M.D.P. Antioxidant activity, total phenolic compounds and flavonoids of mangoes coming from biodynamic, organic and conventional cultivations in three maturation stages. *Br. Food J.* **2011**, *113*, 1103–1113. [CrossRef]

26. Lima, G.P.P.; da Rocha, S.A.; Takaki, M.; Ramos, P.R.R.; Ono, E.O. Comparison of polyamine, phenol and flavonoid contents in plants grown under conventional and organic methods. *Int. J. Food Sci. Technol.* **2008**, *43*, 1838–1843. [CrossRef]

27. Toor, R.K.; Savage, G.P.; Heeb, A. Influence of different types of fertilisers on the major antioxidant components of tomatoes. *J. Food Compos. Anal.* **2006**, *19*, 20–27. [CrossRef]

28. Bavec, M.; Turinek, M.; Grobelnik-Mlakar, S.; Slatnar, A.; Bavec, F. Influence of industrial and alternative farming systems on contents of sugars, organic acids, total phenolic content, and the antioxidant activity of red beet (*Beta vulgaris* L. ssp. *vulgaris Rote Kugel*). *J. Agric. Food Chem.* **2010**, *58*, 11825–11831. [CrossRef] [PubMed]

29. Roussos, P.A.; Gasparatos, D. Apple tree growth and overall fruit quality under organic and conventional orchard management. *Sci. Hortic.* **2009**, *123*, 247–252. [CrossRef]

30. Jakopic, J.; Slatnar, A.; Stampar, F.; Veberic, R.; Simoncic, A. Analysis of selected primary metabolites and phenolic profile of "Golden Delicious" apples from four production systems. *Fruits* **2012**, *67*, 377–386. [CrossRef]

31. Jakopic, J.; Slatnar, A.; Mikulic-Petkovsek, M.; Veberic, R.; Stampar, F.; Bavec, F.; Bavec, M. Effect of different production systems on chemical profiles of dwarf French bean (*Phaseolus vulgaris* L. cv. Top Crop) pods. *J. Agric. Food Chem.* **2013**, *61*, 2392–2399. [CrossRef] [PubMed]

32. Slatnar, A.; Licznar-Malanczuk, M.; Mikulic-Petkovsek, M.; Stampar, F.; Veberic, R. Long-term experiment with orchard floor management systems: Influence on apple yield and chemical composition. *J. Agric. Food Chem.* **2014**, *62*, 4095–4103. [CrossRef] [PubMed]

33. Soares, D.J.; Cavalcante, C.E.B.; Cardoso, T.G.; de Figueiredo, E.A.T.; Maia, G.A.; de Sousa, P.H.M.; de Figueiredo, R.W. Study of the stability of cashew nuts obtained from conventional and organic cultivation. *Semin. Cienc. Agrar.* **2012**, *33*, 1855–1867. [CrossRef]

34. Roth, E.; Berna, A.; Beullens, K.; Yarramraju, S.; Lammertyn, J.; Schenk, A.; Nicolai, B. Postharvest quality of integrated and organically produced apple fruit. *Postharvest Biol. Technol.* **2007**, *45*, 11–19. [CrossRef]

7

Use of Nondestructive Devices to Support Pre- and Postharvest Fruit Management

Guglielmo Costa *, Lorenzo Rocchi, Brian Farneti, Nicola Busatto, Francesco Spinelli and Serena Vidoni

Department of Agricultural Science, Alma Mater Studiorum, University of Bologna, Bologna 40127, Italy; lorenz.rocchi@gmail.com (L.R.); brian.farneti@gmail.com (B.F.); nicola.busatto2@unibo.it (N.B.); francesco.spinelli2@unibo.it (F.S.); serena.vidoni@unibo.it (S.V.)
* Correspondence: guglielmo.costa@unibo.it

Abstract: Fruit quality is greatly affected by the ripening stage at harvest. In order to preserve quality traits, increase product marketability, and extend both the storage time and the shelf life, it is crucial to tailor postharvest strategies to ripening and physiological stages, and these need to be determined precisely. Surveying instruments working with modern technologies such as visible spectrometry can be precise and effective in assessing ripening stage and in grouping fruit in homogeneous classes. This paper reviews results using original nondestructive devices developed at the University of Bologna to define the fruit ripening of several fruit species through a new index (Index of Absorbance Difference (I_{AD})) to compare relationships among fruit ripening stage, fruit quality, and postharvest life. The devices defining the I_{AD} can be used in the field (by the Difference Absorbance (DA)-Meter), at the packinghouse (with the DA-head, a stationary device), and at the cold storage level (with the DA Fruit Logger (DAFL)).

Keywords: fruit ripening; index of absorbance difference (I_{AD}); DA-Meter; DA-Head; DAFL

---

1. Introduction

Fruit quality is a combination of several features that depend on the stage of maturity reached at harvest, postharvest fruit management and disease susceptibility, and storage duration, each influencing general consumer appreciation. In fact, it has been demonstrated that, when fruit are harvested before they reach the proper ripening stage on the tree, fruit quality is poor, as they do not reach their characteristic aroma and flavor, causing, in some situations, disaffection of the consumers for the fruit that do not fulfill their expectations. In this context, the proper choice of the best harvest time plays a crucial role, representing a compromise between the achievement of sufficient quality and the mechanical resistance to harvest and postharvest practices.

Unfortunately, the most common techniques used to determine the ripening stage are often based on destructive methods (°Brix, texture analysis, starch degradation evolution, etc.), and thus they are not applicable to the entire harvest, but only to a small subset of fruits used as a statistical sample. Because the ripening dynamics on-tree can significantly diverge based on the fruit position within the canopy or on the position of the tree in the orchard, even in case of multiple harvest dates performed over a wide harvest window, the fruit batch should be considered heterogeneous [1–5]. A deeper and more precise knowledge of the ripening degree, obtained using a strict protocol and proper devices, could help by segregating the fruit into more uniform groups that each could then be managed with the best storage and marketing strategies. In addition, the possibility of monitoring fruit ripening changes occurring during cold storage could provide critical information for deciding the post-storage temperature that needs to be re-established to guarantee the most appropriate marketing strategy [6]. As a consequence, the ripening stage of fruit at harvest has to be determined with great accuracy following a strict protocol, using proper techniques and devices. However, although fruit quality has always been recognized as a crucial aspect [7,8], only a few quality traits and characteristics are commonly determined at harvest by simple and outdated analyses.

To overcome the problems arising from an imprecise determination of fruit quality, the University of Bologna has focused its efforts in recent years on developing nondestructive devices based on visible/near infrared (VIS/NIR) properties, allowing the definition of fruit ripening and fruit quality attributes by assessing the level of chlorophyll degradation, defined as Index of Absorbance Difference (I_{AD}), and providing a general indication about the fruit ripening stage. The devices have been tested at field and packinghouse levels and in the cold storage room to monitor fruit maturation and ripening evolution from the "field to the fork" level.

In this review, the main results obtained with these nondestructive devices are reported, focusing on the most significant results achieved with different fruit species as far as the relationships among fruit ripening definition, fruit quality enhancement, and disease susceptibility are concerned.

2. The VIS/NIR Devices Used in the Research

The 3 different devices developed for defining fruit ripening as I_{AD} can be used along the supply chain at different levels: (a) in the field with the DA-Meter; (b) in the packinghouse with the DA-Head, a stationary device; and (c) in cold storage with the Difference Absorbance Fruit Logger (DAFL).

2.1. The DA-Meter

The University of Bologna developed and patented three different devices, the DA-Meter, Kiwi-Meter and Cherry-Meter, that have been tested on different fruit species. All three meters do not require any complex calibration and can be used along the productive chain [9]. The first device patented was the DA-Meter (University of Bologna patent n° MO 2005000211) (Figure 1A). The DA-Meter is a portable, user-friendly device for measuring the I_{AD}. The DA-Meter has been tested on several pome and stone fruit cultivars. The I_{AD} is calculated as the difference between the absorbance values at 670 nm and 720 nm, near the chlorophyll a absorbance peak.

Figure 1. The three devices assessing fruit ripening (I_{AD}): (**A**) the DA-Meter at field level; (**B**) the DA-Head at the packinghouse level; and (**C**) the DAFL in cold storage.

The Kiwi-Meter (University of Bologna patent n° PD2009A000081) has been tested on *Actinidia deliciosa* and *Actinidia chinensis* kiwi fruit varieties. The device was developed specifically for kiwifruit, and it differs from the DA-Meter by the wavelengths used: 540 and 640 nm with 800 nm as a reference point. The difference between absorbance at 540 nm versus 800 nm is used for the *Actinidia deliciosa* fruit, and the difference between 640 nm versus 800 nm is used for the *Actinidia chinensis* fruit [10]. In *Actinidia deliciosa*, a higher I_{AD} implies more ripe fruits with a brilliant green colored flesh, while with *Actinidia chinensis*, I_{AD} gradually decreases, staying in a constant range for some time followed by a rapid decrease with the onset of the flesh color break from green to yellow.

The Cherry-Meter is used on cherry (*Prunus avium, Prunus cerasus*) fruit, allowing the determination of I_{AD} with a value range from 0 to 2.6 positively correlated with ripening. The device is made up of six diode LEDs positioned around a photodiode detector. A set of two diode LEDs emit at 560 nm, 640 nm, and 750 nm. The fruit is illuminated alternately with the two monochromatic sources, and for each the amount of light re-emitted by the fruit is measured. Light is detected by a photodiode positioned centrally to a diode crown, and is converted to a digital signal through an analogue to digital converter. I_{AD} is obtained from the differences of the absorbance between two wavelengths for anthocyanin (560 and 640 nm) and the reference value at 750 nm [3].

2.2. The DA-Head

The DA-Head (Figure 1B) is a stationary device that measures the ripening stage as I_{AD} in a similar way to the DA-Meter. This device has been tested on several fruit species (kiwifruit, apricot (*Prunus armeniaca*), pear (*Pyrus communis*), apple (*Malus* × *domestica*), and peach (*Prunus persica*) [5,10,11], although the first and more complete trials have been carried out with the apricot cultivar "La vallée", with the collaboration of the Office Cantonal de Arboriculture of Sion, Switzerland. The device can be considered a pre-commercial prototype [12]. The DA-Head, besides grading fruit on the basis of their size, contemporaneously groups the fruit according to their ripening stage using 2 or 3 remote sensors (DA-heads) which are able to read the I_{AD} of fruit passing on a moving belt. The DA-Heads can be adapted to standard commercial grading machines and operate and select fruit at the same speed of the sorting machine itself, approximately 15 fruit/s.

From the practical point of view, it is a very important achievement that the I_{AD} groups the fruit in homogeneous classes according to their ripening stage, starting in field conditions with the portable DA-Meter and at the packinghouse with the DA-Head.

2.3. The Difference Absorbance Fruit Logger (DAFL)

The last device along the productive chain is the DAFL (Figure 1C). It is a small device positioned over a batch of fruit stored in a cold room that continuously monitors the I_{AD} and the temperature data of each fruit remotely at programmed times (minutes, hours, days, etc.). The data are transmitted via radio signal to a receiver unit connected to a computer, and the information is available through Internet connection [6].

3. Major Accomplishments with the Use of the Devices

3.1. Assessing Proper Harvest Time

The portable DA-Meters (DA-Meter, Kiwi-Meter and Cherry-Meter) are mainly used to monitor the ripening process of fruit still attached to the tree and to assess the proper time to harvest. In particular, in some fruit species (i.e., peach and nectarine) [11,13], the ripening stage can be precisely defined since the I_{AD} and chlorophyll *a* amount in the outer mesocarp are highly correlated. In addition, in stone fruit, and in particular in peach, the I_{AD} correlates with ethylene evolution, the ripening hormone responsible for flesh softening, color development, and sugar accumulation [14,15]. It is worth noting that a trial carried out for several years with different cultivars of peach and nectarine showed that each cultivar has a typical I_{AD} value that remains constant year after year, coincident with the highest ethylene production peak. In addition, the I_{AD} value is strongly correlated with the transcript levels of ripening related genes [13,16]. As a consequence, the determination of the

I_{AD} in nectarines and peach can precisely predict the optimal harvest time [17]. The ripening stage assessed at harvest by the I_{AD} was also recognized and appreciated by consumers who were also able to distinguish fruit of different ripening stages. In a consumer test trial, the most desirable "Stark Red Gold" nectarine fruit were those that reached an I_{AD} value of 0.4–0.5 [18], while "Plus Plus" peach fruit had to reach an I_{AD} value <0.7 to be as desirable (Table 1).

Table 1. "Plus Plus" peach fruit segregated with the DA-stationary machine into 4 ripening classes, the % of the fruit in each class, and consumer likelihood of purchasing each class.

Class	I_{AD}	Fruits (%)	Likelihood of Purchasing (%)		
			Yes	Maybe	No
0	0–0.4	5.5	34.0	57.4	19.4
1	0.4–0.7	22.2	31.9	59.5	23.4
2	0.7–1	61.1	19.1	46.8	23.4
3	>1	11.2	8.5	40.4	21.3

For pome fruit, the I_{AD} was also used to define the proper harvest time of several apple varieties, including "Gala", "Golden Delicious", "Red Delicious", "Fuji", "Granny Smith", and "Pink Lady" [18–20] and of "Abbé Fétel" pear [5]. A consumer test trial performed with "Gala" apples indicated that the most desirable fruit were harvested at an I_{AD} value of 0.6–0.9 [19]. Using "Abbé Fètel" pear, a consumer test trial performed with more than 100 people at the "AGER-Innovapero" Conference (Ferrara, Italy, 18 October 2013) [21] at the end of the storage period showed that the most desirable fruit were those harvested when the I_{AD} had reached value of 1.8–1.9, but when the I_{AD} value was >2 the fruit were not as desirable to consumers.

For cherry, the ripening stage expressed as I_{AD} correlates with anthocyanin content (Figure 2) as well as with other traits normally used to assess fruit quality [3].

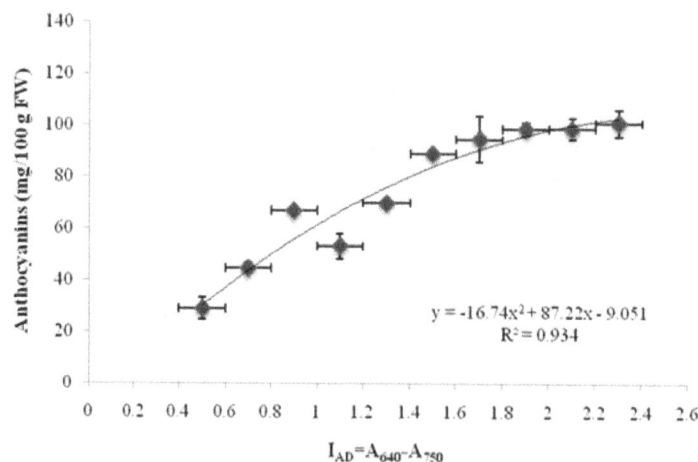

$$y = -16.74x^2 + 87.22x - 9.051$$
$$R^2 = 0.934$$

Figure 2. Correlation of skin anthocyanin content and I_{AD} classes at harvest of "Lala Star" cherry.

With *Actinidia deliciosa* kiwifruit, the robustness of I_{AD} use is represented by the fact that the I_{AD} value significantly correlates with the standard fruit quality trait values (Figure 2) normally used for assessing fruit maturity [10,22]. "Hayward" kiwifruit must be harvested when the soluble solids concentration (SSC) reaches a minimum value of 6.2 °Brix [23]. In some seasons, ripening of *Actinidia deliciosa* fruit, expressed as SSC values, evolves slowly in the last period of fruit growth, complicating the definition of the evolution of fruit ripening. When the Kiwi-meter is used in such circumstances, the I_{AD} might not be fully able to detect small changes in ripening, and readings must be taken with a high accuracy to have robust and suitable data. In contrast, with *Actinidia chinensis* fruit, harvest must be performed when the flesh color changes from green to yellow, a change normally occurring at a hue angle of 103°–105° determined with a colorimeter (Figure 3) [10,24]. However, this determination requires fruit destruction and, as a consequence, can be only performed on a limited number of fruit. The Kiwi-Meter allows a much larger fruit sample, and the I_{AD} results are significantly correlated with hue angle ($R^2 = 0.819$). As a result the ripening stage expressed as I_{AD} is more reliable and repeatable [12,23] on most of the yellow-fleshed cultivars. In addition, the I_{AD} robustness is also underlined by the fact that the value is constant across years as shown in pome and stone fruit [12,13,25].

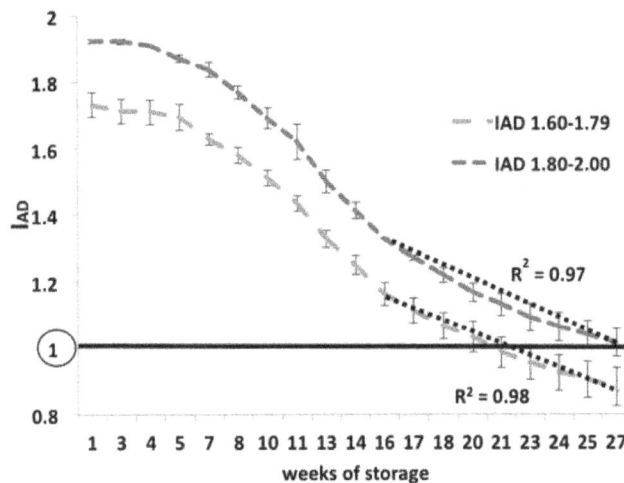

Figure 3. Ripening trends expressed as I_{AD} class during storage of "Abbé Fétel" pear fruit. An I_{AD} value = 1 shown in the figure represents the beginning of skin color change from green to yellow. In the last phase of storage, the I_{AD} change tends to be linear.

3.2. Grouping Fruit in Ripening Homogeneous Classes

The possibility of grouping fruit in homogeneous classes according to their ripening stage is an important use for the I_{AD}, especially for fruit undergoing longer storage periods. Such information can be easily obtained by the I_{AD} value determined at harvest with the DA-Meter on a limited number of fruit, or at the packinghouse with the DA-Head, so that all of the harvested fruit can be grouped into homogeneously-ripening classes. Such trials have been carried out on peach, apricot, apple, pear and kiwifruit. For instance, the DA stationary machine was able to detect within commercial classes of fruit differences that were not previously noticed using standard methods of evaluation, as in the case of classes of "Plus Plus" peaches that were further grouped into 4 subclasses of fruit according to the more precise I_{AD} ripening measure.

Moreover, the differences detected by these new devices, and represented by the I_{AD} index, are economically useful since the consumers are able to recognize the differences in ripening determined by the DA-Meter and appreciate more ripened fruit (Table 1). The same information was also obtained with several cultivars of apricot where the fruit were grouped into classes of uniform ripening with the DA-head device. Notably, consumer appreciation was related to fruit quality/ripening and but was influenced very little by appearance for "Swigold" apricot. In fact, although fruit ripening was quite different among the classes (4 classes characterized by the following I_{AD} values, 0–0.5; 0.5–0.7; 0.7–0.9, and 0.9–2.0, and by the soluble solids content, 14.9; 14.2; 13.2 and 12.5 °Brix, respectively, fruit appearance (skin color and fruit size) in the 4 classes was practically identical (Figure 4) [12], demonstrating that the traditional methodologies, such as colorimeter or caliper measurements, are not suitable for effectively determining fruit physiological stage.

Figure 4. "Swigold" apricot grouped by the DA-head into 4 classes of ripening.

3.3. Fruit Ripening Monitoring in Cold Storage

The Difference Absorbance Fruit Logger (DAFL) (Figure 1C) monitors ripening stage of the fruit maintained in cold room during storage. Trials were carried out with apple, pear and kiwifruit and indicated that fruit collected at higher I_{AD} values at harvest maintained higher values for the entire storage period as compared to fruit harvested at a lower I_{AD}. In pear, the I_{AD} value precisely followed ripening evolution in a sigmoid pattern (Figure 3). In the last phase of storage, the I_{AD} value was linear for both lower and higher initial I_{AD} values, allowing a tentative prediction of the span of time needed to reach the optimal I_{AD} value before removing fruit from cold storage (arbitrarily defined as 1) and driving marketing decisions [6].

3.4. Reducing Postharvest Losses

Several studies have pointed out that fruit disease susceptibility is related to the ripening stage reached by the fruit at harvest [26–28]. In apple, superficial scald susceptibility trials were performed with 3 varieties, "Red Delicious", "Granny Smith", and "Pink Lady", and revealed that when the harvest was performed too early, the fruit were significantly affected by scald. For the susceptible variety "Granny Smith", fruit were harvested at an I_{AD} value of 1.8–2.0 (early ripening, fruit not fully ripe). Scald incidence reached 100% after just two months of storage. In contrast, when fruit were harvested later at a more advanced ripening stage (I_{AD} of 1.6–1.8), fruit were not affected by superficial scald within the first two months of storage (Figure 5) [20].

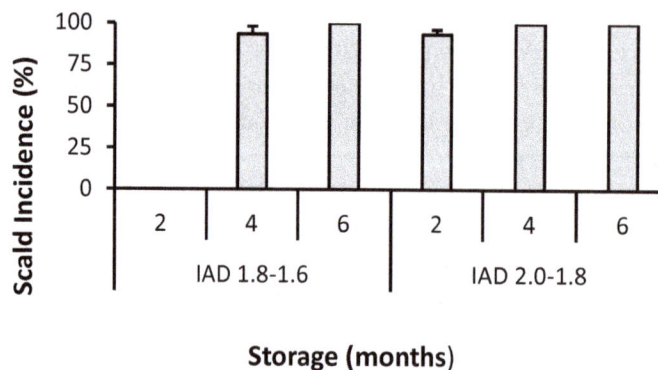

Figure 5. Scald incidence (%) in Granny Smith apples harvested at two different ripening stages determined by I_{AD}. The scald incidence was evaluated on fruit left at room at temperature for 1 week after 2, 4 and 6 months of commercial storage.

Also, in peach, the appearance of brown rot (*Monilia fruticola*) was related to the ripening stage. The damage caused by the fungus was more severe on fruit characterized by an advanced ripening stage [29].

4. Final Remarks

The results of research carried out over the last few years on some of the main temperate fruit species indicated that the I_{AD} represents a powerful and reliable tool for assessing fruit ripening. The I_{AD} devices allow monitoring the evolution of ripening from the field until removal from cold storage. Fruit ripening measured in the field with the DA-Meter establishes the proper harvest time and gives preliminary information about the homogeneity of fruit ripening. Furthermore, at the packinghouse, use of the DA-Head allows grouping of the fruit into homogeneously-ripening classes driving decisions on the best storage strategies and subsequent marketing. Finally, in cold storage, the DAFL allows continuous monitoring of the evolution of fruit ripening.

The I_{AD} can find useful applications in the fruit production chain and represents an essential decision support system tool that can drive pre- and postharvest management of ripening fruit.

Acknowledgments: Research funded by AGER (Project "Agroalimentare e Ricerca"), grant No. 2010–2119.

Conflicts of Interest: The authors declare no conflict of interest.

References

1. Smith, G.S.; Gravett, I.M.; Edwards, C.M.; Curtis, J.P.; Buwalda, J.G. Spatial analysis of the canopy of kiwifruit vines as it relates to the physical, chemical and postharvest attributes of the fruit. *Ann. Bot.* **1994**, *73*, 99–111. [CrossRef]

2. Dallabetta, N.; Costa, F.; Pasqualini, J.; Wehrens, R.; Noferini, M.; Costa, G. The influence of training system on apple fruit quality. *Acta Hortic.* **2013**, *1058*, 55–62. [CrossRef]

3. Nagpala, E.G.L.; Noferini, M.; Farneti, B.; Piccinini, L.; Costa, G. Cherry-Meter: An innovative nondestructive (vis/NIR) device for cherry fruit ripening and quality assessment. In Proceedings of the 7th ISHS Internation Symposium on Cherry, Plasencia, Spain, 23–27 June 2013; ISHS: Leuven, Belgium in press.

4. Bonora, E.; Noferini, M.; Vidoni, S.; Costa, G. Modeling fruit ripening for improving peach homogeneity in planta. *Sci. Hortic.* **2013**, *159*, 166–171. [CrossRef]

5. Vidoni, S.; Rocchi, L.; Donati, I.; Spinelli, F.; Costa, G. Combined use of Planttoon® and IAD to characterize fruit ripening homogeneity in "Abbé Fétel" pears. *Acta Hortic.* **2015**. [CrossRef]

6. Vidoni, S.; Fiori, G.; Rocchi, L.; Spinelli, F.; Musacchi, S.; Costa, G. DAFL: New innovative device to monitor fruit ripening in storage. *Acta Hortic.* **2015**, *1094*, 549–554. [CrossRef]

7. Harman, J.E. Kiwifruit maturity. *Orchard. N. Z.* **1981**, *54*, 126–127.

8. Shewfelt, R.L. What is quality? *Postharvest Biol. Technol.* **1998**, *15*, 197–200. [CrossRef]

9. Costa, G.; Fiori, G.; Torrigiani, P.; Noferini, M. Use of Vis/NIR spectroscopy to assess fruit ripening stage and improve management in post-harvest chain. In *New Trends in Postharvest Management of Fresh Produce I*, Fresh Produce 3 (Special Issue 1) ed.; Sivakumar, D., Ed.; Global Science Book: Ikenobe, Japan, 2009; pp. 1–6.

10. Costa, G.; Bonora, E.; Fiori, G.; Noferini, M. Innovative nondestructive device for fruit quality assessment. *Acta Hortic.* **2011**, *913*, 575–581. [CrossRef]

11. Costa, G.; Bonora, E.; Noferini, M. Modeling system and vis-NIR devices to improve pre- and post-harvest management. *Acta Hortic.* **2015**, *1084*, 809–821. [CrossRef]

12. Costa, G.; Fiori, G.; Rocchi, L.; Vidoni, S.; Berthod, N.; Besse, S.; Knieling, S.; Rossier, J. A prototype of a stationary DA-device to group the apricots in classes of homogeneous ripening. In Proceedings of the ISHS XVI International Symposium on Apricot Breeding and Culture, Shenyang, China, 29 June–3 July 2015.

13. Ziosi, V.; Noferini, M.; Fiori, G.; Tadiello, A.; Trainotti, L.; Casadoro, G.; Costa, G. A new index based on Vis spectroscopy to characterize the progression of ripening in peach fruit. *Postharvest Biol. Technol.* **2008**, *49*, 319–329. [CrossRef]

14. Ruperti, B.; Cattivelli, L.; Pagni, S.; Ramina, A. Ethylene-responsive genes are differentially regulated during abscission, organ senescence and wounding in peach (*Prunuspersica*). *J. Exp. Bot.* **2002**, *53*, 429–437. [CrossRef] [PubMed]

15. Trainotti, L.; Zanin, D.; Casadoro, G. A cell-oriented genomic approach reveals a new and unexpected complexity of the softening in peaches. *J. Exp. Bot.* **2003**, *54*, 1821–1832. [CrossRef] [PubMed]

16. Tadiello, A.; Ziosi, V.; Negri, A.S.; Noferini, M.; Fiori, G.; Busatto, N.; Espen, L.; Costa, G.; Trainotti, L. On the role of ethylene, auxin and a GOLVEN-like peptide hormone in the regulation of peach ripening. *BMC Plant Biol.* **2016**. [CrossRef] [PubMed]

17. Bonora, E.; Noferini, M.; Stefanelli, D.; Costa, G. A new simple modeling approach for the early prediction of harvest date and yield in nectarines. *Sci. Hortic.* **2014**, *172*, 1–9. [CrossRef]

18. Costa, G.; Noferini, M. Use of Nondestructive devices as a decision support system for fruit quality enhancement. *Acta Hortic.* **2013**, *998*, 103–115. [CrossRef]

19. Costamagna, F.; Giordani, L.; Costa, G.; Noferini, M. Use of AD Index to define harvest time and characterize ripening variability at harvest in "Gala"apple. *Acta Hortic.* **2013**, *998*, 117–123. [CrossRef]

20. Farneti, B.; Gutierrez, M.S.; Novak, B.; Busatto, N.; Ravaglia, D.; Spinelli, F.; Costa, G. Use of the index of absorbance difference (IAD) as a tool for tailoring post-harvest 1-MCP application to control apple superficial scald. *Sci. Hortic.* **2015**, *16*, 110–116. [CrossRef]

21. Costa, G.; Rossi, D.; Tamburini, E.; Donegà, V.; Loberti, R. Innovazioni di processo per unapericoltura di qualità. *L'Inf. Agrar.* **2013**, *36*, 44–46. (In Italian)

22. Pellegrino, S.; Costamagna, F.; Noferini, M.; Costa, G. Monitoring of "Hayward" (*Actinidiadeliciosa*) fruit ripening in North-West Italy. *Acta Hortic.* **2011**, *913*, 665–669. [CrossRef]

23. Rocchi, L.; Vidoni, S.; Ceccarelli, A.; Fiori, G.; Costa, G. Use of the DA-index™ for monitoring fruit ripening evolution in *A. chinensis* to precisely assess harvesting time "in planta". *J. Berry Sci.* **2016**. [CrossRef]

24. Costa, G.; Vidoni, S.; Rocchi, L.; Cellini, A.; Buriani, G.; Donati, I.; Spinelli, F. Innovative nondestructive device for fruit quality assessment and early disease diagnosis. *Acta Hortic.* **2015**, *1096*, 69–78. [CrossRef]

25. Costa, G.; Noferini, M.; Fiori, G.; Ziosi, V.; Berthod, N.; Rossier, J. Establishment of the optimal harvest time in Apricot ("Orangered" and "Bergarouge") by means of a new Index based on Vis spectroscopy. *Acta Hortic.* **2010**, *862*, 533–539. [CrossRef]

26. Salzman, R.A.; Tikhonova, I.; Bordelon, B.P.; Hasegawa, P.M.; Bressan, R.A. Coordinate accumulation of antifungal proteins and hexoses constitutes a developmentally controlled defense response during fruit ripening in grape. *Plant Physiol.* **1998**, *117*, 465–472. [CrossRef] [PubMed]

27. Guillén, F.; Castillo, S.; Zapata, P.J.; Martínez-Romero, D.; Valero, D.; Serrano, M. Efficacy of 1-MCP treatment in tomato fruit: 2. Effect of cultivar and ripening stage at harvest. *Postharvest Biol. Technol.* **2006**, *42*, 235–242. [CrossRef]

28. Cantu, D.; Vicente, A.R.; Greve, L.C.; Dewey, F.M.; Bennett, A.B.; Labavitch, J.M.; Powell, A.L.T. The intersection between cell wall disassembly, ripening, and fruit susceptibility to *Botrytis cinerea*. *Proc. Natl. Acad. Sci. USA* **2008**, *105*, 859–864. [CrossRef] [PubMed]

29. Spadoni, A.; Cameldi, I.; Noferini, M.; Bonora, E.; Costa, G.; Mari, M. An innovative use of DA-meter for peach fruit postharvest management. *Sci. Hortic.* **2016**, *201*, 140–144. [CrossRef]

8

Organic Plant Breeding: A Key to Improved Vegetable Yield and Safe Food

Rodel Maghirang *, Maria Emblem Grulla, Gloria Rodulfo, Ivy Jane Madrid
and Maria Cielo Paola Bartolome

Institute of Plant Breeding, College of Agriculture, University of the Philippines Los Baños, Laguna 4031, Philippines; megrulla@gmail.com (M.E.G.); glo_rodulfo@yahoo.com (G.R.); ijwmadrid@gmail.com (I.J.M.); mlbartolome@up.edu.ph (M.C.P.B.)
* Correspondence: vlsection2014@gmail.com or rgmaghr@yahoo.com

Abstract: Most often, organic farming focuses on the improvement of management practices such as nutrient application and pest control, and very seldom deals with variety improvement or breeding. Because it has been dependent on commercially-available varieties developed under conventional high-input methods, traits are expressed resulting in low yields that are commonly attributed to organic farming practices rather than to the adaptability of the cultivar to the system. A research program in the Philippines involving several regions and institutions has pioneered in the evaluation and improvement of varieties through breeding under low-input organic conditions. After making several crosses, pedigree selection, replicated yield and on-farm trials, promising and potential varieties were developed and identified in squash, cucumber, lettuce and yardlong bean. The most promising yield advantages over the respective check varieties ranged up to 47% in squash, 31% in yardlong bean, 42% in lettuce, and 43% in cucumber. Pest and disease resistance were also considered during the selection process, and top performers were moderately to highly resistant. General acceptability in appearance, taste and marketability provided additional selection criteria for considering the top performers and potential varieties. Commercial varieties developed and performing well under conventional high-input methods were mostly not suitable under organic low-input conditions. Hence, breeding under organic low-input conditions is a must to achieve high yield in organic farming systems.

Keywords: low input; squash; cucumber; lettuce; yardlong bean; Philippines

1. Introduction

In the last 14 years, the total organic agricultural land in the world has increased by almost 300%. India has the largest number of organic producers (650,000), Australia has the largest area (17.2 million hectares), the USA has the largest market (24,347 million Euros), and Switzerland has the highest per capita consumption as reported by the International Federation of Organic Agriculture Movements (IFOAM) [1]. It is common practice to approach organic production by replacing application of synthetic chemicals with biologically- or naturally-derived alternatives. Numerous products and methods of cultivation, soil inputs, weed control, nutrient recycling and application, and non-chemical pest control have been developed. Recently, especially in the Asia and Pacific region, efforts have been initiated in breeding varieties for organic production systems. This is an approach not yet been widely exploited by researchers.

Organic varieties or seeds are required in organic production. However, limitations in the availability of organic varieties triggered policy makers to adjust and consider this gap. Conventional F1 varieties from private seed companies are allowed in organic production provided that seeds are produced for at least one generation under an organic system [2,3]. However, previous variety trials show that conventional varieties do not always perform well under organic conditions. This may be because conventional varieties are intentionally developed under optimum conditions but are not adaptable to low input conditions and an organic environment.

Basically, the concept of organic breeding is the same as conventional breeding, with Phenotype (P) = Genotype (G) + Environment (E) + Genotype × Environment (GE), wherein the performance of the variety (P) is dependent on the genetic trait (G), effect(s) of the environment (E), and the interaction between the variety and the environment (GE). However, varieties from conventional breeding may have traits that may be unsuitable for organic production systems, and some important traits for organic farming systems may not be found in the conventional varieties [4]. Breeding under organic conditions may result in improved levels of stress tolerance and disease resistance in the resulting varieties. Hence, the evaluation and improvement of varieties through breeding under organic conditions at minimum levels of input application are essential for the development of the organic sector and for the quality of organic products.

As IFOAM [5] mentioned in a position paper, "organic plant breeding is inevitable." Thus, this research has the objectives of selecting, recommending or developing organic vegetable cultivars such as yardlong bean (*Vigna unguiculata* subsp. *sesquipedalis*), squash (*Cucurbita moschata*), lettuce (*Lactuca sativa*) and cucumber (*Cucumis sativus*) for organic production and consequently to increase vegetable production in organic systems.

2. Experimental Section

2.1. Variety Development

From 1992 to 2002, selected open-pollinated lines and segregating populations of yardlong bean, squash, lettuce and cucumber were subjected to primary evaluation under the DA-BAR funded project "Varietal Evaluation of Selected Vegetables under Organic Conditions" [6]. From these lines and populations, entries with good quality and performance under organic conditions were selected and crosses were made among the selections. The crosses made were evaluated in two sites at the University of the Philippines Los Baños (UPLB), one at the Institute of Plant Breeding (IPB) main compound and another at the PAMANA station along Pili Drive. Pedigree selection was employed under low-input organic conditions.

At the experimental sites, raised beds with dimensions of 1 m × 5 m were prepared for the evaluation of the lines and crosses. Each raised bed was lined with two rows having 20 hills per row and a distance of 50 cm between hills. The evaluation sites were calibrated to be able to select for genotypes with tolerance to weed populations, resistance or tolerance to pests and diseases, and efficiency in nutrient utilization and related stress(es), following Philippine National Standards (PNS) for Organic Agriculture. Organic fertilizers (vermicompost and other compost materials) were applied at a minimal level of about 50%–75% of the nutrition required by the plant, specifically at 2–3 t·ha^{-1}. Fermented plant juice and fermented fruit juice were also applied at specific stages of the plants. Minimal weeding and pest control measures were also practiced. Plastic mulch was used to control weeds, while aromatic pest-repellent plants such as lemon grass and marigold were planted as a means of pest control. Furthermore, flowering plants such as Mexican sunflower and Zinnia were planted to attract pollinators and other beneficial insects, while barrier plants were established around the experimental areas to prevent contamination from neighboring farms. These farm practices were employed for every season of evaluation of the selected lines.

For every generation, individual plants were selected based on general vigor, fruit characteristics, resistance to pests and diseases, and tolerance to environmental stress. Selfing was performed for each individual plant selection, and the selfed seeds obtained were used for the following generation. At the F6 generation, the selected lines were evaluated in an observational trial to determine the top lines to be included in the yield trials. The observational trials were conducted at IPB-UPLB, Palawan Agricultural Center (PAC), DA-Palawan Agricultural Experiment Station (DA-PAES), and the Bureau of Plant Industry—Los Baños National Crop Research and Development Center (Los Baños, Laguna, Philippines).

2.2. Replicated Yield Trials

Replicated yield trials were conducted at PAMANA, Pili Drive, UPLB and DA-PAES. The entries included for each yield trial were the top selected lines and a commercial check variety for each crop. Each trial was carried out using a Randomized Complete Block Design (RCBD) with three replications. The replicated yield trials also followed the aforementioned farm practices employed during the preliminary evaluation of the lines. Data on horticultural and fruit characteristics were gathered from five plant samples per replication. Total yield was obtained and fruits were graded as marketable and non-marketable.

2.3. On-Farm Trials (OFT)

The top performing lines for each crop were further evaluated through on-farm trials in order to assess their acceptability and performance in different locations. The OFT sites were located in some Philippine provinces of Region 4, namely, Laguna, Quezon, Cavite, and Palawan and were all under organic farming conditions. All entries were planted in plots of commercial sizes with at least 20 m² per entry. A commercial hybrid variety which served as the check was also included for each trial. Cultural management followed the practices of the farmer cooperators. At the end of each trial, the cooperators were asked to rank the lines based on their preferences and to identify the lines they preferred the most, as well as the basis of their selection. Notes on the performance of the entries on-farm were taken. Seed production was also performed by the cooperators in order to use their selected lines in the next season.

2.4. Breeder Seed Production

Breeder's seeds of the potential varieties were produced on-station at PAMANA to ensure purity of supply for the following trials and also for variety registration. All top-performing entries across crops were produced following the cultural management practices for organic production. Fruits were allowed to mature prior to extraction of seeds, after which seeds were air dried for 2–3 days and then dried under the sun for another 2–3 days. Completely dried seeds were kept in tightly sealed containers.

2.5. Data Analysis

All data were analyzed by ANOVA using the Statistical Analysis System (SAS) Software (SAS Institute Inc., Cary, NC, USA). Means were compared using Fisher's Least Significant Difference (LSD) test at $p = 0.05$.

3. Results and Discussion

The study resulted in the recommendation of four selections per crop that were vigorous, prolific, early, high yielding and highly acceptable to farmers and the market (Table 1).

3.1. Yardlong Bean

In the development of varieties of yardlong bean, average yields of 22.29 t·ha^{-1} (selection 0801-5-1-1-0), 22.13 t·ha^{-1} (selection 1096-1-1-0-0), 20.29 t/ha (selection 10116-1-1-0-0) and 21.25 t·ha^{-1} (selection 10421-0-0) were achieved. These selections had average yield advantages of 31%, 30%, 25% and 19%, respectively, over the check variety. The selections were highly acceptable in the market due to their medium to long pods. Green podded 1096-1-1-0-0 had an average pod length range of 58cm, light green podded 10421-0-0 was 57 cm, dark green podded 10116-1-1-0-0 was 54 cm, and green podded 0801-5-1-1-0 was 52 cm (Figure 1). Selection 1096-1-1-0-0 had the heaviest pod weight ranging of 22.9–32.7 g, 0801-5-1-1-0-0 ranged from 23–24.4 g, 10116-1-1-0-0 ranged from 23.3–25.7 g, and 10421-0-0 ranged from 23.9–26.3 g.

Table 1. Yield and percent yield advantage above the check yield of potential varieties developed for organic low-input conditions.

Crop/Entry	Yield (t/ha)	Yield Advantage (%)
Yardlong bean		
0801-5-1-1-0	22.29	31.0
1096-1-1-0-0	22.13	30.1
10421-0-0	21.25	24.9
10116-1-1-0-0	20.29	19.3
Check	17.01	
Cucumber		
11622	26.03	43.0
11621	18.87	3.7
11617	18.54	1.8
11624	17.97	−1.3
Check	18.21	
Lettuce		
Le 1103(Looseleaf)	10.96	44.0
Le 0701 (Looseleaf)	6.38	−16.1
Check (Looseleaf)	7.61	
Le 0702 (Cos)	7.74	0.9
Le 1104 (Cos)	6.93	−9.7
Check (Cos)	7.67	
Squash		
10128-1-1	61.03	47.5
1058-1-1	48.03	16.1
1056-1-1	42.83	3.5
10127-1-2	49.67	20.1
Check	41.37	

Figure 1. Yardlong bean top yielders: (**A**) 0801-5-1-1-0; (**B**) 10116-1-1-0-0; (**C**) 10421-0-0 and (**D**) 1096-1-1-0-0.

In Alfonso, Cavite, green-podded yardlong bean, 0801-5-1-1-0 and 10116-1-1-0-0 were sold in the local market. Selection 0801-5-1-1-0 was also chosen for its resistance to bean rust. In Majayjay, Laguna,

10116-1-1-0-0, 10421-0-0 and 0801-5-1-1-0 were preferred for their characteristics, while 1096-1-1-0-0 was not chosen because it did not survive the intermittent rains and waterlogging in the area. The Majayjay cooperator preferred 0801-5-1-1-0 because of its high yield capacity, for good pod qualities such as wider, longer pods, and for its dark green color. In Quezon, the light-green pods of 10421-0-0 were highly preferred. Generally, all cooperators chose all entries because of their moderate to high resistance to common bean diseases and pests, long pods that were stringless and snappy, high yield, and market acceptability.

3.2. Cucumber

For cucumber variety selections, average yields observed were 11622 with 26.03 t·ha^{-1}, 11,621 with 18.87 t·ha^{-1}, 11,617 with 18.54 t·ha^{-1} and 11,624 with 17.97 t·ha^{-1}. Entry 11622 had an average yield advantage over the check variety of 43%, 11621 was 4% more, and aa617 was 1.8% more. Cucumber selection 11,617, which has a light green or bicolor (green with white streaks) fruit skin, had an average fruit weight of 241.6 g and an average length of 14.7 cm; bicolor fruit skin 11621 had a 277.7 g average fruit weight and 16.6 cm average length; light green or bicolor skin 11622 had a 234.4 g average weight and 13.5 cm average length; and, 11,624, which is a bicolored, had a 202.8 g average fruit weight and 11.4 cm average length (Figure 2). All selections flowered on the 26th day after planting. All were thin skinned and sweet.

Figure 2. Cucumber lines selected for their high yield and acceptability: (**A**) 11622; (**B**) 11621; (**C**) 11617 and (**D**) 11624.

Lines 11,621 and 11,624 were the top selections of the farmer-cooperator in Alfonso for their juicy flesh, green and thin skin, and high yield capacity. In its preliminary OFT, line 11,621 was also selected for its vigor and earliness. These cucumber selections were highly accepted by the market around Alfonso and nearby municipalities. In Lucban, Quezon, the cooperator selected 11,621 for its prolific fruiting, thin skin and juicy flesh.

3.3. Lettuce

There are two types of lettuce recommended from the study, the loose-leaf and the Cos or Romaine type (Figure 3). Two selections were identified per lettuce type. Loose-leaf type Le 1103 had the highest average yield of 10.96 t·ha^{-1} with average yield advantage over the check of 44%, and Le 0701 yielded 6.38 t·ha^{-1}. While Le 0701 had a lower average yield, it was highly acceptable. Le 0702, on the other hand, was the top yielder of the Cos type with a 7.74 t·ha^{-1} and a yield advantage of 0.9%. Loose-leaf Le 1103 has light green color with a purple tinge, is curly and spreading, while Le 0701 is light green, curly and also has spreading leaves. The Cos-type Le 0702 has green, closed-heads with slightly curly leaf blades, and is upright, while Le 1103 is green, has a flat-head and is semi-upright. These selections were also late bolting, highly uniform and vigorous.

Figure 3. Loose-leaf (**A**,**B**) and cos-type lettuce selections (**C**,**D**).

Consumers in the markets in Alfonso, and Tagaytay, Cavite, preferred loose-leaf types—Le 07-01 and Le 11-03 and Cos-type Le 07-02. The cooperator-farmers chose these selections because of their late-bolting trait, too. In Lucban, Quezon, Le 0701, a loose-leaf type, was selected for its late-bolting, high yield, and vigor, while Le 1104 was preferred for its resistance or tolerance to stem rot that was observed in the area. Le 1104 was also high yielding and late bolting. The cooperator in Silang, Cavite, chose Le 0701 for loose-leaf and Le 0702 for a Cos-type for their high yield. In Majayjay, Laguna, the cooperator chose loose-leaf Le 0701 and Cos type Le 0702 due to their high yield, acceptability based on their taste and appearance, and their high marketability. They were also late bolters and vigorous when grown on regular plots or in vertical gardens.

3.4. Squash

Squash selection 10128-1-1 was the top yielder from two preliminary yield trials in DA-PAES. Average total yield of this selection was 61.03 t·ha^{-1} with a yield advantage of 47.5% above the check variety, followed by line 1058-1-1 with 48.03 t·ha^{-1} (a 16% yield advantage) and line 1056-1-1 with 42.83 t·ha^{-1} (a 3.5% yield advantage). These three top yielders are the round types with varying sizes. The last squash selection was 10127-1-2 which is a long-necked or cacao-type squash that yielded an average 49.67 t·ha^{-1} (a yield advantage of 20%). Line 10128-1-1 has yellow orange flesh with average length and diameter of 14 cm and 15 cm, respectively, while 1058-1-1 has yellow flesh with an average fruit length and diameter of 10 cm and 13 cm, respectively, and 1056-1-1 also has yellow orange flesh with average length of 9.5 cm and average fruit diameter of 13 cm. Lastly, cacao type 10127-1-2 has a 33.5 cm average fruit length and a 10 cm average diameter. The flesh color is also yellow orange (Figure 4).

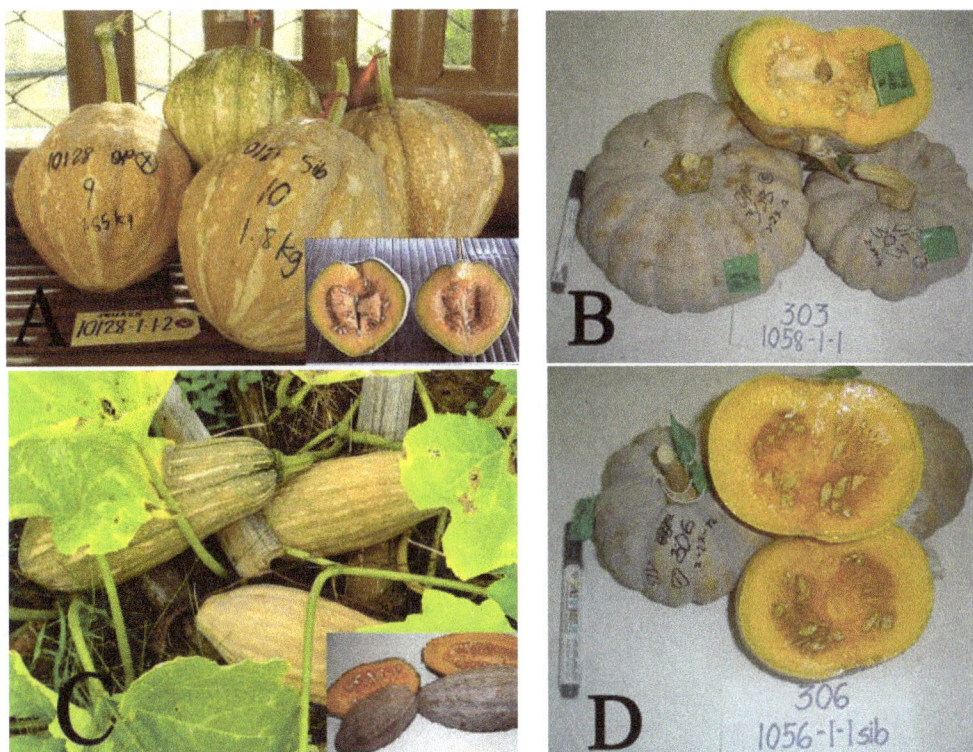

Figure 4. Round-type squash (**A**) 10128-1-1; (**B**) 1058-1-1; (**D**) 1056-1-1; and cacao-type squash (**C**) 10127-1-2 selections.

In Alfonso, Cavite, the farmer-cooperator selected the solo type 10128-1-1 for its size, shape and fruit yield capacity. Organic market-goers in Lucban, Quezon, preferred line 10128-1-1 because of its size which is suitable for a one-serving dish. It was also selected by the cooperator-producer because of its heart-shaped fruits and its low number of seeds. In Tayabas, Quezon, lines 10127-1-2 and 1056-1-1 were preferred for their prolific yield. Another cooperator located in Brgy. Isugod, Quezon, Palawan, selected line 10128-1-1 for its highly acceptable fruit.

4. Conclusions

Commercial varieties from conventional production systems do not always perform best under organic systems. Therefore, it is inevitable that breeding vegetable varieties adaptable for organic low-input conditions will be needed to achieve a high yield in organic farming. From the present studies, varieties developed under organic conditions were observed to be at par or yield more than commercial varieties commonly grown for the market. These lines have also shown resilience under adverse conditions such as high disease incidence and environmental aberrations along with high consumer acceptability. Thus, these varieties developed under organic agricultural systems show promise for answering the need for diversity, sustainability and efficiency in crop production.

Acknowledgments: This research was conducted in collaboration with Department of Agriculture-Palawan Agriculture Experiment Station (DA-PAES) and Bureau of Plant Industry- LBNCRDC, and supported and funded by the Philippine Council for Agriculture, Aquatic and Natural Resources Research and Development (PCAARRD). The authors would like to deeply recognize and express gratitude to the research team for their dedication and commitment.

Author Contributions: Rodel Maghirang served as the overall team leader. He was responsible for the conceptualization of the research and was also involved in the monitoring of the trials, interpretation of data, and critical revision of the article. Maria Emblem Grulla, Maria Cielo Paola Bartolome, Ivy Jane Madrid, and Gloria Rodulfo were all involved in the conduct of the trials, data collection and analysis, and drafting of the manuscript.

Conflicts of Interest: The authors declare no conflict of interest.

References

1. IFOAM. *Consolidated Annual Report of IFOAM—Organics International*; IFOAM-Organics International Head Office: Bonn, Germany, 2015.
2. Ratanawaraha, C.; Ellis, W.; Panyakul, V. *Organic Agri-Business: A Status Quo Report for Thailand 2007*; Thai-German Programme for Enterprise Competitiveness, Sustainable Agriculture Foundation and GreenNet Foundation: Bangkok, Thailand, 2007.
3. Lammerts van Bueren, E.T. Organic Plant Breeding and Propagation: Concepts and Strategies. Ph.D. Thesis, Wageningen University, Wageningen, The Netherlands, 2002.
4. Legzina, L.; Skrabule, I. *Plant Breeding for Organic Farming: Current Status and Problems in Europe*; Compendium ENVIRFOOD: Talsi, Latvia, 2005.
5. IFOAM. *IFOAM Position on the Use of Organic Seed and Plant Propagation Material in Organic Agriculture*; IFOAM Head Office: Bonn, Germany, 2011.
6. Maghirang, R.G.; Rodulfo, G.S.; Enicola, E.E.; Candelaria, R. Organic Breeding and Seed Production in Selected Vegetables. In Proceedings of the Organic Seed Preconference of the 17th IFOAM Organic World Congress, Namyangju, Korea, 26–27 September 2011.

9

Effects of Drying and Blanching on the Retention of Bioactive Compounds in Ginger and Turmeric

Haozhe Gan [†], Erin Charters [†], Robert Driscoll [†] and George Srzednicki [*]

School of Chemical Engineering, University of New South Wales, Sydney 2052, Australia;
z3365252@student.unsw.edu.au (H.G.); z3374995@student.unsw.edu.au (E.C.); r.driscoll@unsw.edu.au (R.D.)
[*] Correspondence: g.srzednicki@unsw.edu.au
[†] These co-authors graduated from the university and approved their participation in this research.

Abstract: Ginger and turmeric, members of the *Zingiberaceae* family, are widely used for their pungent and aromatic flavour in foods and also for their medicinal properties. Both crops are often grown by smallholders in mountain areas on rich former forest soils with no need for fertilizers and pesticides, fulfilling de facto the conditions of organic agriculture. They are consumed fresh or dried. Drying is often performed without taking into account the content of bioactive compounds in the dried product. Various bioactive compounds have been identified in their rhizomes, and their content affects the price of the dried product. Hence, this study focused on the effects of drying treatments and blanching on the retention of bioactive compounds in the dried products. The bioactive compounds in ginger rhizome (*Zingiber officinale* Roscoe) are gingerols (particularly 6-gingerol). The drying treatments that were applied to fresh ginger included constant and also changing temperature conditions. Due to the short drying time, 60 °C was the optimal drying temperature to retain 6-gingerol. However, the changing temperature conditions significantly improved the retention of 6-gingerol. As for blanching, it had a significant negative effect on 6-gingerol retention. Turmeric (*Curcuma longa*) is known for its bright yellow colour and pharmacological properties due to curcumin, a phenolic compound. Drying was performed under constant conditions at 38 °C, 48 °C, 57 °C and 64 °C and a relative humidity of 20% and 40%. Drying at 57 °C with a lower relative humidity was the best drying treatment, yielding the highest amount of curcumin among non-blanched samples. Blanching for 15 min exhibited the highest curcumin yield while blanching for 5 min and 30 min did not have much effect. The findings of this study will benefit the industry in terms of improved quality control and cost reduction.

Keywords: ginger rhizome (*Zingiber officinale* Roscoe); air drying; changing conditions; 6-gingerol; turmeric rhizome (*Curcuma longa*); blanching; curcumin

1. Introduction

Ginger (*Zingiber officinale* Roscoe) and turmeric (*Curcuma longa*) belong to the *Zingiberaceae* family. Both species are spices that are cultivated and consumed in subtropical and tropical regions all around the world. They are erect perennial herbs that are grown for their rhizomes. A rhizome is an underground plant stem that is capable of producing shoots and roots of a new plant. Rhizomes of both plants are consumed in many subtropical and tropical countries such as China, India, Jamaica and Australia. They are consumed fresh or dried. Ginger is often consumed as a condiment and served with food or in beverages as flavoured tea or soft drink, whereas turmeric is often used in dried form in curries. Both are highly regarded in culinary preparations due to their characteristic smell and flavour produced by volatile components and pungent components. Other important applications are in pharmacology, and their medicinal properties have been known for centuries. The moisture content (MC) of fresh ginger is between 83%–94% on a wet basis [1,2]. Polyphenol compounds including 10-gingerol, 8-gingerol, 6-gingerol, and their derivatives, are detected in the roots of ginger, and extracts have been proven to have high antioxidant activity [3,4] and anti-inflammatory effect [5].

The MC of fresh turmeric is about 83%–87% [6]. According to Parthasarathy et al. [7] turmeric contains 6.3% protein, 5.1% fat, 3.5% minerals, and 69.4% carbohydrates. The essential oil has 1% α-phellandrene, 0.6% sabinene, 1% cineol, 0.5% borneol, 25% zingiberene and 53% sesquiterpenes. The components of turmeric can vary amongst cultivars. In addition, the content of curcumin depends on location of growth. Curcumin is the component of turmeric responsible for its colour and all its medicinal properties. It has a molecular formula of $C_{21}H_{20}O_6$ and its structure has been identified as diferuloylmethane. It is insoluble in water and soluble in ethanol and acetone. It makes up 2.5%–6% of the rhizome. Curcumin and its two related demethoxy compounds, demethoxycurcumin and bisdemethoxycurcumin, are known as curcuminoids. These components have been identified as antioxidants. Cyclocurcumin is a newly identified curcuminoid isolated from the fraction of turmeric found to be active as a nematicide. New phenolics have been identified as well, which are antioxidants and anti-inflammatory.

Among the bioactive compounds of interest in the rhizomes of these two plants are gingerols in ginger and curcumin in turmeric. The content of these compounds in the dried product generally determines the price of the commodity. Drying is an important processing step in the postharvest handling of ginger and turmeric as it reduces their MC and water activity, hence reducing microbial activity and chemical reactions which decrease deterioration and increase shelf life. It also reduces the size of products, which decreases storage and transportation costs. Yet, some adverse effects on product quality caused by drying cannot be ignored which include the loss of volatile aromatic compounds, decrease of antioxidant activity, and degradation of valuable nutrient content [8]. Also, the formation of some new components can take place as a result of thermal reactions [9].

Blanching in hot water after harvesting is a traditional process that removes the raw odour and improves drying time. There are conflicting views on the effect of blanching on bioactive compounds in ginger and turmeric. Thuwapanichayanan et al. [10] found that blanched ginger powder had lower total phenolic content and antioxidant activity compared to the untreated sample due to the loss of phenolics into the hot water during blanching. Some studies reported that high temperatures, such as experienced in blanching, cause thermal degradation of curcumin [11], while other studies have shown that blanching protects the bioactive ingredients from the effects of drying [12]. Therefore, this study investigated the effects of blanching time and drying conditions on the concentration of the key bioactive compounds, 6-gingerol in dried ginger, and curcumin in dried turmeric.

2. Materials and Methods

2.1. Ginger and Turmeric Samples

Mature, fresh ginger samples of Australian origin were obtained from the Buderim Ginger Pty Ltd in Yandina, Queensland, Australia. Fresh turmeric was procured as rhizomes from Earthcare Enterprises, Maleny, Queensland, Australia.

2.2. Sample Preparation

Ginger and turmeric samples were stored at −18 °C in a refrigerator after procurement. Frozen samples were defrosted at 4 °C for 24 h, and placed at ambient temperature for 12 h before use. Defrosted rhizomes were peeled and sliced to 4 mm thickness. In all drying treatments, in addition to untreated sliced samples, three batches of samples were blanched in a 70 °C water bath for 5, 15, or 30 min, respectively.

2.3. Drying Treatments

2.3.1. Dryer

Air drying of rhizome slices was conducted in a cabinet dryer (developed by the University of New South Wales Food Engineering Research Group) (Figure 1). The dryer consisted of a temperature and relative humidity (RH) controller, a steam injection unit, a fan, and a drying chamber. The weight of samples was about 200 g for each batch. The samples were spread in a single layer on a stainless steel tray placed in the drying chamber. A load cell connected to a computer was placed below the tray for automatic recording of weight. The drying air flow was parallel to the tray surface with a constant air velocity of 0.75 ± 0.03 m·s^{-1}. The temperature and RH were monitored online using a data logger (Datataker DT50).

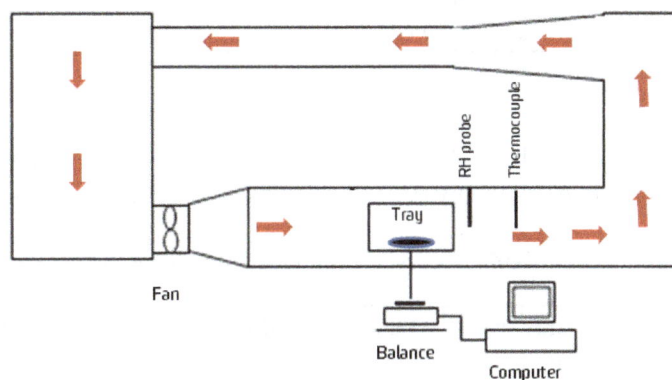

Figure 1. Schematic of the laboratory-scale cabinet dryer used in the experiments.

2.3.2. Drying Conditions

Ginger

Under constant conditions the drying run was stopped after a constant sample weight was reached. Under changing conditions, the change to the next condition took place when the sample weight fell to 60 g. Then, the samples were dried until they reached a constant weight. The weights of samples were recorded online at 15 min intervals. Initial and final MCs of the ginger slices were determined in a vacuum oven according to AOAC standards [13], i.e., at 70 ± 1 °C under 13.3 kPa pressure for 24 h. MC determination was the average of triplicate samples. The results were expressed as means with standard deviation. Comparison of means was done with ANOVA using EXCEL™ (Microsoft Corp., Redmond, WA, USA). The ginger samples drying treatments are in Table 1.

Table 1. Drying conditions of ginger [2].

S/N	Strategy	Temperature (°C)	Relative Humidity (%)	Blanching (min)
1	Temperature control only	40	9–12	0
2		40	9–12	5
3		40	9–12	15
4		40	9–12	30
5		50	6–8	0

Table 1. *Cont.*

S/N	Strategy	Temperature (°C)	Relative Humidity (%)	Blanching (min)
		Temperature & RH Control		
6	Constant conditions	30	30	0
7		40	10	0
8		40	30	0
9		50	25	0
10		60	15	0
11	Changing conditions	40 then 60	30 then 15	0
12		60 then 40	15 then 30	0
13		40 then 30	10 then 30	0
14		30 then 40	30 then 10	0
15		50 then 60	25 then 15	0
16		60 then 50	15 then 25	0

Turmeric

The turmeric samples were subjected to drying treatments as shown in Table 2. Turmeric samples were dried under constant conditions only. MC was determined in the same way as in ginger [13].

Table 2. Drying conditions turmeric.

S/N	Temperature (°C)	Relative Humidity (%)	Blanching (min)
1	40	20	0
2	40	40	0
3	50	20	0
4	50	20	5
5	50	20	15
6	50	20	30
7	60	20	0
8	60	40	0
9	70	20	0

2.4. Determination of Bioactive Compounds

2.4.1. Gingerol

Extraction

The 6-gingerol (≫98% purity) standard was obtained from Sigma-Aldrich (St. Louis, MO, USA). Methanol was of HPLC grade. Water for HPLC analysis was purified with a Milli-Q system (Merck, KGaA, Darmstadt, Germany). A stock solution of 6-gingerol in methanol (5.0 mg/mL) was prepared to produce a series of dilutions. The dilutions were prepared from the stock solution by dilution with 10% methanol and 90% water. For establishing standard curves, the solutions were prepared containing 5, 10, 20, 40, 60, 80 and 100 µg/mL, respectively. All solutions were stored in amber glass bottles at 4 °C before use.

Dried and fresh ginger ware pulverized and passed through a 40 mesh (0.42 mm) sieve before extraction. A powder sample (1 g) from each treatment was dissolved in 25 mL methanol and sonicated for 30 min. The mixtures were centrifuged at 10,000 rpm for 10 min and supernatant was filtered through Whatman filter paper (No. 1). Then, it was diluted with water until the final solvent ratio was 10% methanol and 90% water. All the extracts were kept at 4 °C. Extracts of ginger were filtered through a 0.45 µm nylon filter into an Agilent amber vial with cap and a wide opening for HPLC analysis.

Instrumentation and Chromatographic Conditions

A Prominence LC-20AD HPLC system (Shimadzu, Kyoto, Japan) was used in this study. The separation of the extract was conducted using a C18 column, 3.5 μm, 2.1 × 150 mm. Water (A) and methanol (B) constituted the mobile phase which was used for separation. The gradient elution had the following profile: 0–5 min, 50%–60% B; 5–9 min, 60% B; 9–14 min, 60%–95% B; 14–15 min, 95% B; 15–16 min, 95%–50% B; 16–25 min, 50% B. The injection volume was 40 μL and the flow rate was 0.18 mL/min. The UV detector (0~1000 nm) wavelength was set at 281 nm and the column temperature was maintained at 30 °C.

2.4.2. Curcumin

Extraction

Dried samples were ground in a Pulverisette, type 14,202 laboratory mill (Fritsch, Idar-Oberstein, Germany) at 15,000 rpm with a 1 mm mesh screen and stored in glass flasks. The powdered samples (10 g) were extracted with dichloromethane using a Soxhlet extractor in a water bath at 70 °C for 1.5 h. The extract was concentrated in a rotary evaporator, then diluted in ethanol. The curcumin standard (Sigma-Aldrich) was used for identification and quantitation. A calibration curve was produced using the following concentrations of the standard: 30, 100, 300, 700 and 1000 ng/mL.

Instrumentation and Chromatographic Conditions

The same HPLC as for 6-gingerol was used for curcumin detection. The separation of the extract was performed using a C18 column, 2 μm, 2.1 × 150 mm. Formic acid (0.1%) (A) and acetonitrile (B) constituted the mobile phase which was used for separation. The gradient elution profile is shown in Table 3. The injection volume was 5 μL. The detector wavelength was set at 360 nm.

Table 3. The gradient elution profile.

Step	Time (min)	A%	B%	Flow Rate (mL/min)
0	0.00	99.5	0.5	6.0
1	0.34	99.5	0.5	6.0
2	1.40	45.0	55.0	6.0
3	4.50	40.0	60.0	6.0
4	4.80	0.5	99.5	7.5
5	6.60	0.5	99.5	7.5
6	7.00	99.5	0.5	7.5
7	10.00	99.5	0.5	7.5

3. Results

3.1. Drying

3.1.1. Ginger

The MC of fresh rhizomes was between 536%–904% on a dry basis (db) (Table 4). A higher temperature and lower RH of the drying air led to lower a MC (i.e., higher moisture loss) of the final product. Treatments with higher RH in the dryer produced a higher MC in the final product. The results of changing conditions indicated that treatments with mild temperature and humidity conditions at the initial stage led to a higher MC of the dried sample, while initial harsh conditions caused a lower MC. Thus, the initial conditions of air drying are a primary factor determining final MC of the product. The results of blanching as a pre-treatment of ginger slices (Table 4) showed that blanched slices were generally less hygroscopic than non-blanched slices. The MCs of fresh ginger after 15 and 30 min were on average 383% db lower than those of fresh ginger without blanching. This could be due to gelatinisation of starch molecules which may have blocked sorption sites and reduced water adsorption. A less hygroscopic polymer network may have formed around the blanched surfaces, thus disfavoring moisture adsorption.

3.1.2. Turmeric

The final MCs decreased with an increase in temperature, a decrease in relative humidity and a shorter blanching time (Table 5).

Table 4. Moisture content (MC) of samples subjected to different drying treatments.

Treatment	Fresh MC (% db)	Dried MC (% db)
40 °C/9–12% RH/No blanching	904 [b] ± 3	11.9 [f] ± 0.2
40 °C/9–12% RH/5 min blanching	922 [a] ± 5	10.7 [g] ± 0.3
40 °C/9–12% RH/15 min blanching	523 [i] ± 2	9.7 [g] ± 0.2
40 °C/9–12% RH/30 min blanching	519 [i] ± 1	10.3 [g] ± 0.2
50 °C/6%–8% RH	935 [a] ± 1	8.5 [h] ± 0.3
60 °C/15% RH	536 [h] ± 4	7.5 [i] ± 0.2
50 °C/25% RH	604 [f] ± 2	11.0 [f] ± 0.2
40 °C/30% RH	924 [a] ± 3	18.4 [c] ± 0.2
40 °C/10% RH	666 [e] ± 3	10.2 [g] ± 0.3
30 °C/30% RH	587 [f] ± 5	21.6 [a] ± 0.2
60 °C/15% RH to 50 °C/25% RH	761 [d] ± 1	13.8 [e] ± 0.2
50 °C/25% RH to 60 °C/15% RH	865 [c] ± 1	17.8 [d] ± 0.2
60 °C/15% RH to 40 °C/30% RH	596 [f] ± 4	20.2 [b] ± 0.2
40 °C/30% RH to 60 °C/15% RH	566 [g] ± 4	20.5 [b] ± 0.2
40 °C/10% RH to 30 °C/30% RH	634 [e] ± 3	11.8 [c] ± 0.2
30 °C/30% RH to 40 °C/10% RH	671 [b] ± 3	19.4 [f] ± 0.2

Values are mean ± standard deviation of triplicate analyses. In the same column, values followed by the same letter are not significantly different ($p \leq 0.05$).

Table 5. Moisture characteristics of turmeric slices subjected to different drying treatments.

Treatment	Fresh MC (% db)	Dried MC (% db)
40 °C/20% RH	190 [g] ± 3	7.4 [b] ± 0.2
40 °C/40% RH	335 [e] ± 3	10.5 [a] ± 0.3
50 °C/20% RH	377 [d] ± 3	7.0 [c] ± 0.1
50 °C/20% RH/5 min blanching	530 [a] ± 3	5.7 [d] ± 0.2
50 °C/20% RH/15 min blanching	420 [c] ± 3	7.0 [c] ± 0.3
50 °C/20% RH/30 min blanching	490 [b] ± 2	7.9 [b] ± 0.3
60 °C/20% RH	274 [f] ± 3	5.2 [e] ± 0.2
60 °C/40% RH	256 [f] ± 2	7.6 [b] ± 0.3
70 °C/ 20% RH	325 [e] ± 4	6.1 [d] ± 0.1

Values are mean ± standard deviation of triplicate analyses. In the same column, values followed by the same letter are not significantly different ($p \leq 0.05$).

3.1.3. Retention of 6-Gingerol in Ginger

The peak of 6-gingerol is shown in Figure 2a. A linear regression between retention area and standard concentration was calculated and a coefficient of determination (R^2) of 0.9788 was obtained showing a good linear relationship. Under constant drying conditions, the content of 6-gingerol decreased with increasing temperature and RH. Under changing conditions, it appeared that with specific combinations of drying conditions, the concentration of 6-gingerol significantly increased when compared with constant conditions (Table 6).

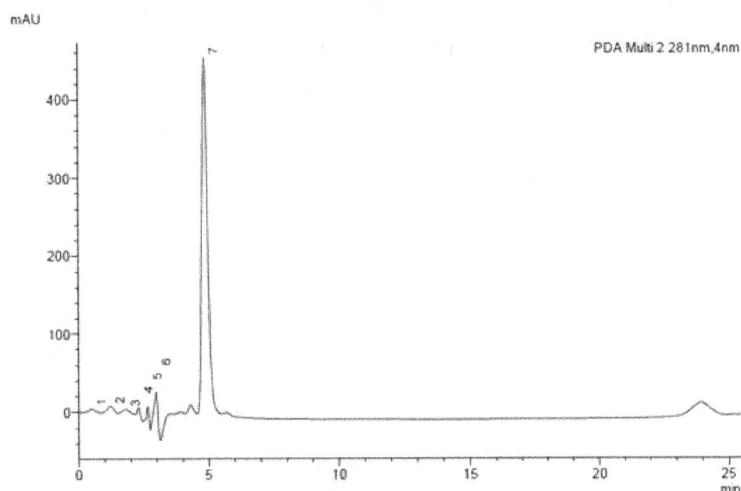

(a)

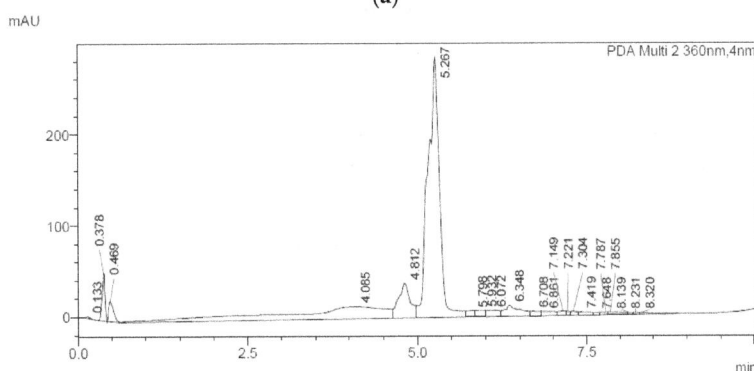

(b)

Figure 2. (a) Peak of 6-gingerol (peak #7); (b) Peak of curcumin from a drying treatment of 60 °C 20% RH.

Table 6. Effects of different drying treatments on 6-gingerol concentration in dried samples.

Treatment	Retention Area	6-Gingerol Concentration (%)	Drying Time (min)
40 °C/9%–12% RH	3572383	1.273 [e] ± 0.063	343 [d] ± 3
40 °C/9%–12% RH/5 min blanching	2289876	0.773 [i] ± 0.005	353 [d] ± 3
40 °C/9%–12% RH/15 min blanching	1196832	0.348 [j] ± 0.012	353 [d] ± 3
40 °C/9%–12% RH/30 min blanching	1059796	0.294 [j] ± 0.005	831 [a] ± 4
50 °C/6%–9% RH	3037359	1.064 [h] ± 0.010	289 [e] ± 3
60 °C/15% RH	2991478	1.047 [h] ± 0.015	177 [h] ± 3
50 °C/25% RH	3306548	1.169 [g] ± 0.011	233 [g] ± 4
40 °C/30% RH	3659751	1.307 [d] ± 0.067	297 [e] ± 3
40 °C/10% RH	3702015	1.323 [d] ± 0.005	257 [f] ± 2
30 °C/30% RH	3512811	1.250 [f] ± 0.050	489 [b] ± 3
60 °C/15% RH to 50 °C/25% RH	4155793	1.500 [b] ± 0.036	185 [h] ± 4
50 °C/25% RH to 60 °C/15% RH	4007405	1.442 [c] ± 0.050	203 [h] ± 3
60 °C/15% RH to 40 °C/30% RH	3532484	1.257 [e] ± 0.005	193 [h] ± 3
40 °C/30% RH to 60 °C/15% RH	3675330	1.313 [d] ± 0.054	257 [f] ± 2
30 °C/30% RH to 40 °C/10% RH	4470837	1.623 [a] ± 0.025	425 [c] ± 2
40 °C/10% RH to 30 °C/30% RH	4458393	1.618 [a] ± 0.025	305 [e] ± 3

Values are mean ± standard deviation of triplicate analyses. In the same column, values followed by the same letter are not significantly different ($p \leq 0.05$).

The combined treatments of 60 °C 15% RH and 50 °C 25% RH resulted in 0.35% more 6-gingerol retention vs. either of the constant conditions. Mild conditions as initial treatment led to higher 6-gingerol concentration. Blanching treatment reduced 6-gingerol retention. After 5 min of blanching, almost half of the 6-gingerol was lost vs. samples without blanching, and it decreased further with 15 and 30 min of blanching, possibly due to the high temperature of blanching.

3.1.4. Retention of Curcumin in Turmeric

The chromatogram indicated a curcumin peak at 5.267 min (Figure 2b). The greatest curcumin content (area under the curve) with samples dried at 50 °C 20% RH and blanched for 15 min. The greatest content for the non-blanched samples was at 60 °C 20% RH which had 82.2% of the content of the highest samples (Table 7).

Table 7. Effects of different drying treatments on curcumin concentration in dried samples.

Treatment	Area under the Curve	Normalised Concentration (%)
40 °C/20% RH	3308511	73.7
40 °C/40% RH	2646929	58.9
50 °C/20% RH	3447014	76.8
50 °C/20% RH/5 min blanching	3506873	78.1
50 °C/20% RH/15 min blanching	4490862	100.0
50 °C/20% RH/30 min blanching	3375932	75.2
60 °C/20% RH	3691512	82.2
60 °C/40% RH	2953344	65.8
70 °C/20% RH	3327375	74.1

4. Conclusions

The following conclusions were drawn from the study:

- Increasing drying temperature decreased 6-gingerol content in the dried product.
- Changing drying conditions significantly increased the concentration of 6-gingerol.
- Blanching treatment had a significant negative effect on 6-gingerol retention.
- Increasing drying temperature at low RH tended to increase curcumin retention.
- Blanching for 15 min favoured curcumin retention in the dried product.

Author Contributions: Haozhe Gan conducted the research on drying of ginger, conceived and designed the experiments, performed the experiments and analysed the data. Erin Charters conducted the research on drying of turmeric, conceived and designed the experiments, performed the experiments and analysed the data. George Srzednicki and Robert Driscoll wrote the paper.

Conflicts of Interest: The authors declare no conflict of interest.

References

1. Phoungchandang, S.; Saentaweesuk, S. Effect of two stage, tray and heat pump assisted-dehumidified drying on drying characteristics and qualities of dried ginger. *Food Bioprod. Process.* **2010**, *89*, 429–437. [CrossRef]
2. Gan, H. Effects of Air Drying Treatments on the Concentration of the Key Bioactive Compound 6-Gingerol in Ginger. Master's Thesis, School of Chemical Engineering, The University of New South Wales, Sydney, Australia, 15 December 2013.
3. Stoilova, I.; Krastanov, A.; Stoyanova, A.; Denev, P.; Gargova, S. Antioxidant activity of a ginger extract (*Zingiber officinale*). *Food Chem.* **2007**, *102*, 764–770. [CrossRef]
4. Sakulnarmrat, K.; Srzednicki, G.; Konczak, I. Antioxidant, enzyme inhibitory and antiproliferative activity of polyphenolic-rich fraction of commercial dry ginger powder. *Int. J. Food Sci. Tech.* 2015. [CrossRef]
5. Dugasani, S.; Pichika, M.R.; Nadarajah, V.D.; Balijepalli, M.K.; Tandra, S.; Korlakunta, J.N. Comparative antioxidant and anti-inflammatory effects of 6-gingerol, 8-gingerol, 10-gingerol and 6-shogaol. *J. Ethnopharmacol.* **2010**, *127*, 515–520. [CrossRef] [PubMed]

6.   Charters, E. The Effects of Drying and Blanching on Curcumin Yields of Root Turmeric. Bachelor's Thesis, School of Chemical Engineering, The University of New South Wales, Sydney, Australia, 31 October 2014.

7.   Parthasarathy, V.A.; Chempakam, B.; Zachariah, T.J. *Chemistry of Spices*; Centre for Agriculture and Bioscience International (CABI): Wallingford, CO, USA, 2008.

8.   Chan, E.; Lim, Y.Y.; Wong, S.K.; Lim, K.K.; Tan, S.P.; Lianto, F.S.; Yong, M.Y. Effects of different drying methods on the antioxidant properties of leaves and tea of ginger species. *Food Chem.* **2009**, *113*, 166–172. [CrossRef]

9.   Chen, C.C.; Ho, C.T. Volatile compounds in ginger oil generated by thermal treatment. *ACS Symp. Ser.* **1989**, *409*, 366–375.

10.   Thuwapanichayanan, R.; Phowong, C.; Jaisut, D.; Štencl, J. Effects of pretreatments and drying temperatures on drying characteristics, antioxidant properties and color of ginger slice. *Acta Univ. Agric. Silv. Mendel. Brun.* **2014**, *62*, 1125–1134. [CrossRef]

11.   Chen, L.; Bai, G.; Yang, S.; Yang, R.; Zhao, G.; Xu, C.; Leung, W. Encapsulation of curcumin in recombinant human H-chain ferritin increases it water-solubility and stability. *Food Res. Int.* **2014**, *62*, 1147–1153. [CrossRef]

12.   Blasco, M.; Garcia-Perez, J.V.; Bon, J.; Carreres, J.E.; Mulet, A. Effect of blanching and air flow rate on turmeric drying. *Food Sci. Technol. Int.* **2006**, *12*, 315–323. [CrossRef]

13.   Association of Official Analytical Chemists (AOAC). *AOAC Official Methods of Analysis of AOAC International. AOAC Official Method 934.06, Moisture in Dried Fruits*; Association of Official Analytical Chemists (AOAC): Rockville, MD, USA, 1997.

10

Combined Effects of Fertilizer, Irrigation, and Paclobutrazol on Yield and Fruit Quality of Mango

Babul C. Sarker [1],*, Mohammad A. Rahim [2] and Douglas D. Archbold [3]

[1]   Principal Scientific Officer, Pomology Division, Horticulture Research Centre, Bangladesh Agricultural Research Institute, Joydebpur, Gazipur 1701, Bangladesh

[2]   Department of Horticulture, Bangladesh Agricultural University, Mymensingh 2202, Bangladesh; marahim1956@yahoo.com

[3]   Department of Horticulture, University of Kentucky, Lexington, KY 40546-0091, USA; darchbol@uky.edu

*   Correspondence: bsarker_64@yahoo.com

Abstract: Combinations of fertilizer rates, foliar N sprays, irrigation practices, and paclobutrazol were studied to determine how much they could alter and/or improve mango (*Mangifera indica* L.) growth, flowering, and yield. Two treatment combinations derived from several years of prior studies of individual practices were compared: one combination was comprised of the best (BT) individual practices from the prior studies and included three applications of fertilizer, a 4% KNO_3 spray application before flowering, paclobutrazol at 7.5 g/L, and weekly irrigation, and the other combination was comprised of the next best (NB) individual practices including two applications of the same amount of fertilizer, a 4% urea spray before flowering, paclobutrazol at 10.0 g/L, and biweekly irrigation. Both combinations significantly reduced terminal shoot growth and leaves per terminal shoot, advanced the date of flowering and harvest, increased panicle number, length and secondary branching, increased fruit set, fruit number at harvest, fruit size, and yield, with BT producing larger fruit and a greater yield than NB. Although both combinations produced fruit with higher quality than the control, the BT combination produced fruit with the higher total soluble solids, reducing, non-reducing, and total sugar content, and vitamin C content than the NB combination. Both BT and NB combinations of the optimums identified in the prior studies were successful at advancing bloom and harvest and increasing yield more than any of the optimum individual components alone, by 14-fold more than untreated trees for the BT combination, suggesting there were additive, if not synergistic, effects on mango. Further studies are warranted to assess the sustainability of these effects over longer periods of time, and to ascertain if the effects occur across mango cultivars and production environments.

Keywords: paclobutrazol; soil drenching; panicle emergence; flowering; peel–pulp ratio

1. Introduction

Mango (*Mangifera indica* L.) is a delicious fruit which belongs to the family Anacardiaceae. The principal mango producing countries are India, China, Thailand, Indonesia, and Pakistan [1]. In Bangladesh, which ranks ninth, about 242,605 tons of fruit are produced from an area of 51,012 ha, with an average yield of 4.75 tons per ha [2], although it falls short of fulfilling national demand. Irregular flowering, low fruit set, as well as fruit retention leading to low yield, fruit of poor quality, and a short harvest period are the main hindrances to increasing mango availability. Optimizing and integrating various key management practices such as fertilizer application, irrigation, foliar sprays of KNO_3 or urea, and soil drench application of paclobutrazol might extend the availability period and increase yield and quality. Many mango growers in Bangladesh do not employ these management practices on an annual basis, or have few guidelines for optimizing them, and need a set of recommendations to increase and sustain production.

Yearly soil application of N, P, and K has markedly increased the number of mango fruit per plant, fruit weight, yield, and fruit quality [3–7]. Effective rates of N, P and K have ranged from 0.4 to 0.8 kg per tree. Foliar KNO_3 applications have also advanced flowering and harvest date, increased yield, and reduced alternate bearing in mango [8,9]. Maximum yield was found by foliar urea application [10]. Potassium nitrate, especially in combination with urea, produced good results on flowering and yield parameters [11]. Irrigation of mango trees can be beneficial, as it has been shown to reduce fruit drop resulting in satisfactory production [12]. However, in Bangladesh, bearing trees are often irrigated only after fruit set and thereafter at fortnight intervals, but not all growers practice this.

Soil application of the growth retardant paclobutrazol was reported to reduce tree growth and induce precocious flowering, fruit set, fruit retention, and increase yield in bearing mango trees [13,14]. In addition, improvements in fruit quality parameters, such as ascorbic acid, total sugar, reducing sugar, and total soluble solids content, have also been observed [14–17].

A series of studies assessing strategies of soil and foliar fertilization, irrigation, and soil application of paclobutrazol were carried out to identify the best treatments for optimizing mango yield, fruit quality, and harvest period, and to be used to develop management recommendations. Cow manure, widely available to mango growers in Bangladesh, was studied as a soil fertilizer. It comprised 13.8 g N/kg, 3.2 g P/kg, 0.36 g K/kg, 0.13 g S/kg, and 0.02 g Zn/kg, and was applied at 12.5 to 37.5 kg per tree in 1, 2, or 3 separate installments over several months [18]. The highest rate, split into three applications, produced the highest number of fruit, total yield, and fruit quality, and delayed the start of harvest by 11 days. To determine if foliar nutrient application was beneficial, KNO_3 at 4%, 6% and 8% (w/v), and urea at 2% and 4% (w/v), were applied before flowering on 15 November 2007. KNO_3 at 4% exhibited the highest yield, and fruit vitamin C and total soluble solids (TSS) content, followed by urea at 4% [19], with both advancing the start of harvest by 5 days. Regular irrigation starting at fruit set prevented fruit drop and improved fruit in size and quality [20]. To determine which irrigation strategies had significant impacts on fruit yield and quality, trees were irrigated on 15 October and 15 November 2007 and then from fruit set to maturity at 7, 14, or 21 day intervals versus no irrigation. Plants irrigated twice on 15 October and 15 November 2007 and then from fruit set to maturity at 7 and 14 day intervals produced about two-fold higher yields (22.2 and 21.6 kg/plant, respectively) and the best quality [21], and delayed the start of harvest by 7 days. Paclobutrazol applied at 7.5 g/L on 15 October 2007 produced the highest number of fruits as well as greatest yield per plant fruit weight, and had the best quality, compared to lower and higher rates and a December application, with harvest starting about 15 days earlier. A subsequent study indicated that a July application of paclobutrazol at 7.5 g/L or 10.0 g/L significantly advanced harvest by 22 days over control and produced 4.7-fold higher yield and bigger fruit over control.

After having assessed the cultural practices individually, combinations of the optimum individual practices were assessed to determine if they were additive or synergistic in their effects on tree growth and components of yield. Thus, an integrated experiment with split applications of fertilizer, KNO_3 and urea sprays, periodic irrigation, and soil drench applications of paclobutrazol was performed, combining the best individual practices from the previous studies as one treatment, and the second best practices as another treatment, to determine effects on mango yield and fruit quality.

2. Materials and Methods

2.1. Location

The experiment was carried out at the Bangladesh Agricultural University (BAU) Germplasm Centre, Department of Horticulture, Mymensingh, Bangladesh which is located at $24°26'$ latitude and $90°15'$ longitude with an altitude of 8.3 m above sea level. Biochemical analyses were carried out in the Department of Biochemistry of Bangladesh Agricultural University (BAU), Mymensingh, Bangladesh.

2.2. Soil and Climate

The soil at the Germplasm Centre is sandy loam which belongs to the Old Brahmaputra Flood Plain Alluvial Tract [22]. The selected soil samples of the experimental area were analyzed at the Laboratory of the Soil Science Division, Bangladesh Agricultural Research Institute, Joydebpur, Gazipur. Soil pH was 6.5 and 6.3 at a depth of 0–15 cm and 15–30 cm, respectively. The total %N, available P, and available K were 0.056, 38 µg/mL and 0.10 meq/100 mL at a soil depth of 0–15 cm, respectively, and 0.036, 13 µg/mL and 0.07 meq/100 mL at a soil depth of 15–30 cm, respectively. The experimental site enjoys a sub-tropical climate characterized by heavy rainfall (means of 361 mm/month and 2166 mm total), high temperature (mean 28.34 °C), a mean of 161 h/month of sun during April to September, and scant rainfall (mean 79 mm/month and 555 mm total), low temperature (22.6 °C) and a mean of 187 h/month of sun during October to March (Source: Weather Yard, Department of Irrigation and Water Management, BAU, Mymensingh, Bangladesh, 2005).

2.3. Experimental Material

The cultivar used in the study, Amrapali (BARI Mango-3), is precocious, medium dwarf, annual, and prolific in bearing and has good fruit quality [23]. The age of the plants was 11 years at the initiation of experiment. Plant spacing was 5 m × 5 m.

2.4. Experimental Design

The experiment was laid out in a Randomized Complete Block Design (RCBD) with 3 replications. The treatments were: (1) the best (BT) individual treatments of fertilizer (37.5 kg cow dung with 518.43 g N, 120 g P, 187.5 g K, 67.45 g S, and 8.09 g Zn per plant) applied in three installments, a 4% KNO_3 spray application on 15 November 2007 approximately 1.5 to 2 months prior to anticipated flowering, irrigation on 15 October and 15 November 2007 and then continued from fruit set (6 March 2008) to maturity at 7 day intervals, soil drench application of paclobutrazol at 7.5 g/L of active ingredient on 15 July 2007 (all of the best treatments from prior studies); (2) the 2nd or next best (NB) individual treatments of fertilizer (same as above) applied in two installments, a 4% urea spray application on 15 November 2007 approximately 1.5 to 2 months prior to anticipated flowering, irrigation on 15 October and 15 November 2007 and then continued from fruit set to maturity at 14 day intervals, and a soil drench application of paclobutrazol at 10 g/L of active ingredient on 15 July 2007; and (3) untreated control.

2.5. Application of Fertilizer and Paclobutrazol

Fertilizer was applied in two or three applications. The first application was the total amount of cow dung, triple superphosphate (TSP), gypsum, and zinc sulfate applied on 15 September. If in two applications, $\frac{1}{2}$ of the urea and muriate of potash (MoP) were applied on 15 September and the remainder was applied on 15 March. If in three applications, $\frac{1}{4}$ of the total amounts of urea and MoP were applied on 15 March and again at the 3rd application on 15 May. Very light irrigation was provided at each time of fertilizer application.

Paclobutrazol (Syngenta, India) was prepared from a 25% (w/v) stock solution for final concentrations of 7.5 and 10 g/L of active ingredient. Treatments were applied as a soil drench in which 10 small holes (10–15 cm depth) were prepared in the soil around the collar region of the trees just inside the fertilizer ring [24]. One liter of the prepared solutions was poured into the holes of each tree, and the soil was reworked after application. Only water was applied to the control plants.

2.6. Tree Growth and Yield Measurements

The length and number leaves of 10 randomly-selected terminal shoots at flowering on 20 December 2007 were measured. The individual leaf area of 50 leaves was measured, taking 5 from each of the 10 above-selected shoots, using a leaf area meter and was expressed as cm^2. Starting at the first appearance of a panicle, the number of panicles per plant was counted at 10 day intervals up to completion of panicle emergence. Ten panicles were randomly selected from each treatment on 18 February 2008, and the length and number of secondary branches per panicle was recorded. The initial number of fruit of each panicle, and the fruit retained per panicle were counted at 10 day intervals starting from the pea stage up to harvest starting from 6 March 2008. The average date of harvest (when one or two ripe fruit dropped from the tree, all the fruit were harvested at one time), number of fruit/tree, fruit weight, and total yield/tree was recorded.

2.7. Fruit Quality Measurements

After harvest, 10 randomly-selected fruit per tree were ripened at room temperature, and fruit quality was determined. Each fruit was peeled, the stone removed, and each was weighed along with the remaining pulp. The edible (pulp) portion, stone-to-pulp ratio, peel-to-pulp ratio, shelf-life (the difference between the harvest date and the date up to which the fruit remained edible was considered as shelf life), total soluble solids (TSS), titratable acidity, vitamin C, dry matter, and reducing, non-reducing, and total sugar content. TSS of the pulp was measured using a hand refractometer. The titratable acidity of mango pulp was determined [25] using 0.1 N NaOH solution. Vitamin C content was determined [26]. Reducing sugar content of mango pulp was determined by the dinitrosalicylic acid method [27]. The total sugar content of mango pulp was colorimetrically determined by the anthrone method [28].

2.8. Statistical Analysis

Treatment means were analyzed by analysis of variance and separated by Fisher's Least Significant Difference (LSD) test at $p = 0.05$ [29].

3. Results

3.1. Tree Growth

Both BT and NB had similar effects of reducing terminal shoot length and increasing panicle length and the number of secondary branches per panicle (Table 1). The NB treatment had the fewest leaves per terminal shoot, although the BT treatment also reduced terminal shoot length. The NB treatment only had the smallest leaf dimensions and area per leaf. Interestingly, the BT combination had the widest panicles.

Table 1. Effects of combinations of growth regulator and management practices on leaf, shoot and panicle characters of mango. Treatments were the combinations of the best and next best treatments from prior studies, as defined in the text.

Treatment Combination	Length of Terminal Shoot (cm)	Leaves per Terminal Shoot	Leaf			Length of Panicle (cm)	Width of Panicle (cm)	Secondary Branches per Panicle
			Length (cm)	Width (cm)	Area (cm^2)			
Control	16.3 a [z]	12.5 a	22.6 a	4.5 a	63.1 a	22.9 b	17.1 b	24.8 b
Best	14.2 b	9.3 c	21.1 a	4.4 a	60.9 a	28.1 a	19.1 a	32.7 a
Next Best	12.8 b	8.4 b	19.2 c	3.8 b	52.3 b	26.9 a	17.5 b	31.7 a
LSD (0.05)	1.5	0.8	1.89	0.4	5.8	1.6	1.3	3.3

[z] Means followed by different letters within columns significantly differ by Fisher's LSD at $p = 0.05$.

3.2. Emergence, Number of Panicles, Fruit Set, and Fruit Retention

The first panicle emerged 49 and 44 days earlier for the BT and NB treatment combinations, respectively, than the control (Table 2). The BT combination produced more panicles per plant from emergence until equaled by the NB combination on the final measurement date.

Initial fruit set was greater for both treatments than the control (Table 3). From 26 March to harvest, the BT combination had 2-fold greater fruit retained per panicle followed by the NB combination than the control.

Table 2. Effects of combinations of growth regulator and management practices on emergence and number of panicles per plant of mango. Treatments were the combinations of the best and next best treatments from prior studies, as defined in the text.

Treatment Combination	Date of Appearance of First Panicle	Number of Panicles per Plant at					
		30 December 2007	9 January 2008	19 January 2008	29 January 2008	8 February 2008	18 February 2008
Control	7 February 2008	0 c [z]	0 c	0 c	0 c	6.8 b	32.0 b
Best	20 December 2007	15.0 a	48.3 a	96.0 a	148.8 a	189.5 a	197.7 a
Next Best	25 December 2007	9.4 b	30.0 b	88.0 b	134.3 b	180.2 a	191.6 a
LSD (0.05)	-	1.4	3.9	7.1	10.1	12.9	16.0

[z] Means followed by different letters within columns significantly differ by Fisher's LSD at $p = 0.05$.

Table 3. Effects of combinations of growth regulator and management practices on fruit set and fruit retention per panicle. Treatments were the combinations of the best and next best treatments from prior studies, as defined in the text.

Treatments	Fruit Set per Panicle	Number of Fruit Retained per Panicle in 2008 at								
		16 March	26 March	5 April	15 April	25 April	5 May	15 May	25 May	Harvest
Control	6.54 b [z]	5.87 b	2.65 c	1.75 b	1.45 c	1.17 c	1.10 c	1.00 c	0.90 c	0.82 c
Best	10.83 a	10.42 a	5.25 a	4.50 a	3.53 a	2.43 a	2.22 a	2.15 a	2.10 a	2.05 a
Next Best	10.17 a	9.62 a	3.42 b	2.87 ab	1.97 b	1.67 b	1.67 b	1.65 b	1.62 b	1.62 b
LSD (0.05)	1.34	1.09	0.62	1.64	0.32	0.34	0.20	0.13	0.17	0.18

[z] Means followed by different letters within columns significantly differ by Fisher's LSD at $p = 0.05$.

3.3. Harvest, Fruit Yield, and Fruit Characters

The harvest date was three or more weeks earlier for the treatment combinations than the control (Table 4). The BT combination produced the most and largest fruit, and total yield (Figure 1), followed by the NB combination, and both were significantly greater than the control. In fact, the BT combination produced 8-fold more fruit, and fruit weight was 61% greater, than the control. Fruit dimensions for both combinations were greater than the control but did not differ. The BT combination produced a higher edible portion than the control, with the NB combination intermediate between them. The stone:pulp ratio and peel:pulp ratio was lowest in the BT combination, followed by the NB combination, and was highest in the control. Both treatment combinations increased shelf life compared to controls.

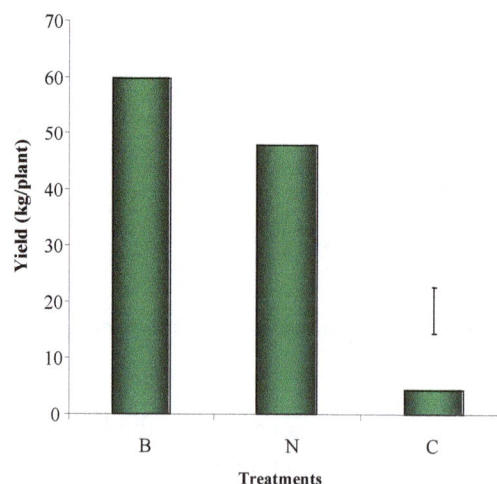

Figure 1. Effects of combinations of growth regulator and management practices on yield per plant of mango. Treatments were the combinations of the best (BT) and next best (NB) treatments from prior studies, as defined in the text. Vertical bar represents LSD at $p = 0.05$.

Table 4. Effects of combinations of growth regulator and management practices on number of fruit and fruit characters of mango. Treatments were the combinations of the best and next best treatments from prior studies, as defined in the text.

Treatment	Date of Harvest	Number of Fruit per Plant	Fruit Weight (g)	Fruit Length (cm)	Fruit Breadth (cm)	Fruit Thickness (cm)	Edible Portion (%)	Stone:pulp Ratio	Peel:pulp Ratio	Shelf Life (days)
Control	1 July 2008	23 c [z]	191.0 c	9.1 b	6.4 b	5.9 b	63.9 b	0.3 a	0.3 a	8.7 b
Best	4 June 2008	197 a	307.8 a	10.7 a	7.5 a	6.6 a	69.5 a	0.2 b	0.2 b	9.8 a
Next Best	6 June 2008	175 b	278.8 b	10.3 a	7.4 a	6.5 a	67.0 ab	0.2 b	0.2 b	9.6 a
LSD (0.05)		16	23.2	1.0	0.4	0.4	4.0	<0.1	<0.1	0.7

[z] Means followed by different letters within columns significantly differ by Fisher's LSD at $p = 0.05$.

3.4. Fruit Quality

The BT combination exhibited the highest TSS, but NB did not differ from the control (Table 5). The NB combination had a higher pulp pH than the control, with the BT combination similar to both. Both combination treatments had lower titratable acidity and moisture content, and higher vitamin C content, than the control, but did not differ from one another. The BT combination produced higher dry matter content, and reducing sugar, non-reducing sugar, and total sugar content than the NB combination, which was greater than the control.

4. Discussion

Both BT and NB combinations significantly impacted mango vegetative growth, flowering, and fruiting. While both were expected to differ from untreated trees, the primary question was whether they differed from one another, as one was comprised of the best individual treatments from prior studies while the other was comprised of the next best treatments, as the names indicate. The treatments differed in their components in some important ways. Both had equal levels of fertilizer applied, but it was applied in three installments in BT and only two in NB. KNO_3 was foliar-applied in BT, while urea was used at the same concentration in NB. For the NB combination, trees were less frequently irrigated from fruit set to maturity and had a higher rate of paclobutrazol included than the BT combination. While each component had effects on vegetative growth, flowering, and/or fruiting individually, it was not known how they might additively and/or synergistically interact to affect these traits.

Both combination treatments reduced vegetative growth compared to the control (Table 1). Our prior study had shown that paclobutrazol, regardless of concentration, caused a marked reduction in terminal shoot length, leaf number per terminal shoot, and leaf area compared to a control, but that reductions were at a maximum at a high concentration (10.0 g/L) [30]. However, the NB combination in this study had a higher paclobutrazol concentration than the BT, but it did not reduce terminal shoot growth more; although leaves per shoot, and leaf length, width, and area, were reduced more. The reduction in mango growth by paclobutrazol may be due to enhanced total phenolic content of terminal buds and alteration of the phloem-to-xylem ratio of the stem, altering assimilate partitioning and patterns of nutrient supply [31]. Soil-applied paclobutrazol reduced gibberellin production and growth in new mango shoots [32,33]. Less frequent irrigation reduced mango terminal shoot growth in our prior work [21] but both irrigation levels produced the same growth in the present study, and both grew less than the unirrigated control, so irrigation level was not a factor affecting growth.

Earlier flowering and greater panicle production were promoted by both BT and NB treatments (Table 2), although BT flowered earlier and increased panicle number in the first weeks of flowering more than NB. The reduction in shoot growth may have increased total available carbohydrates, and a higher C:N ratio in the shoots favors flower bud initiation in mango [34]. The flowering response was also likely due to the foliar KNO_3 and urea applications [19]. Others have shown similar responses to KNO_3 treatment [8,35]. The response to KNO_3 may be mediated, not by increasing N in the tissue, but by promoting ethylene biosynthesis which affects floral induction in mango [36]. Paclobutrazol also induced earlier flowering and greater panicle production in our earlier study [30]. In contrast, more frequent irrigation caused a slight delay in date of flowering [21].

Initial fruit set per panicle was similar in both treatments, but the number of fruit retained by BT was greater than by NB from early in the fruiting period until harvest (Table 4), as previously observed for paclobutrazol treatment [30], KNO_3 and urea treatments [19], and in response to irrigation [21]. The same response was noticed to paclobutrazol, and found that treated trees had higher reserves which enhanced fruit set compared to low reserves and low fruit set in untreated trees [17].

Both combination treatments advanced harvest date by 24–26 or more days in the present work (Table 4). Paclobutrazol alone can cause this, as found by [37] and in our prior work, although only by 10 to 15 days [30]. KNO_3 and urea treatment also advanced harvest date although only by 5 days [19], but irrigation delayed it by 7 days [21]. Interestingly, the differences in harvest dates were less, 27–29 days, than the differences in the first bloom dates, 44–49 days.

The BT combination increased yield by 14-fold and the NB combination by 11-fold over control trees, with BT yield significantly greater than NB yield (Figure 1). These increases far exceeded the response to each treatment component alone [18,19,21,30], when yield increases ranged from 2-fold to 7-fold greater than control trees. The present results are a strong indication that the combinations provided additive impacts on yield.

The increased total yield was the result of effects on both fruit per plant and mean fruit weight. All treatments alone also increased fruit number and yield [18,19,21,30]. The increased fruit weight in the combination treatments might be due to an increased rate of photosynthesis and an increase in chlorophyll content. Paclobutrazol increased leaf water use efficiency by increasing the rate of photosynthesis for a given level of leaf stomatal conductance or transpiration [38]. An increase in fruit weight and yield per tree from paclobutrazol has been reported in several studies with different mango varieties [13,17,39–41], from KNO_3 [42] and urea application [10], and from increased fertilizer application [43].

Both treatments exhibited better physico-chemical characteristics (pulp percentage, stone percentage, peel percentage, total soluble solids, titratable acidity, ascorbic acid, total reducing and non-reducing sugar) compared to the control, similar to the reports by others with the individual and/or combined treatments [4,14,15,44].

5. Conclusions

This paper represents the culmination of a set of studies on the effects of individual management practices in fertilization, foliar N application, irrigation, and paclobutrazol application to identify the optimum for each alone. Both combinations significantly reduced shoot growth, likely due to the presence of the growth retardant paclobutrazol, and shifted the bloom and harvest dates to earlier times. Although both combinations produced more fruit that also had higher quality than the control, the BT combination produced a greater yield and fruit with a higher total soluble solids, reducing, non-reducing, and total sugar content, and vitamin C content than the NB combination. The BT and NB combinations of the optimums identified in the prior studies were successful at advancing bloom and harvest and increasing yield more than any of the optimum individual componentsalone (Table 6), suggesting there were additive, if not synergistic, effects on mango. Although the present results are clear, indicating that implementation of these combinations of treatments could increase yield and ensuing economic returns very significantly over use of a single treatment, there are important considerations that need to be addressed. Given the low replication number and single year of data with a single cultivar, it is not known if similar responses will occur on broader plantings of "Amrapali" (BARI Mango-3) or other mango cultivars, or under different edaphic or environmentalconditions. In addition, the long-term sustainability of the responses to the treatment combinations should be studied, as well as the possible carryover of effects of the treatments, especially to soil-applied nutrients and to paclobutrazol, into the year(s) after application. This might be a concern if there was a cumulative effect from several years of application. Nonetheless, the "packages" of treatments appeared superior to any one treatment alone, and warrant further study to answer these additional issues.

Table 5. Effects of combinations of growth regulator and management practices on quality attributes of mango. Treatments were the combinations of the best and next best treatments from prior studies, as defined in the text.

Treatments	TSS (%)	pH	Titratable Acidity (%)	Vitamin C (mg/100 g pulp)	Moisture Content (%)	Dry Matter Content (%)	Reducing Sugars (%)	Non-Reducing Sugars (%)	Total Sugars (%)
Control	24.8 b [z]	5.74 b	0.23 a	29.4 b	81.9 a	17.9 c	4.8 c	12.8 c	17.5 c
Best	28.7 a	5.89 ab	0.18 b	35.9 a	77.7 b	22.3 a	5.4 a	14.5 a	19.9 a
Next Best	26.2 b	5.95 a	0.18 b	34.8 a	79.0 b	21.0 b	5.2 b	13.7 b	18.9 b
LSD (0.05)	1.9	0.16	0.01	1.2	1.4	1.1	0.2	0.6	0.7

[z] Means followed by different letters within columns significantly differ by Fisher's LSD at $p = 0.05$.

Table 6. Summary of major effects of the individual factors and their combination on key yield traits. The top individual treatment of each pair was considered the best and was included in the Best combination, and the next best treatment was included in the Next Best combination.

Treatment	Advance (−) or Delay (+) Relative to Control in Date of First Panicle	Date of First Harvest	Final Fruit per Panicle	Fruit Weight (g)	Yield vs. Control
Fertilizer in 3 applications [z]	−3	+11	1.87	217	5.6X more
Fertilizer in 2 applications	−3	+5	1.74	201	4.6X more
Control	0	0	0.62	161	-
4% KNO₃ [y]	−18	−5	1.63	193	2.5X more
4% Urea	−14	−5	1.37	203	2.2X more
Control	0	0	0.73	175	-
Irrigation at 7 day intervals [x]	+2	+10	1.47	183	1.7X more
Irrigation at 14 day intervals	+2	+10	1.17	187	1.7X more
Control	0	0	0.57	169	-
Paclobutrazol 7.5 g/L [w]	−21	−22	2.00	304	4.7X more
Paclobutrazol 10.0 g/L	−21	−22	1.70	261	3.4X more
Control	0	0	0.63	178	-
Best	−44	−25	2.05	308	14X more
Next Best	−44	−25	1.62	279	11X more
Control	0	0	0.82	191	-

[z] From [18]; [y] From [19]; [x] From [21]; [w] From [45].

Acknowledgments: The author is grateful to the authorities at the Bangladesh Agricultural Research Institute (BARI) for awarding the scholarship and deputation to complete the research work which led to the Ph.D. The author extends his special thanks to the authorities of the Bangladesh Agricultural University (BAU) Germplasm Centre, Department of Horticulture, Bangladesh Agricultural University, Mymensingh, for providing logistical support during the entire period of the field experiments.

Author Contributions: Babul C. Sarker designed and carried out the experiment, analyzed data and prepared the manuscript. Mohammad A. Rahim contributed through supervision and providing necessary suggestions during the research. Douglas D. Archbold contributed in organization and presentation of the data and the manuscript, and in writing the final draft. All the authors read and approved the manuscript.

Conflicts of Interest: The authors declare no conflict of interest.

References

1. FAOSTAT. Food and Agriculture Organization of the United Nations, Statistics Division. 2016. Available online: http://faostat3.fao.org/home/E (accessed on 20 June 2016).

2. Bangladesh Bureau of Statistics, Statistics Division, Planning Division, Ministry of Planning. *Yearbook of Agricultural Statistics of Bangladesh*; Government of the People's Republic of Bangladesh: Dhaka, Bangladesh, 2005; p. 90.

3. Reddy, Y.T.N.; Kurian, R.M.; Kohli, R.R.; Gorakh, S.; Singh, G. Effect of nitrogen, phosphorus and potassium on growth, yield and fruit quality of "Totapuri" mango (*Mangifera indica*). *Indian J. Agric. Sci.* **2000**, *70*, 475–478.

4. Satapathy, S.K.; Banik, B.C. Studies on nutritional requirement of mango cv. Amrapali. *Orissa J. Hortic.* **2002**, *30*, 59–63.

5. Sharma, R.C.; Mahajan, B.V.C.; Dhillon, B.S.; Azad, A.S. Studies on the fertilizer requirements of mango cv. Dashehari in sub-mountainous region of Punjab. *Indian J. Agric. Res.* **2000**, *34*, 209–210.

6. Suriyapananont, V.; Subhadrabandhu, S. Fertilizer trials on mangoes (*Mangifera indica* L.) var. Nam Dok Mai in Thailand. *Acta Hortic.* **1992**, *321*, 529–534. [CrossRef]

7. Zhou, X.C.; Liu, G.J.; Yao, J.W.; Ai, S.Y.; Yao, L.X.; Zhou, X.C.; Liu, G.J.; Yao, J.W.; Ai, S.Y.; Yao, L.X. Balanced fertilization on mango in Southern China. *Better Crops Int.* **2001**, *15*, 16–20.

8. Khattab, M.M.; Haseeb, G.M.; Shaban, A.E.; Arafa, M.A. Effect of paclobutrazol and potassium nitrate on flowering and fruiting of Ewais and Sidik mango trees. *Bull. Fac. Agric.* **2006**, *57*, 107–123.

9. Sergent, E.; Ferrari, D.; Leal, F. Effects of potassium nitrate and Paclobutrazol on flowering induction and yield of mango (*Mangifera indica* L.) cv. Haden. *Acta Hortic.* **1997**, *455*, 180–187. [CrossRef]

10. Gupta, R.K.; Brahmachari, V.S. Effect of foliar application of urea, potassium nitrate and NAA on fruit retention, yield and quality of mango cv. Bombai. *Orissa J. Hortic.* **2004**, *32*, 7–9.

11. Yeshitela, T.; Robbertse, P.J.; Stassen, P.J.C. Potassium nitrate and urea sprays affect flowering and yields of "Tommy Atkins" (*Mangifera indica*) mango in Ethiopia. *S. Afr. J. Plant Soil* **2005**, *22*, 28–32. [CrossRef]

12. Hossain, A.K.M.A. *Production Technology of Mango*; Horticulture Research Centre, BARI: Gazipur, Bangladesh, 1994; p. 122.

13. Kulkarni, V.J. Chemical control of tree and promotion of flowering and fruiting in mango (*Mangifera indica* L.) using paclobutrazol. *J. Hortic. Sci.* **1988**, *63*, 557–566. [CrossRef]

14. Singh, S.; Singh, A.K. Regulation of shoot growth and flowering in mango cv. Gulab Khas by paclobutrazol. *Ann. Agric. Res.* **2006**, *27*, 4–8.

15. Karuna, K.; Mankar, A.; Singh, J. Effect of urea and growth substances on yield and physico-chemical characteristics of mango. *Hortic. J.* **2005**, *18*, 131–133.

16. Vijayalakshmi, D.; Srinivasan, P.S. Improving the quality attributes of "off" year Alphonso mango through chemicals and growth regulators. *Orissa J. Hortic.* **2000**, *28*, 31–33.

17. Yeshitela, T.; Robbertse, P.J.; Stassen, P.J.C. Paclobutrazol suppressed vegetative growth and improved yield as well as fruit quality of "Tommy Atkins" mango (*Mangifera indica*) in Ethiopia. *N. Z. J. Crop Hortic. Sci.* **2004**, *32*, 281–293. [CrossRef]

18. Sarker, B.C.; Rahim, M.A. Effects of doses and splits of fertilizer application on harvesting time, yield and quality of mango cv. Amrapali. *Bangladesh J. Agric. Res.* **2012**, *37*, 279–293. [CrossRef]

19. Sarker, B.C.; Rahim, M.A. Yield and quality of mango (*Mangifera indica* L.) as influenced by foliar application of potassium nitrate and urea. *Bangladesh J. Agric. Res.* **2013**, *38*, 145–154. [CrossRef]

20. Singh, L.B. *The Mango Botany, Cultivation and Utilization*; Leonard Hill Ltd.: London, UK, 1968; p. 230.

21. Sarker, B.C.; Rahim, M.A. Effect of irrigation frequency on harvesting time, yield and quality of mango. *J. Bangladesh Soc. Agric. Sci. Technol.* **2010**, *7*, 11–16.

22. United Nations Development Programme (UNDP). *Land Resource Appraisal of Bangladesh for Agricultural Development Report 2: Agroecological Regions of Bangladesh*; Food And Agriculture Organization (FAO): Rome, Italy, 1988; p. 577.

23. Singh, R.N. *Mango*; Indian Council of Agricultural Research: New Delhi, India, 1996; p. 134.

24. Burondkar, M.M.; Gunjate, R.T. Control of vegetative growth and induction of regular and early cropping in "Alphonso" mango with paclobutrazol. *Acta Hortic.* **1993**, *341*, 206–215. [CrossRef]

25. Rangana, S. *Manual of Analysis of Fruit and Vegetable Products*; Tata McGraw-Hill Pub. Co. Ltd.: New Delhi, India, 1979; p. 634.

26. Plummer, D.T. *An Introduction to Practical Biochemistry*; Tata McGraw-Hill Pub. Co. Ltd.: New Delhi, India, 1971; p. 229.

27. Miller, G.L. Use of dinitro salicylic acid reagent for determination of reducing sugar. *Anal. Chem.* **1972**, *31*, 426–428. [CrossRef]

28. Jayaraman, J. *Laboratory Manual in Biochemistry*; Wiley Eastern Ltd.: New Delhi, India, 1981; p. 62.

29. Gomez, K.A.; Gomez, A.A. *Statistical Procedures for Agricultural Research*; John Wiley and Sons: New York, NY, USA, 1984; p. 680.

30. Sarker, B.C.; Rahim, M.A. Vegetative growth, harvesting time, yield and quality of mango (*Mangifera indica* L.) as influenced by soil drench application of paclobutrazol. *Bangladesh J. Agric. Res.* **2012**, *37*, 335–348. [CrossRef]

31. Kurian, R.M.; Iyer, C.P.A. Stem anatomical characteristics in relation to tree vigour in mango (*Mangifera indica* L.). *Sci. Hortic.* **1992**, *50*, 245–253. [CrossRef]

32. Cardenas, K.; Rojas, E. Effect of paclobutrazol and nitrates of potassium and calcium on the development of the mango "Tommy Atkins". *Bioagro* **2003**, *15*, 83–90.

33. Ram, S. Hormonal physiology of flowering in "Dashehari" mango. *J. Appl. Hortic.* **1999**, *1*, 84–88.

34. Jogdande, N.D.; Choudhari, K.G. Seasonal changes in auxin content and its role in flowering of mango (*Mangifera indica* L.). *Orissa J. Hortic.* **2001**, *29*, 10–12.

35. Dalal, S.R.; Gonge, V.S.; Jadhao, B.J.; Jogdande, N.D. Effect of chemical on flowering and fruit yield of mango cv. Pairy. *Int. J. Agric. Sci.* **2005**, *1*, 24–25.

36. Mosqueda-Vazquez, R.; Avila-Resendiz, C. Floral induction of mango with KNO_3 applications and its inhibition by $AgNO_3$ or $CoCl_2$ application. *Hortic. Mex.* **1985**, *1*, 93–101.

37. Singh, D.B.; Ranganath, H.R. Induction of regular and early fruiting in mango by paclobutrazol under tropical humid climate. *Indian J. Hortic.* **2006**, *63*, 248–250.

38. Quinlan, J.D. New approaches to the control of fruit tree forms and size. *Acta Hortic.* **1981**, *120*, 95–105. [CrossRef]

39. Singh, Z.; Dhillon, B.S. Effect of paclobutrazol on floral malformation, yield and quality of mango (*Mangifera indica* L.). *Acta Hortic.* **1992**, *296*, 51–53. [CrossRef]

40. Shinde, A.K.; Waghmare, G.M.; Wagh, R.G.; Burondkar, M.M. Effect of dose and time of paclobutrazol application on flowering and yield of mango. *Indian J. Plant Physiol.* **2000**, *5*, 82–84.

41. Anbu, S.; Parthiban, S.; Rajangam, J.; Thangaraj, T. Induction of off-season flowering in mango (*Mangifera indica* L.) using paclobutrazol. *South Indian Hortic.* **2002**, *49*, 384–385.

42. Oosthuyse, S.A. Effect of KNO_3 sprays to flowering mango trees on fruit retention, fruit size, tree yield and fruit quality. *Acta Hortic.* **1997**, *455*, 359–366. [CrossRef]

43. Feungchan, S.; Yimsawat, T.; Chindaprasert, S.; Hongsbhanich, N.; Daito, H. The effect of the fertilizer application interval on the mango. *Kaen Kaset Khon Kaen Agric. J.* **1989**, *17*, 100–105.

44. Ghosh, S.N.; Chattopadhyay, N. Foliar application of urea on yield and physico-chemical composition of mango fruits cv. Himsagar under rainfed condition. *Hortic. J.* **1999**, *12*, 21–24.

45. Sarker, B.C.; Rahim, M.A.; Pomology Division, Horticulture Research Centre, Bangladesh Agricultural Research Institute, Joydebpur, Gazipur 1701, Bangladesh. Unpublished Data, 2012.

11

Effect of Organic Production Systems on Quality and Postharvest Performance of Horticultural Produce

Francesco Giovanni Ceglie [1,2], **Maria Luisa Amodio** [2] **and Giancarlo Colelli** [2,*]

[1] Organic Agriculture Department, Mediterranean Agronomic Institute of Bari, CIHEAM-IAMB, Via Ceglie, 9, Valenzano, BA 70010, Italy; ceglie@iamb.it

[2] Department of Science of Agriculture, Food and Environment, University of Foggia, Via Napoli 25, Foggia, FG 71122, Italy; marialuisa.amodio@unifg.it

* Correspondence: giancarlo.colelli@unifg.it

Abstract: Organic standards include a well-defined set of practices and a list of technical tools that are permitted by regulation. Organic products are mainly purchased for their safety and absence of synthetic pesticide residues. Furthermore, a diet based on organic products claims to provide health benefits due to the high nutritional value compounds that are more concentrated in organic products compared to conventional ones. As the scientific basis of the differences between organically- and conventionally-grown fruits and vegetables is under debate, some of the published work, together with some recent unpublished results, will be covered in the present review. In addition, the effect of different approaches to organic horticultural production will be described. Many studies have confirmed lower nitrate content, especially in leafy vegetables, and higher antioxidant compounds in organically-grown fruits in comparison to conventional ones. A recent study reported organic kiwifruit as higher in ascorbic acid and total phenol content than conventional kiwifruit. These differences were maintained throughout cold storage. Similarly, in organic grapes, antioxidant-related compounds were significantly higher than in conventionally-grown grapes. Analogous results were obtained with organic strawberries grown in protected conditions. However, conventional products usually result in higher moisture content, and this should be taken into account to confirm the differences on a dry matter basis. Possible explanations for the effects of organic farming practices on nutritional quality and postharvest performance of fresh produce are the following: (i) organic amendments provide a high input of exogenous organic matter and of nutrients for a long period; in contrast, mineral fertilizers, allowed only in conventional farming systems, are highly concentrated in nutrients that are directly available for root uptake in a shorter time period; (ii) the use of synthetic pesticides (only possible in conventional agriculture) slows down defence mechanisms against pathogens, with the consequence of favoring primary metabolism; (iii) cultural practices may result in different plant composition and nutritional quality, which in turn influence cold storage performance of the products as these differences, both in fertility and pest management, affect the allocation of secondary plant metabolites (such as ascorbic acid and phenolic compounds).

Keywords: shelf life; secondary metabolites; oxidative stress; strawberry; tomato; nutrition

1. Introduction

Organic standards include a well-defined set of practices and a list of technical tools that are permitted by regulations (*i.e.*, Reg n.889/08 in UE and the National Organic Program in U.S.). A diet based on organic products claims to provide health benefits due to the higher concentration of nutritional compounds compared to conventional ones, and the absence of pesticide residues [1]. The present challenge of feeding the world requires new strategies to ensure food security which is surely based on food availability and access, but also on food safety and nutritional quality. Organic production systems may be a way to ensure the sustainability of production, allowing preservation of natural resources for present and future generations, while providing a high quality and long shelf life of the product [2]. Despite the importance of this issue, few comparative studies have focused on postharvest aspects, and the scientific basis of the possible difference between organically- and conventionally-managed produce is still debated [3]. Some published studies, together with some recent unpublished results, will be covered in the present review. In particular, comparison studies that were conducted under similar environmental conditions with respect to climate, soil characteristics and availability of nutrients, have been considered for the effect of diverse pre-harvest practices on postharvest performance.

2. Diverse Plant Response to Different Systems of Production

The overall quality characteristics of the plant parts which are used as food are the result of the interactions among genotype, environmental conditions, cultural practices and postharvest handling and processing techniques. Any environmental factor that varies from the ideal potentially generates a stress which may either promote or inhibit specific plant responses. Such stress may lead to oxidative stress that produces reactive oxygen species (ROS) in the chloroplasts (free radicals like superoxide and molecular forms like singlet oxygen) and in the mitochondria (mainly superoxide). Ascorbic acid, glutathione, phenolic compounds, and alkaloids, among the antioxidants, are non-enzymatic defenses used by the plant to cope with this dangerous situation [4]. In general, an organic production system is a stressful system largely due to an insufficient supply of mineral nitrogen throughout the crop cycle, fostering phenolics and ROS which are natural defense substances. For organic plants, the content of secondary metabolites, that also include antioxidants, have been positively correlated with these natural defense substances [5]. However, antioxidant biosynthesis requires plant resource reallocation from primary metabolism to secondary metabolite production. As a consequence, cultural practices, both organic and conventional, may result in different plant composition and nutritional quality at harvest due to different biotic and/or abiotic stress conditions, which, in turn, influence the concentration of secondary metabolites.

3. Quality and Postharvest Performances of Organic and Conventional Fruit and Vegetables

In the scientific literature, different studies have claimed opposite results. Woëse *et al.* [6] and Smith Spangler *et al.* [3] estimated large but not statistically significant differences in quality between organic and conventionally-managed crops, which led them to conclude that the two production systems had similar qualitative characteristics. However, in the last decade, other studies have contributed to enlarge the dataset for this comparison. Lairon [7] confirmed the absence of pesticide residues in organic food (97% of samples had no detectable level of pesticide) and reported a lower content of nitrogen in organic vegetables in comparison to conventional ones. Similarly, Baranski *et al.* [8] concluded that the concentration of total nitrogen was 10% lower in organic compared to conventional crops (nitrate 30% and nitrite 87% lower) [9]. The low percentage of nitrogen forms was correlated with a high concentration of secondary metabolites (such as phenols and vitamins, which do not contain nitrogen). Several studies have confirmed elevated levels of secondary metabolites in organic carrots [10], sweet peppers [11], and tomatoes [12]. Other studies have reported opposite results with tomatoes [13] and carrots [14]. Plant secondary metabolites are important for their role in enhancing human health [15], and, as a result, much attention has been dedicated to this family of compounds in the recent comparison studies. Luthria [16] found similar phenolic acid content in eggplant cultivated with organic or conventional techniques. Chassy *et al.* [17] reported higher values of antioxidant activity in organic tomato and bell peppers, while D'Evoli *et al.* [18] found that total polyphenol content was higher in conventional plums, but ascorbic acid, α-tocoferol, and β-carotene were higher in organic samples. Furthermore, the majority of studies comparing the polyphenol content of plant products from different farming systems indicated significantly higher concentrations of these substances in organic fruits and vegetables compared to conventional ones. Lairon [7] highlighted the high level of antioxidants, minerals, and dry matter in organically-grown products. However, conventional products usually result in higher water content, and this should be taken into account with the differences confirmed in terms of dry matter. Bourn and Prescott [19] observed higher dry matter content in organically-grown carrots and spinach in comparison with the conventional ones. Ceglie *et al.* [20] and Conti *et al.* [21] found a higher dry matter content in organic strawberries. Dry matter at harvest has been proposed as a promising predictor of post-harvest soluble solids in apples, also affecting the relative amount of mass that could be lost during cold storage [22]. Woëse *et al.* [6] reported no clear trend in the level of total sugar content in organic vegetables compared with conventional ones.

Different types and amounts of secondary metabolites at harvest may affect antioxidant changes during the cold storage period and determine the diverse postharvest performance of the products. Despite the relevance of the issue, few comparative studies have focused on such postharvest issues. Differences during cold storage in incidence of physiological disorders, soluble solids content, firmness, and mineral content have been reported [23,24]. After ethylene treatment, organic banana showed faster peel color changes, and lower gravimetric water balance and pulp/peel ratio, and impedance in comparison to conventionally-grown banana [25]. During simulated marketing conditions, organoleptic characteristics and resistance to deterioration were higher for organic strawberries in comparison to the conventional fruit [26]. Hasey *et al.* [27] reported higher soluble solids content and firmness in organically-grown kiwifruit compared with conventionally-grown kiwifruit. In contrast, Benge *et al.* [28] reported that conventional kiwifruit showed higher soluble solids content and similar softening behavior and decay than organic ones, analyzed at the same firmness stage. The levels of calcium in kiwifruit were negatively correlated with the incidence of soft patches. Moreover, higher levels of Ca as well as of NO_3^-, Mg, Fe, and Zn were observed at harvest and during 25 days of cold storage at 10 °C for conventional "Meyer" lemons (*Citrus meyeri* Tan.) in comparison with fruits from an organic orchard [29].

4. Postharvest Performance of Organically and Conventionally Grown Kiwifruits

Amodio *et al.* [30] compared postharvest performance of organic and conventional "Hayward" kiwifruit harvested at the same maturity stage. Fruit shape and peel characteristics before cold storage, maturity indices (CO_2 and C_2H_4 production, firmness, color, soluble solids content, and acidity) and compounds associated with flavor and nutritional quality (minerals, sugars and organic acids, ascorbic acid, total phenols, and antioxidant activity) were determined at 0, 35, 72, 90, and 120 days of cold storage at 0 °C, and after 1 week of shelf-life simulation at 20 °C. At harvest, organic and conventional kiwifruit had similar soluble solids content. Conventional kiwifruit had a higher firmness and L* value, and a lower hue angle and chromaticity, resulting in a lighter green color when compared with the organic kiwifruit. During cold storage, soluble solids content increased more in conventional than in organic kiwifruit. Concerning nutritional compounds, ascorbic acid, and total phenols were more concentrated in organic than conventional kiwifruit, resulting in a higher antioxidant activity. These differences were maintained throughout the cold storage period. The two production systems resulted in different morphological attributes since organic kiwifruits exhibited a larger total and columella area, smaller flesh area, more spherical shape, and thicker skin compared to conventional kiwifruit. Also, the levels of the main mineral constituents were higher in organic than in conventional kiwifruit.

5. Postharvest Quality of Organic and Conventional White Table Grapes

The effects of organic *versus* conventional production of table grapes grown at two locations were evaluated at harvest and during 14 days of cold storage at 0 °C (Amodio *et al.*, unpublished). Respiration rate, firmness, color, soluble solids content, acidity, organic acids, ascorbic acid, total phenols, antioxidant activity, flavonols, and peel characteristics were determined at harvest and during cold storage. Only from one location, phenolic compounds were significantly higher in the organic table grapes (505.1 $\pm$ 52.4 mg gallic acid/100 g) than in the conventional ones (370 $\pm$ 58 mg gallic acid/100 g). Similarly, antioxidant activity was higher in organic grapes (1211 $\pm$ 134 mg Trolox/100 g) than in conventional ones (763 $\pm$ 98 mg Trolox/100 g). This difference was observed until the sixth day of cold storage. Vitamin C in organic and conventional table grapes was similar at harvest and throughout cold storage. Storability was greatly affected by the agricultural system resulting in a shorter shelf-life for organic grapes, which scored the highest values for firmness and appearance.

6. Organic *vs.* Conventional Products: Comparative Analysis in a "Cul De Sac"

Organic and conventional production systems are comprised of a number of very diverse approaches: sequences of crop rotation, sources and quality of inputs, timing and doses of water and fertilizers, weed management and cover crop use, and pathogen and pest control. Organic and conventional farming cannot indeed be defined as two isolated clusters of agronomic practices. On one hand, it is possible to have high or moderate input intensity conventional farming systems. An example of the latter is the restricted use of chemical inputs by private standards (*i.e.*, for commercial reasons), such as for "zero pesticide residues" labels. In contrast, organic farming relies on more complex agro-ecosystem management that includes leguminous crops and organic matter-based amendments, although it may also be managed with a simple input substitution based on the replacement of chemical fertilizers and pesticides with organic-allowed inputs. In the scientific literature, this process in known as "conventionalization" of the organic farming [31] which may be summarized as the development of organic farming practices that might not be sustainable but that are not excluded by organic regulations [32]. In particular, organic greenhouse production is a clear example of this. The highly intensive production levels recorded in greenhouse systems requires high availability of nutrients to sustain crop growth. As a consequence, organic growers simplify crop rotation, excluding cover crops, and use easily soluble organic fertilizers. These issues should be taken into account in studies that aim to compare the quality and post-harvest performances of organic *vs.* conventional farming systems. Comparative analyses have to face the heterogeneity of the circumstances; observed differences and/or similarities would not necessarily be related to certification systems of the products because organically-certified products might have been produced by a "conventionalized" organic system and vice-versa. If organic-conventional system comparisons implemented the same agronomic practices and only varied in the type of nutrient or pesticide input, this would not be an organic *vs.* conventional system comparison but a comparison of organic *vs.* conventional inputs [33]. In this respect, there is room to begin comparative analyses among different approaches to organic methods of production in order to appreciate, within the greater heterogeneous cluster of organic practices, the impact on the quality of an organic conventionalized system *vs.* other organic systems based on agro-ecological practices. Such comparison studies should be considered as investigations of the effects of pre-harvest practices on post-harvest quality and product shelf life. In this regard, two research studies are reported below as preliminary results of this promising research approach.

7. Effects of Different Systems of Organic Production on Quality and Post-Harvest of Tomato

Tomato, a climacteric fruit, represents a significant source of folate, vitamin C, polyphenols, and other antioxidants [34]. A conventionalized organic production system based on organic commercial fertilizers was compared with two organic production systems based on agro-ecological practices represented by (i) animal manure amendments and dead mulches of cover crops; or (ii) on-farm compost amended green manuring of cover crops. For the three production systems, tomato fruit respiration rate, morphological, physical, sensorial characteristics, and nutritional compound content were monitored. It was observed that dehydroascorbic acid was significantly higher in the conventionalized system for tomato fruit harvested at breaker and pink stages. However, this difference disappeared after 10 days of cold storage at 15 °C. Organic agro-ecological systems were able to obtain a similar tomato yield and fruit quality as the "conventionalized" organic system which used off-farm inputs only. This was confirmed both at harvest and during cold storage. Further information and results of this experiment have been reported [35]. It is worth noting that when the environmental conditions are similar in terms of climate, soil characteristics, availability of water, and soluble nutrients for the roots, any differences in the cultural practices (even relevant ones such as cover crop mixtures *vs.* bare soil, organic dead mulch *vs.* plastic mulch, manure *vs.* commercial organic fertilizers) did not affect crop quality and postharvest performance.

8. Effects of Different Systems of Organic Production on Quality and Post-Harvest of Strawberry

Ceglie *et al.* [20] compared quality characteristics and postharvest performance of organically- and conventionally-produced strawberry fruit. Three organic farming systems (two "agro-ecological" and one "conventionalized") were compared to a conventional system. The production system affected the quality parameters of the strawberries cultivated in unheated tunnels. Conventionally-grown strawberry fruit were greater in diameter, chroma values, and firmness compared to the organically-produced fruit regardless of the specific organic system. Vitamin C, malic acid, tartaric acid, fructose, and glucose content were higher in the organic strawberries compared to conventional fruit. In fact, organically-grown strawberries received the highest evaluation for sweetness and aroma. Similar results were obtained for different varieties in the same geographic area (South of Italy) as reported by Conti *et al.* [21]. During the entire period of cold storage titratable acidity, citric acid, and total phenol values presented a similar decreasing trend from all of the production systems. However, the two organic agro-ecological systems differed in some aspects from both the "conventionalized" organic and the conventional system of production. At harvest, ascorbic acid and sucrose concentration were higher in both the organic agro-ecological than in the other two systems. Furthermore, during the entire cold storage period, tartaric acid and total phenols were higher in fruit from the organic agro-ecological systems than in either the conventional or the "conventionalized" organic systems. Thus, a "conventionalized" organic system may be halfway between conventional farming and organic-agro-ecological systems both in terms of quality and in terms of environmental sustainability. In this regard, Reganold *et al.* [36] linked the high quality of organic strawberry fruit with the high capability and stress resilience of organically-managed soil. Further studies on other commodities and with a longer-term assessment perspective are necessary to individuate sets of cultural practices applicable under the organic regulation which may enhance the quality of organic produce.

9. Conclusions

This review represents a small contribution to the wider picture of the quality of horticulture produce resulting from systems of production. Organic production systems have the objective of including a rational use of natural resources with high quality and shelf life performance. In many cases, conventional farming systems have achieved such high results. Nowadays, organic agriculture is increasing in terms of area of production and number of operators. A wide set of solutions have been proposed which, although valid with respect to organic certification standards, still need scientific assessment concerning claims of sustainability and high quality production. More in-depth analyses may relate the organic *vs.* conventional comparison to the more general issue of pre-harvest effects on postharvest performance of crops. In this respect, the balance between primary and secondary metabolic pathways seem to be an important aspect resulting from the complex interaction of genotype, environment, and agricultural practices which lead to differences in quality and postharvest performance of fresh fruits and vegetables. The need to improve the quality of food available in the world in a sustainable way should orient research on agricultural practices to increase the nutritional composition of fresh fruits and vegetables and to enhance the shelf life.

Author Contributions: This work was a product of the combined effort of all of the authors. The authors equally contributed to the literature reviewing, to the manuscript writing and revisions.

Conflicts of Interest: The authors declare no conflict of interest.

References

1. Oates, L.; Cohen, M.; Braun, L.; Schembri, A.; Taskova, R. Reduction in urinary organophosphate pesticide metabolites in adults after a week-long organic diet. *Environ. Res.* **2014**, *132*, 105–111. [CrossRef] [PubMed]

2. Rembiałkowska, E. Quality of plant products from organic agriculture. *J. Sci. Food Agric.* **2007**, *87*, 2757–2762. [CrossRef]

3. Smith Spangler, C.; Brandeau, M.L.; Hunter, G.E.; Bavinger, J.C.; Pearson, M.; Eschbach, P.J.; Sundaram, V.; Liu, H.; Schirmer, P.; Stave, C.; *et al.* Are organic foods safer or healthier than conventional alternatives? A systematic review. *Ann. Intern. Med.* **2012**, *157*, 348–366. [CrossRef] [PubMed]

4. Gill, S.S.; Tuteja, N. Reactive oxygen species and antioxidant machinery in abiotic stress tolerance in crop plants. *Plant Physiol. Bioch.* **2010**, *48*, 909–930. [CrossRef] [PubMed]

5. Winter, C.K.; Davis, S.F. Organic foods. *J. Food Sci.* **2006**, *71*, R117–R124. [CrossRef]

6. Woëse, D.; Lange, C.; Boess, K.; Bogl, W. A comparison of organically and conventionally grown foods—Results of a review of the relevant literature. *J. Sci. Food Agric.* **1997**, *74*, 281–293. [CrossRef]

7. Lairon, D. Nutritional quality and safety of organic food. A review. *Agron. Sustain. Dev.* **2010**, *30*, 33–41. [CrossRef]

8. Barański, M.; Srednicka-Tober, D.; Volakakis, N.; Seal, C.; Sanderson, R.; Stewart, G.B.; Leifert, C. Higher antioxidant and lower cadmium concentrations and lower incidence of pesticide residues in organically grown crops: A systematic literature review and meta-analyses. *Br. J. Nutr.* **2014**, *112*, 794–811. [CrossRef] [PubMed]

9. Caruso, G.; Conti, S.; La Rocca, G. Influence of crop cycle and nitrogen fertilizer form on yield and nitrate content in different species of vegetables. *Adv. Hortic. Sci.* **2011**, *25*, 81–89.

10. Sikora, M.; Hallmann, E.; Rembiałkowska, E. The content of bioactive compounds in carrots from organic and conventional production in the context of health prevention. *Rocz. Państw. Zakł. Hig.* **2009**, *60*, 217–220. [PubMed]

11. Del Amor, F.M.; Serrano-Martinez, A.; Fortea, I.; Nunez-Delicado, E. Differential effect of organic cultivation on the levels of phenolics, peroxidase and capsidiol in sweet peppers. *J. Sci. Food Agric.* **2008**, *88*, 770–777. [CrossRef]

12. Pieper, J.R.; Barrett, D.M. Effects of organic and conventional production systems on quality and nutritional parameters of processing tomatoes. *J. Sci. Food Agric.* **2009**, *89*, 177–194. [CrossRef]

13. Rossi, F.; Godani, F.; Bertuzzi, T.; Trevisan, M.; Ferrari, F.; Gatti, S. Health-promoting substances and heavy metal content in tomatoes grown with different farming techniques. *Eur. J. Nutr.* **2008**, *47*, 266–272. [CrossRef] [PubMed]

14. Krejčová, A.; Návesník, J.; Jičínská, J.; Černohorský, T. An elemental analysis of conventionally, organically and self-grown carrots. *Food Chem.* **2016**, *192*, 242–249. [CrossRef] [PubMed]

15. Lundegårdh, B.; Mårtensson, A. Organically produced plant foods—Evidence of health benefits. *Acta Agric. Scand. B* **2003**, *53*, 3–15. [CrossRef]

16. Luthria, D.; Singh, A.P.; Wilson, T.; Vorsa, N.; Banuelos, G.S.; Vinyard, B.T. Influence of conventional and organic agricultural practices on the phenolic content in eggplant pulp: Plant-to-plant variation. *Food Chem.* **2010**, *121*, 406–411. [CrossRef]

17. Chassy, A.W.; Bui, L.; Renaud, E.N.C.; Van Horn, M.; Mitchell, A.E. Three-year comparison of the content of antioxidant micronutrients and several quality characeeristics in organic and conventionally managed tomatoes and bell peppers. *J. Agric. Food Chem.* **2006**, *54*, 8244–8252. [CrossRef] [PubMed]

18. D'Evoli, L.; Tarozzi, A.; Hrelia, P.; Lucarini, M.; Cocchiola, M.; Gabrielli, P.; Franco, F.; Morroni, F.; Catelli-Forti, G.; Lombardi Boccia, G. Influence of cultivation system on bioactive molecules synthesis in strawberries: Spin-off on antioxidant and antiproliferative activity. *J. Food Sci.* **2010**, *75*, 94–99. [CrossRef] [PubMed]

19. Bourn, D.; Prescott, J. A comparison of the nutritional value, sensory qualities, and food safety of organically and conventionally produced foods. *Crit. Rev. Food Sci. Nutr.* **2002**, *42*, 1–34. [CrossRef] [PubMed]

20. Ceglie, F.G.; Mimiola, G.; Dechiara, M.L.; Amodio, M.L.; Colelli, G. A comparative study of quality and post-harvest performances of organically and conventionally grown strawberry (Fragaria × ananassa DUCH. "Festival"). In Proceedings of the 2nd Congress of the Italian Network for Research in Organic Farming—RIRAB, Rome, Italy, 11–13 June 2014.

21. Conti, S.; Villari, G.; Faugno, S.; Melchionna, G.; Somma, S.; Caruso, G. Effects of organic *vs.* conventional farming system on yield and quality of strawberry grown as an annual or biennial crop in southern Italy. *Sci. Hort.* **2014**, *180*, 63–71. [CrossRef]

22. McGlone, V.A.; Jordan, R.B.; Seelye, R.; Clark, C.J. Dry-matter—A better predictor of the post-storage soluble solids in apples? *Postharvest Biol. Technol.* **2003**, *28*, 431–435. [CrossRef]

23. DeEll, R.; Prange, K. Postharvest physiological disorders, diseases and mineral concentrationsof organically and conventionally grown McIntosh and Cortland apples. *Can. J. Plant Sci.* **1993**, *73*, 223–230. [CrossRef]

24. Weibel, F.P.; Treutter, D.; Graf, U.; Haseli, A. Sensory and health related fruit quality of organic apples. A comparative field study over three years using conventional and holistic methods to assess fruit quality. In Proceedings of the 11th International Conference on Cultivation Technique and Ohytopatological Problems in Organic Fruit Growing, Weinsberg, Germany, 3–5 February 2004; pp. 185–195.

25. Nyanjage, M.O.; Wainwright, H.; Bishop, C.F.H.; Cullum, F.J. A comparative study on the ripening and mineral content of organically and conventionally grown Cavendish bananas. *Biol. Agric. Hortic.* **2001**, *18*, 221–234. [CrossRef]

26. Cayuela, A.; Vidueira, M.; Albi, A.; Gutierrez, F. Influence of the ecological cultivation of strawberries (Fragaria × ananassa cv. Chandler) on the quality of the fruit and on their capacity for conservation. *J. Agric. Food Chem.* **1997**, *45*, 1736. [CrossRef]

27. Hasey, J.K.; Johnson, R.S.; Meyer, R.D.; Klonsky, K. An organic *versus* conventional farming system in kiwifruit. *Acta Hort.* **1997**, *444*, 223–228. [CrossRef]

28. Benge, J.R.; Banks, N.H.; Tillmann, R.; Nihal de Silva, H. Pairwise comparison of the storage potential of kiwifruit from organic and conventional production system. *N.Z. J. Crop Hort. Sci.* **2000**, *28*, 147–152. [CrossRef]

29. Uckoo, R.M.; Jayaprakasha, G.K.; Patil, B.S. Phytochemical analysis of organic and conventionally cultivated Meyer lemons (*Citrus meyeri* Tan.) during refrigerated storage. *J. Food Compos. Anal.* **2015**, *42*, 63–70. [CrossRef]

30. Amodio, M.L.; Colelli, G.; Hasey, J.K.; Kader, A.A. A comparative study of composition and postharvest performance of organically and conventionally grown kiwifruits. *J. Sci. Food Agric.* **2007**, *87*, 1228–1236. [CrossRef]

31. Goldberger, J.R. Conventionalization, civic engagement, and the sustainability of organic agriculture. *J. Rural Stud.* **2011**, *27*, 288–296. [CrossRef]

32. De Wit, J.; Verhoog, H. Organic values and the conventionalization of organic agriculture. *NJAS-Wagen. J. Life Sci.* **2007**, *54*, 449–462. [CrossRef]

33. Seufert, V.; Ramankutty, N.; Foley, J.A. Comparing the yields of organic and conventional agriculture. *Nature* **2012**, *485*, 229–232. [CrossRef] [PubMed]

34. Charanjeet, K.; George, B.; Deepa, N.; Singh, B.; Kapoor, H. Antioxidant status of fresh and processed tomato. *J. Food Sci. Technol.* **2004**, *41*, 479–486.

35. Ceglie, F.G.; Muhadri, L.; Piazzolla, F.; Martinez-Hernandez, G.B.; Amodio, M.L.; Colelli, G. Quality and postharvest performance of organically grown tomato (*Lycopersicon. Esculentum* L. "Marmande") under unheated tunnel in Mediterranean climate. *Acta Hort.* **2015**, *1079*, 487–494. [CrossRef]

36. Reganold, J.P.; Andrews, P.K.; Reeve, J.R.; Carpenter-Boggs, L.; Schadt, C.W.; Alldredge, J.R.; Ross, C.F.; Davies, N.M.; Zhou, J. Fruit and soil quality of organic and conventional strawberry agro-ecosystems. *PLoS ONE* **2010**, *5*, e12346. [CrossRef]

12

Vacuum Packaging Controlled Crown Rot of Organically-Grown Balangon (*Musa acuminata* AAA Group) Banana

Elda Esguerra *, Dormita Del Carmen, Roxanne Delos Reyes and Ryan Anthony Lualhati

Crop Science Cluster, Postharvest and Seed Sciences Division, Postharvest Horticulture Training and Research Center (PHTRC), College of Agriculture, University of the Philippines Los Baños (UPLB), Laguna 4031, Philippines; drdcarmen@yahoo.com (D.D.C.); roxanne.edr@gmail.com (R.D.R.); raolualhati@gmail.com (R.A.L.)
* Correspondence: elda_esguerra@yahoo.com

Abstract: Balangon bananas take about 23 to 28 days from harvest to reach Japan since the fruit have to be assembled from small and scattered farms, hence the problems of premature ripening and crown rot. The effectiveness of vacuum packaging in retarding ripening and in controlling crown rot has not been documented for organically-grown Balangon bananas. Balangon bananas harvested from farms in Don Severino Benedicto, Negros Occidental, Phillipines, were washed three times in tap water, then packed (wet packing) in a 13-kg capacity corrugated fibreboard carton lined with 0.05 mm thick low density polyethylene (LDPE) bag, and vacuum-packed using an ordinary vacuum cleaner. Bananas treated with 1% sodium bicarbonate were also subjected to vacuum packing. Packaged bananas were then loaded in refrigerated vans (13.0–13.5 °C), transported to Manila and then to the UPLB-PHTRC laboratory for simulated domestic and international shipments which took about 25 days from harvest until the bananas reached Japan. Bananas were then taken out of the sealed LDPE, allowed to equilibrate at 18 °C, treated with 2500 µL/L ethephon, and held at 23 °C for ripening. During the 25-day holding at 13.0–13.5 °C, bananas that were vacuum-packaged remained green. In the control (not vacuum-packaged), a few fingers in each hand started to ripen. The most significant effect of vacuum packaging in combination with 13 °C storage was the control of crown rot, particularly when bananas started to ripen. With vacuum packaging, the incidence of crown rot at the ripe stage was 2.8% compared with 55.7% in the control. Sodium bicarbonate did not control crown rot alone, nor contribute to the reduction caused by packaging and vacuum associated with the control of decay was the high visual quality rating of the fruit. Extended storage under vacuum-packed conditions did not significantly affect the physico-chemical and sensory attributes of bananas at the ripe stage.

Keywords: organic banana; crown rot; vacuum packaging; sodium bicarbonate

1. Introduction

Banana is one of the major sources of income and foreign exchange earnings which accounts for about 9% of total agricultural exports of the Philippines. The major export variety is Cavendish, and is mainly handled by multinational corporations. The other variety that is exported and has found a niche market in Japan as a "non-chemical" banana is the Bungulan (*Musa acuminata*, AAA group), popularly known as Balangon [1]. Balangon accounts for about 47% of the export volume of small banana holdings. Balangon bananas for export are grown by clusters of farmers who follow a protocol for product quality and safety. However, even with this quality system, problems still occur when the fruit reaches Japan. It was reported that losses of about 6%–15% of the fruit occur due to crown rot and anthracnose (Alter Trade Japan, 2012 personal communication) and mechanical damage. Since Balangon bananas are grown organically, and fruit come from different production areas where pre- and postharvest management practices differ, it is expected that quality problems will occur. Moreover, the long interval from harvest to final distribution to market cooperatives in Japan, which takes about 21–28 days, ultimately results in quality deterioration and decay. In Balangon banana, Alvindia et al. [2] reported that the most active crown rot pathogens were *Colletotrichum musae*, *Fusarium verticillioides*, *Lasiodiplodia theobromae*, and *Thielaviopsis paradoxa*. These fungi infect the crown through fresh wounds created during dehanding and trimming.

Modified atmosphere packaging (MAP) is an ideal preservation technique and is known to have great potential in extending the postharvest life of fruits and vegetables [3]. Vacuum packaging which is an active form of MAP has been shown to retard ripening of bananas even with extended holding [4,5]. Vacuum packaging is currently being practiced in conventionally-grown Cavendish bananas exported to the Middle East wherein sea shipment takes about 18–21 days from the port of Davao, Philippines. In the case of these conventionally-grown bananas, pretreatment with fungicide after harvest is the supplementary treatment for crown rot control. For organic Balangon bananas, synthetic fungicides are not allowed, hence the need for alternative supplementary treatments to MAP that are compatible with organic standards. Alvindia [6] recommended supplementary treatments of generally regarded as safe (GRAS) compounds like sodium bicarbonate for the control of crown rot of bananas. His earlier studies showed a 92% reduction in decay using sodium bicarbonate in combination with a biological control agent [7]. This study was conducted with the following objectives: (a) to control the incidence and severity of crown rot of Balangon bananas during 25-day storage under MAP at 13.0–13.5 °C and during poststorage at 23 °C; (b) to determine the physico-chemical changes at the ripe stage of Balangon bananas subjected to MAP; and (c) to assess the sensory acceptability at the ripe stage of previously-stored bananas under MAP.

2. Experimental Section

Green but commercially mature, organically-grown Balangon bananas were harvested from farms in Don Salvador Benedicto, Negros Occidental, Philippines, and transported to the central packinghouse facility in Bacolod City. Dehanded bananas were washed three (3) times (in wash tanks) with running water. During the first washing, the crown was trimmed and the damaged fingers were removed. Hands of bananas were then divided into clusters consisting of 2–3 clusters per hand.

2.1. Vacuum Packaging Treatment

Vacuum packaging as an active form of MAP was employed in the study. The banana corrugated fibreboard carton (13-kg capacity) was lined with non-perforated 0.05 mm thick, low density polyethylene (LDPE) bag prior to packing of bananas that were still wet (wet packing as practiced commercially). To prevent abrasion, polystyrene liners were inserted in between the layer and clusters of bananas. Each carton weighing about 13.0 kg contained 18 clusters of bananas. Vacuum was applied for about 10 s using an ordinary vacuum cleaner to evacuate the air inside the LDPE bag. The bags were sealed by making a knot (through hand twisting) and tied with rubber bands. For the combination treatment of vacuum packaging and 1% sodium bicarbonate, baking soda (BS) which was composed mainly of sodium bicarbonate was used. The banana clusters were dipped for one (1) min in a freshly-prepared 1% solution of BS, then packed wet in LDPE bags and subjected to vacuum packing. To determine the efficacy of sodium bicarbonate in controlling crown rot when used as a single treatment, clusters of bananas dipped in 1% BS were packed wet in boxes without LDPE liner. The control consisted of banana clusters washed only in water (as currently practiced by the company) and packed wet in cartons without LDPE liner.

2.2. Inter-Island Domestic Transport and Simulated Export Shipment to Japan

Packed cartons of bananas were temporarily kept under ambient condition in the company packinghouse for about 12 h and were then loaded in a reefer van (13.5 °C) together with other bananas for export and transported to Manila. Holding in the domestic reefer vans lasted for 14 days rather than the normal 7 days since the bananas could not be loaded in the international reefer van due to port congestion in Manila. A simulated sea shipment export to Japan was conducted at the PHTRC-UPLB laboratory wherein experimental cartons of bananas were stored at 13.0 °C for 11 days representing the 5-day sea shipment and 6-d temporary storage upon arrival in Japan.

2.3. Induction of Ripening after MAP

On the 26th day from harvest, bananas were removed from vacuum packaging and exposed to air at 18 °C for induction of ripening. Once the pulp temperature reached 18 °C, bananas were dipped for 30 s in 2500 µL/L ethephon (2-chloroethyl phosphonic acid, 46% a.i.) as ethylene source insuring that the crown and the pedicel were not dipped in the solution. The temperature of the ripening room was set at 23 °C to simulate the condition during retail distribution in Japan.

2.4. Data Gathered

The levels of oxygen (O_2), carbon dioxide (CO_2) and ethylene (C_2H_4) in boxes of bananas subjected to vacuum packaging were determined at different points during the domestic inter-island and export sea shipment to Japan as follows:

DAH 7: 7 days after MA packaging, arrival of bananas at Alter Trade Corp. warehouse
DAH 8: 8 days after MA packaging, arrival at PHTRC-UPLB laboratory
DAH 14: 14 days after MA packaging, shipment departure for Japan
DAH 19: 19 days after MA packaging; arrival of bananas in Japan
DAH 26: 26 days after MA packaging, opening of the MA-packed bananas for initiation of ripening

Strips of electrical tape were placed onto each LDPE bag, and headspace samples were taken through the tape using a 1.0 mL disposable syringe. The O_2 and CO_2 that accumulated in the headspace were determined using a gas chromatograph (Shimadzu 8A GC) fitted with a thermal conductivity detector (TCD), and for C_2H_4, a flame ionization detector (FID, Shimadzu 12A GC). Upon opening of the vacuum-packaged bananas and during ripening, the incidence and severity of crown rot were monitored in each box following a rating of 0 to 7 [6] as follows (Table 1):

Table 1. Crown rot index used in assessing the severity of infection.

Crown Rot Index	Description
0	no discoloration or mycelial growth on the crown
1	discoloration or mycelial growth limited on the surface of the crown
2	discoloration or mycelial growth less than 10% of crown area
3	11%–40% discoloration or mycelial growth on crown area
4	41%–70% discoloration or mycelial growth on crown area
5	71%–100% discoloration or mycelial growth on crown area
6	Discoloration or mycelial growth advanced to finger stalks
7	Finger stalk rot occurs causing the fingers to drop when handled

The change in the visual quality of the fruit was monitored daily at 23 °C using the following rating: 8, 9 (excellent, field fresh), 6, 7 (very good, defects fair), 4, 5 (good, defects moderate), 3 (limit of marketability), and 1, 2 (poor, limit of edibility). Defects attributed to deteriorative changes after harvest were considered to include discoloration, crown rot, anthracnose, and shriveling. Firmness was determined on two paired sides at the middle portion of the finger using a fruit pressure tester (digital force gauge SX Series Model 2256, Aikoh Engineering Co., Ltd., Osaka, Japan) and the value was expressed as kg-force. Total soluble solids (TSS) content was determined after homogenizing 10 g pulp in 20 mL water and drops of clear extract was placed in an Atago digital refractometer. Titratable acidity (TA) was determined from the extract titrated with 0.1 N NaOH to a faint pink color with phenolphthalein as indicator. To determine if the different treatments affected the internal quality of the fruit, bananas were subjected to sensory evaluation at the ripe stage with 10 sensory panelists. A 9-point Hedonic rating scale was used where 9 represents the most favorable response.

2.5. Experimental Design

The experiment was laid out in a Completely Randomized Design (CRD) consisting of three replicates per treatment with each box representing a replicate. Each box consisted of 18 clusters of bananas. Analysis of Variance (ANOVA) was done using the Statistical Analysis System (SAS Institute Inc., Cary, NC, USA), and means were compared using the Least Significance Difference (LSD) at 5% level.

3. Results and Discussion

3.1. Gas Levels during Storage

Vacuum packaging resulted in a decrease in O_2 and an increase in CO_2 levels (Figure 1A,B). Pesis et al. [4] reported that after vacuum packaging, O_2 levels retained in the bag ranged from 0.3% to 3% depending on the permeability of the plastic film. Oxygen levels in vacuum packs decreased to 4%–7% while that of the MAP+BS packs ranged from 4%–5% which was lower than MAP alone although differences were not significant (Figure 1A). In the case of CO_2, very low levels (1.08%–2.67%) accumulated in the vacuum packs indicative of dramatic retardation of respiration rate brought about by the modification of the atmosphere. One of the responses of crops to low O_2 is the reduction in the rate of respiration [4,8]; hence CO_2, which is a product of respiration, did not accumulate in the MA packs. Ethylene did not accumulate to levels that would induce ripening of bananas kept under MAP. Throughout the 25-day storage of bananas under vacuum packaging, C_2H_4 concentrations remained low at 0.08–0.17 µL/L.

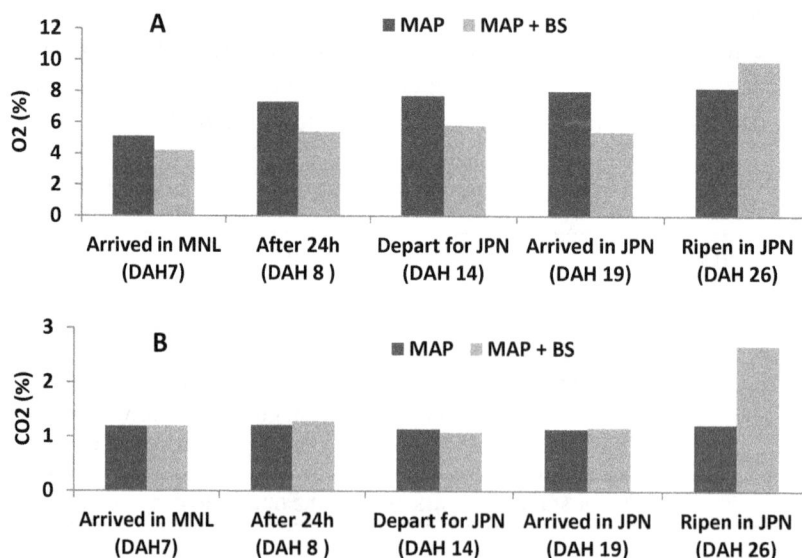

Figure 1. Oxygen (**A**) and carbon dioxide (**B**) levels in MA packs of bananas measured at different handling steps from harvest (start of MAP) until bananas reached Japan (simulated sea shipment). Each mean was obtained from three (3) cartons containing 18 clusters of bananas. Differences between treatments were not significant at 5% level, LSD.

3.2. Incidence and Severity of Crown Rot

Modified atmosphere packaging through its active form of vacuum packaging significantly controlled the incidence of crown rot during the 25-day holding under MAP and during ripening in air at 23 °C (Figure 2A). The supplementary treatment of 1% sodium bicarbonate applied in the form of BS did not contribute to the control of crown rot as suggested by the earlier studies of Alvindia [7]. Vacuum packaging in combination with 1% BS resulted in 5.6% incidence of crown rot in contrast to 2.8% in vacuum packaging alone, although differences between treatments were not significant. Crown rot was observed only on the 3rd day of ripening at 23 °C (post-storage) when bananas would already have been sold to consumers in Japan. Severity of crown rot was likewise very low at 0.2 (Figure 2B), and was observed only as surface molds on the crown which can be readily removed by slight trimming of the crown prior to retail packaging. The earlier proposal of Ke and Kader [9] that MAP can replace the use of postharvest chemicals like fungicides and insecticides for disease and insect control, respectively, was apparent in this study. Kader [10] further recommended that MAP can be a component of integrated pest management in fresh produce since low O_2 and high CO_2 might have fungistatic effects.

On the other hand, treatment with 1% BS alone (non-vacuum packaged) did not control crown rot during storage at 13.0–13.5 °C or during ripening in air at 23 °C. Upon removal from cold storage, incidence of crown rot was already high at 16.7% and progressively increased during ripening, reaching 88.9% on the 5th day (Figure 2A). In the control (non-vacuum packed, no BS), crown rot was absent upon removal from storage at 13 °C and became evident only on the 3rd day of ripening at 23 °C. On the 5th day, incidence of crown rot reached 50% and was lower than bananas subjected to 1% BS alone. The increase in the severity of crown rot during ripening of the control and 1% BS-treated fruits closely followed the increase in incidence (Figure 2B). On the 3rd and 5th day of ripening at 23 °C, the severity rating of crown rot in 1% BS alone reached 3.8 (41%–70% discoloration and mycelial growth on crown area), and decay had spread in the pedicel of some fingers in a cluster. The control fruit had a lower crown rot severity rating than the 1% BS-treated fruit, although mycelial growth on the crown was already prominent.

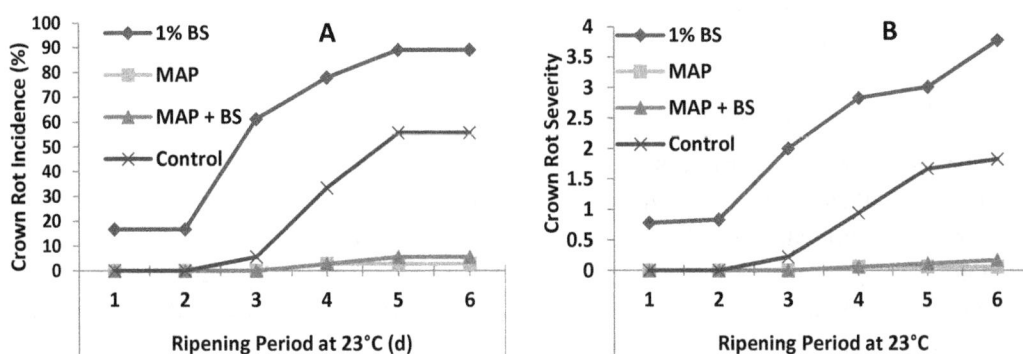

Figure 2. Incidence (**A**) and severity rating (**B**) of crown rot of Balangon bananas upon removal from MAP (day 1) and during ripening in air at 23 °C. Bananas were previously kept under MAP for 25 days at 13.0–13.5 °C simulating domestic interisland and export sea shipment to Japan. Each point represents the mean of three replicates per treatment with each replicate consisting of 18 clusters of bananas.

3.3. Rate of Ripening

Upon removal from MAP at 13.0–13.5 °C storage, bananas were still green with one to two fingers in some clusters exhibiting traces of yellow. Generally, however, ripening of Balangon bananas were retarded during the 25-day storage at 13.0–13.5 °C regardless of whether they were held under MAP or not. The prolonged green life of bananas is important in terms of export shipment to Japan where only green or unripe bananas are allowed entry. When bananas were induced to ripen with ethephon and held in air at 23 °C, the fruit ripened normally attaining the table ripe stage on the 5th day. There were no significant differences in the rate of ripening among treatments indicating that bananas held under MAP were exposed only to mild O_2 stress such that upon return to air, normal metabolism ensued. Mild O_2 stress does not result in injury and can increase longevity and maintain product quality [8,11].

3.4. Visual Quality Rating (VQR)

Due to the low incidence and severity of crown rot of vacuum-packaged Balangon bananas, the decrease in the visual quality or external appearance of the fruit was low during ripening at 23 °C (Figure 3). On the 6th day when bananas were at full yellow peel color, VQR of vacuum-packed banana was 6.2 where fruit were still highly marketable. On the other hand, bananas treated with 1% BS alone and not subjected to vacuum packing exhibited the fastest decline in VQR attributed mainly to the occurrence of moderate to severe crown rot with the pedicel of some fingers in a cluster exhibiting decay. The significantly low VQR of the control fruit was also attributed to crown rot.

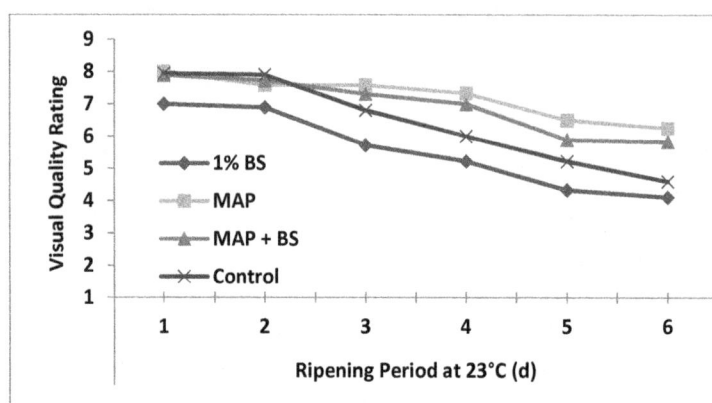

Figure 3. Change in the visual quality rating of Balangon bananas during ripening in air at 23 °C. Bananas were previously kept under MAP for 25 days at 13.0–13.5 °C simulating domestic inter-island and export sea shipment to Japan. Each point represents the mean of three replicates per treatment with each replicate consisting of 18 clusters of bananas.

3.5. Physico-Chemical Changes at The Ripe Stage

Significant difference in firmness at the ripe stage were obtained only between the control and the previously MAP-stored bananas where the control fruits had the higher firmness value (Table 2). The TSS content did not vary significantly among treatments, ranging from 21.2 to 23.5 Brix. Similarly, TA did not vary significantly among treatments. The lack of significant differences in TSS and TA values among treatments indicated that the mild O_2 stress under MAP did not affect the internal quality attributes of the fruit as has been reported by [8].

Table 2. Physico-chemical characteristics of Balangon bananas at the ripe stage [1].

Treatment	Firmness (kg-Force)	TSS (Brix)	TA (%)
1% Baking soda (BS)	0.71 ab	23.4 NS	0.24 NS
MAP	0.66 b	23.5	0.23
MAP + BS	0.71 ab	21.9	0.24
Control (No VP)	0.76 a	21.2	0.25

[1] Balangon bananas were previously stored under MAP for 25 days at 13.0–13.5 °C then withdrawn from MAP and induced to ripen with 2500 µL/L ethephon and ripened at 23 °C. Mean separation within columns for each characteristic at 5% LSD. NS indicates no significant difference among means. Each data represents the mean obtained from 3 replicates per treatment consisting of 3 fruits per replicate.

3.6. Sensory Evaluation at the Ripe Stage

The low O_2 and relatively high CO_2 with MAP storage at 13.0–13.5 °C did not result in anaerobiosis that would lead to development of off-odor and off-flavor at the ripe stage. Bananas when subjected to sensory evaluation were acceptable to the panelists whether fruits were subjected to MAP or not (Table 3). The creamy-white pulp color and aroma did not vary significantly among treatments. Significant differences were obtained in the following sensory attributes: sweetness, balance of sweetness and sourness, flavor, smoothness, and overall acceptability. In all these attributes, MAP-stored and bananas subjected to 1% BS alone had higher scores than the control.

Table 3. Sensory scores of Balangon bananas at the ripe stage [1].

Sensory Attribute	Sensory Score/Treatment			
	1% BS	MAP	MAP + BS	Control
Pulp color	7.4 NS	7.4	7.5	7.2
Aroma	7.3 NS	7.5	7.5	7.1
Sweetness	7.5 a	6.3 ab	7.1 a	6.0 b
Sourness	6.7 NS	6.8	7.2	6.5
Balance of sweetness & sourness	7.5 a	6.5 ab	6.8 a	5.9 b
Flavor	7.8 a	6.7 ab	7.0 ab	6.3 b
Smoothness	8.0 a	7.3 a	7.6 a	6.3 b
Firmness	7.5 NS	7.3	7.5	6.8
Overall acceptability	7.9 a	7.1 a	7.2 a	6.5 b

[1] Balangon bananas were previously stored for 25 days simulating refrigerated (13.0–13.3 °C) sea shipment from the packinghouse in Bacolod City to the port of Japan then ripened in air at 23 °C. Mean separation within rows for each sensory parameter at 5% LSD. NS indicates no significant difference among means. Each data represents the mean score of 10 sensory panelists. Sensory scores followed: 9 = like extremely, 8 = like very much, 7 = like moderately, 6 = like slightly, 5 = neither like nor dislike, 4 = dislike slightly, 3 = dislike moderately, 2 = dislike very much, 1 = dislike extremely.

4. Conclusions and Recommendation

Modified atmosphere packaging in its active form as vacuum packaging dramatically reduced the incidence of crown rot of organically-grown Balangon bananas by 94%–97% when used alone, and by 89%–94% when a supplementary treatment with 1% BS was used, indicating that BS can be eliminated. Treatment with 1% BS alone did not control crown rot. Ripening was retarded during the 25-day simulated domestic inter-island and export sea shipment to Japan at 13.0–13.5 °C. The ripening behavior of bananas proceeded normally upon removal from MAP and subsequent exposure to air at 23 °C. The physico-chemical changes in fruit firmness, TSS and TA were not affected, nor were the desirable sensory attributes of bananas at the ripe stage indicating that the mild O_2 and CO_2 stress during storage under MAP did not affect the quality of fruits. These results showed the potential for MAP to reduce losses in Balangon bananas (estimated at 22%) due to delays in shipment to Japan.

Acknowledgments: The authors gratefully acknowledge the project funding from the Department of Agriculture-Bureau of Agricultural Research (DA-BAR), the collaboration of Alter Trade Corporation (ATC), and the staff of PHTRC-UPLB for their assistance during sensory evaluation. Acknowledgement is due to Melody Orogo for the statistical analysis.

Author Contributions: Elda Esguerra is responsible for the inception of the experiment, data to be gathered and writing the paper, Dormita Del Carmen for the experimental design and coordination with Alter Trade Corporation, Roxanne Delos Reyes and Ryan Anthony Lualhati for the data gathering and analysis.

Conflicts of Interest: The authors declare no conflict of interest.

References

1.   Valmayor, V.V.; Espino, R.R.C.; Pascua, O.C. *The Wild and Cultivated Bananas of the Philippines*; Philippine Agriculture and Resources Research Foundation, Inc.: Los Baños, Philippines, 2002; p. 242.

2. Alvindia, D.G.; Kobayashi, T.; Yaguchi, Y.; Natsuaki, K.T. Pathogenicity of fungi isolated from non-chemical bananas. *Jpn. J. Trop. Agric.* **2002**, *44*, 215–223.

3. Thompson, A.K. *Controlled Atmosphere Storage of Fruits and Vegetables*; CAB International: New York, NY, USA, 1998.

4. Pesis, E.; Ben Arie, R.; Feygenberg, O.; Villamizar, F. Ripening of ethylene-pretreated bananas is retarded using modified atmosphere and vacuum packaging. *HortScience* **2005**, *40*, 726–731.

5. Kudachikar, V.B.; Kulkani, S.G.; Keshava Prakash, M.N. Effect of modified atmosphere packaging on quality and shelf life of "Robusta" banana stored at low temperature. *J. Food Sci. Technol.* **2011**, *3*, 319–324. [CrossRef] [PubMed]

6. Alvindia, D.G. Revisiting hot water treatments in controlling crown rot of banana *cv*. Buñgulan. *Crop Prot.* **2012**, *33*, 59–64. [CrossRef]

7. Alvindia, D.G. Sodium bicarbonate enhances the efficacy of *Trichoderma harzianum* DGA01 in controlling crown rot of banana. *J. Gen. Plant Pathol.* **2004**, *79*, 136–144. [CrossRef]

8. Kader, A.A. Regulation of fruit physiology by controlled/modified atmospheres. *Acta Hortic.* **1995**, 59–61. [CrossRef]

9. Ke, D.; Kader, A.A. Potential of controlled atmosphere for insect disinfestation of fruits and vegetables. *Postharvest News Inf.* **1992**, *3*, 31–37.

10. Kader, A.A. The future of modified atmosphere research. *Acta Hortic.* **2010**, *857*, 213–217. [CrossRef]

11. Abdul-Rahaman, A.; Bishop, C. Evaluating the effects of biodegradable and conventional modified atmosphere packaging on the shelf life of organic Cavendish bananas. *J. Postharvest Technol.* **2013**, *1*, 29–35.

13

The Influence of Crop Habitat and Control Strategies on Pepper Viruses in Andalusia (Spain)

Almudena Simón, Carmen García, Fernando Pascual, Leticia Ruiz and Dirk Janssen *

Instituto Andaluz de Investigación y Formación Agraria, Pesquera, Alimentaria y de la Producción Ecológica (IFAPA), Centro La Mojonera, Camino de San Nicolas 1, La Mojonera 04745, Spain; almudenasimonmartinez@gmail.com (A.S.); mariac.garcia.g@juntadeandalucia.es (C.G.); fervirologia@hotmail.com (F.P.); leticia@thetemplatestore.com (L.R.)
* Correspondence: dirk.janssen@juntadeandalucia.es

Abstract: Andalusia, southern Spain, is a major horticultural production region within the Mediterranean, where over 10,000 ha are dedicated to the production of pepper (*Capsicum annuum* L.). Approximately two-thirds of the area dedicated to this crop is in a greenhouse and the remaining one-third is comprised of open field crops. Using pepper as a model, we identified and compared the major diseases caused by viruses in the different geographic regions and agronomic systems within the region. Symptomatic samples were collected during 2009 and analyzed by ELISA and RT-PCR for the presence of *Tomato spotted wilt virus* (TSWV), *Cucumber mosaic virus* (CMV), *Tomato mosaic virus* (ToMV), *Pepper mild mottle virus* (PMMoV), *Potato virus Y* (PVY), *Tobacco mild green mosaic virus* (TMGMV), *Tomato chlorosis virus* (ToCV) and *Parietaria mottle virus* (PMoV). Contingency table analysis showed a significant relationship between the presence of major diseases caused by viruses in pepper crops and the different agrosystems in terms of location (inland versus coastal), disease control management (chemical versus integrated), cropping system (open field versus greenhouse), and virus-resistant versus susceptible cultivars. Pepper crops in plastic-covered greenhouses were predominantly associated with arthropod-transmitted virus diseases, such as TSWV. CMV was predominant in provinces located inland, and PMoV was found independent of the agrosystem, disease control methods, or geographic location.

Keywords: *Capsicum annuum; cucumber mosaic virus; parietaria mottle virus; potato virus Y;* tobamovirus; *tomato spotted wilt virus;* greenhouse

1. Introduction

Increasing numbers of virus species cause diseases in horticulture crops worldwide. These crops are grown in different geographic locations, each with their own environmental and agro-technical features. Consequently, there exists an infinite range of agrosystems developed to reduce the effect of pests and diseases and to improve agronomic productivity. These agrosystems could be considered ecological communities in which plant pathogens have their own dynamics, defined for example by host suitability, climate, and vectors [1]. In the case of plant viruses that are transmitted by vectors, their management can involve conventional (chemical) control and/or integrated (including biological) control [2]. The use of virus-resistant cultivars constitutes an important means to control virus diseases [3]. Greenhouses and other protected horticultural environments allowing year-round production offer some physical protection against arthropods and diseases. An element determining the success of control can be the existence of arthropod pest and disease reservoirs surrounding the greenhouses, associated with wild vegetation, or with open-field cultivated crops [4]. This also holds for plant viruses, and examples exist where the incidence in outdoor crops during seasonal periods is believed to affect that in nearby greenhouses during off-season production such as in Mediterranean horticultural production where summer open-field and greenhouse winter crops often overlap [5].

Pepper (*Capsicum annuum* L.) production in Spain ranks fifth on a global scale. The Region of Andalusia (southern Spain) has over 10,900 ha dedicated to pepper, and most is cultivated in greenhouses (about 7500 ha) which include various "parral" type and multi-tunnel houses located along the coast of the province of Almeria. The remaining 3400 ha are produced outdoors in the remaining seven provinces [6]. Greenhouse-grown cultivars are planted mostly for fresh consumption and open-field crops are for the food processing industry. Modern horticulture is performed in highly intensive farming systems, but often production suffers heavy losses from arthropod pests and diseases [7], and in the case of pepper, these losses are due to disease caused by plant viruses. Some of these viruses are contact and seed-transmitted, including *Tomato mosaic virus* (ToMV, [8]), *Pepper mild mottle virus* (PMMoV, [9]) and *Tobacco mild green mosaic virus* (TMGMV, [10]), and some are aphid transmitted, including *Cucumber mosaic virus* (CMV, [11]) and *Potato virus Y* (PVY, [11]), and have been present in the region and crop species for many decades. Others are considered emerging viruses and include the western flower thrips *Frankliniella occidentalis*-transmitted *Tomato spotted wilt tospovirus* (TSWV, [12]), *Bemisia tabaci*-transmitted *Tomato chlorosis crinivirus* (ToCV, [13]) and the pollen-transmitted *Parietaria mottle ilarvirus* (PMoV, [14,15]).

Here, relationships between the presence of major diseases caused by viruses in pepper crops and the different agrosystems in terms of location (inland versus coastal), disease control management (chemical versus integrated), cropping system (open field versus greenhouse), and virus-resistant versus susceptible cultivars, are examined.

2. Materials and Methods

2.1. Sample Collections

During 2010, samples were collected from crops located in 71 towns and villages in Andalusia, distributed among 18 different pepper producing regions within the provinces of Almeria, Granada, Malaga, Cadiz, Huelva, Seville, Cordoba and Jaen (Figure 1). The numbers of samples collected per ha of crop surface corresponded to at least 2.5% of the relative crop surface dedicated to pepper in each of the eight provinces, according to available statistical information [6] (Table 1). The sampled crops were characterized and labelled according to the administrative location (province), geographic origin as either coastal (H1, C1, M1, G1, A1 and A2, in Figure 1) or inland (S1, M2, M3, Co1, Co2, J1, J2, G2, G3, G4, A3 and A4), the type of farm (crops grown in a greenhouse or in the open field), the plant health control strategy applied (arthropod pest control based on chemical treatments, or on integrated pest control which includes biocontrols, as determined after visual inspection of the crops and interviewing the farmers), and the crop cycles (crops planted in autumn or in spring).

Table 1. Area (ha) of pepper crops under cultivation in Andalusia during 2009 [6] and the numbers of samples collected for the present study.

Province	Open Field	Protected	Total	Samples
Almeria	73	7432	7505	186
Cadiz	902	–	902	27
Cordoba	593	–	593	15
Granada	264	306	570	22
Huelva	155	4	159	4
Jaen	264	–	264	8
Malaga	156	590	745	24
Seville	183	25	208	11
Andalusia	2598	8357	10,946	297

Figure 1. Map of Andalusia with pepper producing areas where virus surveys were undertaken during the year 2010. The main towns located within the sampling zones were (A1) Nijar, Almeria, (A2) El Ejido, Roquetas de Mar, Adra, (A3) Laujar, Sta. Cruz (A4) Tijola, (G1) Albuñol, Motril, (G2) Alhama de Granada, Zafarraya, (G3) Guadix, Darro, Baza, (G4) Cullar, Puente de Don Fabrique, (M1) Torrox, Velez Malaga, (M2) Coin Alhaurin de la Torre, Ronda, (M3) Antequera, Archidona, (C1) San Lucar de Barrameda, Chipiona, Vejer de la Frontera, (J1) Cazorla, Úbeda, (J2) Jaen, Bailen, (Co1) Córdoba, Puerto Genil, Santaella, (Co2) El Viso, Hinojosa del Duque, (S1) Burguillos, Carmona, La Rinconada, and (H1) Hinojosa, Moguer, Almonte.

Sampled pepper plants belonged to one of a wide range of cultivar types such as Lamuyo, Bell Pepper, Kappya, Italian Sweet Pepper, Chili pepper, and snack-sized peppers, and were from either local or commercial cultivars, including varieties from the following seed companies: De Ruiter Seeds, Enza Zaden España, Nunhems Spain S.A, Rijk Zwaan Iberica S.A., Semillas Fitó, Syngenta Seeds, Tozer Seeds Iberica, Western Seed 2000 SL, Zeraim Ibérica S.A., and Zeta Seeds. Samples consisted of fruit and leaves from plants displaying symptoms suggesting virus infection, including leaf mosaic, ring and leaf spotting, leaf deformation, leaf puckering, and/or plant stunting. All samples were kept at 4 °C and were analyzed 2–5 days after collection.

2.2. Virus Detection

All samples were analyzed for the presence of TSWV, CMV, ToMV, PMMoV, TMGMV and PVY using ELISA and specific antisera from Loewe Biochemica GmbH (Sauerlach, Germany) following the manufacturer's instructions. PMoV and ToCV were detected using RT-PCR: total RNA was purified from 0.2 g leaf tissue using Trizol reagent (Invitrogen, Carlsbad, CA, USA) according to manufacturer's instructions. RT-PCR reactions were developed with the Titan One Tube RT-PCR system (Roche) employing total RNA preparations from pepper and following manufacturer's instructions. PMoV was amplified using the following primers as described (5′-GATGTTGCCGCCGACGATTCTA-3′ and 5′-TTTTCCCACAACCCGAACAC-3′), designed after Genbank Sequence Accession No. AY496068. These primers produce an amplicon of the expected size (475 bp) [14]. ToCV was detected with the primers 5′-TTTGTTCCTCTTTGGGTTTC-3′ and 5′-ATGGGTTTTCTGATGATAAT-3′, designed to amplify a 708-bp fragment from the ToCV genome [9]. The PCR conditions to generate PMoV and ToCV amplicons were 95 °C for 3 min, 35 cycles of denaturation for 20 s at 95 °C, annealing for 30 s at 55 °C and extension for 40 s at 72 °C followed by one final extension cycle for 8 min at 72 °C.

2.3. Statistics

From the number of samples with viruses, contingency tables were constructed, and the Fisher's exact test for equality of distributions was used to investigate whether the resulting proportions could be explained based on different factors, such as geographic origin, climate, crops producing, as well as arthropod pest and disease control systems.

3. Results

Table 1 summarizes the numbers of pepper samples collected and their geographic location. All of the samples were negative for ToCV. TSWV was the virus found most frequently (71% of samples), followed by CMV and PMoV with similar frequencies of 17% and 16.5%, respectively. ToMV, PMMoV and PVY were less frequently detected: 4.71% of samples contained one of the tobamoviruses (ToMV, PMMoV or TMGMV), and only 1.35% of samples had PVY. All of these virus species are commonly found in Andalusia although not all viruses were detected in all provinces.

From the collected samples, 268 (90.23%) proved to be infected by at least one of the virus species analyzed. In samples collected from most of the Andalusian provinces, at least two viruses were detected, either as simple or as double infections (Table 2). In the coastal provinces of Almeria, Granada, Malaga and Cadiz, TSWV was the most prevalent virus (88.71%, 63.64%, 54.17% and 92.59% respectively). Whereas the percentage of CMV in Almeria was only 2.69%, it was 22.73% in Granada, and 54.17% in Malaga. In contrast, the percentage of PMoV was similar in Almeria (23.66%) and Granada (27.27%). That of ToMV was only 4.17% in Malaga. The situation is different for the inland provinces of Seville, Cordoba and Jaen where CMV was the prevalent virus, detected in 81.82% of samples from Seville, in 73.33% of samples from Cordoba and in 63.64% of the samples in the province of Jaen. Interestingly, PVY and PMMoV were detected in 36.40% and 27.27%, respectively, of samples from Seville (Table 2).

Table 2. Percentages of collected samples with single and mixed infections of viruses detected from pepper crops of the eight provinces of Andalusia.

	Province							
	Almeria	Granada	Malaga	Cadiz	Huelva	Seville	Cordoba	Jaen
Viruses	%							
One Infection								
TSWV	68.98	54.54	41.67	92.59	0	0	0	0
CMV	1.6	0	41.67	0	0	9.09	73.33	62.5
PMoV	4.81	0	0	0	0	0	0	0
Two Infections								
TSWV + CMV	0.53	0	8.33	0	0	0	0	12.5
TSWV + PMoV	17.65	4.54	0	0	0	0	0	0
CMV + PMoV	0	18.18	0	0	0	0	0	0
CMV + TMGMV	0	0	0	0	0	36.36	0	12.5
Three Infections								
TSWV + CMV + ToMV	0	0	4.17	0	0	0	0	0
TSWV + PMoV + CMV	0.53	4.54	0	0	0	0	0	0
TSWV + PMoV + TMGMV	0.53	0	0	0	0	0	0	0
CMV + PVY + TMGMV	0	0	0	0	0	9.09	0	0
Four Infections								
CMV + PVY + PMMoV + TMGMV	0	0	0	0	0	27.27	0	0
No Virus	5.35	18.18	7.17	7.41	100	18.18	26.67	12.5

TSWV, *Tomato spotted wilt virus*; ToMV, *Tomato mosaic virus*; PMMoV, *Pepper mild mottle virus*; TMGMV, *Tobacco mild green mosaic virus*; PMoV, *Parietaria mottle virus*; CMV, *Cucumber mosaic virus*; PVY, *Potato virus Y*.

Of all samples 71.38% were from cultivars that were resistant to one or several of the viruses studied, whereas 22.90% were from cultivars without any virus resistance (Table 3). Virus resistant cultivars were almost completely restricted to the greenhouses located along the coastal provinces (Almeria, Granada, Malaga and Cadiz). Although cultivars resistant to CMV were commercially available, none of the collected samples had CMV-resistance genes. At present, there are no commercial cultivars that are resistant to PMoV. As for all of the inland provinces (Seville, Cordoba and Jaen), the collected samples were from plants that had no resistance gene. In regions where PVY, ToMV and PMMoV were found, no cultivars resistant to tobamovirus were commonly used (Table 3).

Table 3. Percentages of collected samples with and without resistance to the viruses studied in the pepper crops from the eight provinces of Andalusia.

	Province							
	Almeria	Granada	Malaga	Cadiz	Huelva	Seville	Cordoba	Jaen
Virus-Resistance					%			
TSWV	4.84	9.09	0	0	0	0	0	0
PVY	4.84	0	8.33	0	0	0	0	0
ToMV	10.75	36.36	0	0	100	0	0	0
ToMV + PMMoV	17.74	4.54	0	74.07	0	0	0	0
TSWV + ToMV	13.44	0	0	0	0	0	0	0
TSWV + ToMV + PVY	5.38	0	0	0	0	0	0	0
TSWV + ToMV + PMMoV	22.58	9.09	0	0	0	0	0	0
ToMV + PMMoV + PVY	9.14	0	0	0	0	0	0	0
TSWV + ToMV + PMMoV + PVY	4.30	0	0	0	0	0	0	0
Local cultivars, without resistance	3.76	22.73	91.67	0	0	100	100	100
Unknown cultivars	3.22	18.18	0	25.93	0	0	0	0

TSWV, *Tomato spotted wilt virus*; ToMV, *Tomato mosaic virus*; PMMoV, *Pepper mild mottle virus*; TMGMV, *Tobacco mild green mosaic virus*; PMoV, *Parietaria mottle virus*; CMV, *Cucumber mosaic virus*; PVY, *Potato virus Y*.

All of the descriptive data linked to the pepper samples, geographic origin, climate, other crops being produced, as well as arthropod pest and disease control systems, were added to contingency tables and subjected to the Fisher's exact test for equality of distributions in order to investigate whether the resulting proportions could be explained based on these diverse factors. The results (Table 4) show that most viruses were specifically detected according to location, cropping system, disease control strategy, and crop cycles ($\alpha < 0.05$). Only PMoV was independent of whether the crops were coastal or inland located, open field or greenhouse-grown, and autumn-winter or spring-summer crop cycles. Tobamoviruses were independent only for the crop cycle period.

Table 4. Table of the relationships [a] and statistical significance among viruses detected during a survey of pepper crops and the different geographic locations, farm types, pest and disease control methods, and crop cycles.

	CMV + PVY	ToMV + PMMoV + TMGMV	PMoV	TSWV
Number of samples	54	14	49	211
Location: coastal versus inland	<0.01 **,[b]	0.03 **	1.70 ns	59.71 **
Farm type: greenhouse versus open field	<0.01 **	0.02 **	2.35 ns	263.90 **
Pest/Disease control: chemical versus integrated	38.60 **	0.29 *	39.00 **	0.04 **
Crop cycle: spring versus autumn	81.34 **	0.59 ns	0.59 ns	0.02 **

[a] Expressed as relative proportions; [b] ns = not significant, $\alpha \geq 0.05$; * $\alpha < 0.05$; ** $\alpha < 0.0001$ (Fisher's exact test for equality of distributions).

4. Discussion

In Andalusia, horticulture crops are produced in two cropping agrosystems [16,17]. One type is located in the inland regions with a sub-continental climate, having cold winters and hot summers, and in which crops are grown outdoors using cultivars without resistance to pathogens. In some cases, seeds are extracted and retained for the next crop. While obtained from selected plants and apparently healthy fruit, they lack preventive disease treatments. In this type of farm, the crop cycle takes place during the warmer months of the year. Fruit yields suffer considerable quality losses from viruses and other diseases, and they are largely sold locally. The second system is located along the coastal areas of the provinces of Almería, Granada, Malaga and Cadiz (Figure 1). The latter province hosted over 1000 ha of cut flower production since the 1980's, but that has gradually been replaced by vegetable crops over the last ten years. In Malaga, subtropical cultures predominate. All of these areas have mild winter climates that permit greenhouse production during autumn-winter and even during spring. Arthropod pest control is based on integrated management and includes cultural practices, mechanical insect exclusions (including insect screens on lateral and roof windows), and biological control [18]. On these farms, high quality commercial seeds are used that are F1 hybrids that have been chemically treated and contain resistance genes against diseases [19]. The yields obtained are generally high and the produce is sold on the national and European market [20].

In contrast to other parts of the world, the most prevalent viruses in the Mediterranean over the past 20 years have been mostly restricted to two tobamoviruses, several aphid-transmitted potyviruses, CMV, and thrips-transmitted TSWV [21]. The relatively constant plant health situation means that pepper could also serve as a model to compare pathological situations within and between production areas. The main viruses in pepper from France are CMV, PVY, TMV, PMMoV and TSWV [22]. In our study, CMV was present in all provinces except Cadiz (Table 2). This virus is non-persistently transmitted by aphids, mostly of the species *Aphis gosspii* and *Myzus persicae*. CMV has a very broad host range, including member species of the *Solanaceae, Cucurbitacae, Leguminosae* and many spontaneous plant species [23]. This wide host range leads to extensive virus reservoirs close to horticultural crops. In Almeria, most greenhouses have insect screens of 10×20 threads/cm^2 that exclude aphids that reside outside the greenhouse. However, in Malaga and Granada, most of the farms produce open field crops where CMV reaches higher incidences.

We found that, within Andalusia, open field pepper crops suffered mostly from seed-transmitted viruses and those that are aphid-transmitted such as CMV (Tables 2 and 4). Similar results were obtained for field-grown pepper in Turkey with PVY and TMV in double infections [24], or with PVY and ToMV in Ethiopia [25], with PVY and CMV in Serbia [26], and in tomatoes in Valencia (Spain) where PVY and ToMV were the predominant viruses [5]. The inland climatic conditions of the Andalusian regions where these field pepper crops were located did not favor the presence of *F. occidentalis* thrips and the transmission of TSWV. During the summer, high temperatures (>40 °C) and low humidity (~20%), which can produce a drastic reduction in the populations of thrips [27,28], are common. On the other hand, crops produced in greenhouses without these climate extremes had high incidences of diseases caused by viruses such as TSWV. These crops were physically protected against invading aphids using insect screens, and therefore viruses such as CMV were rarely in evidence.

Populations of viruses carried in seed or transmitted by vectors to field tomato crops can determine levels of infection of greenhouse-grown tomato during a consecutive season [5]. In particular, arthropods cross boundaries between different habitats [29], and this could affect the transmission of viruses as they may be vectors, and influence the response of their predators and parasitoids. For pepper crops we would expect pathogen spillover might take place, defined as when the epidemics in a host population are driven not by transmission within that population but by transmission from a reservoir population [30]. However, from our work, most virus infections appeared specific to location, cropping system, disease control strategy and crop cycles (Table 4). Probably, the recent large-scale application of integrated pest management in pepper greenhouses explains the significant association between the presence of TSWV and the type of disease management [18]. The use of cultivar resistance against tobamoviruses has proven effective as well. These viruses are rarely found in greenhouses from Almeria where resistance has been used since 2000 [31] (Table 2). The presence of tobamoviruses was related to the location and type of farms as well as the type of plant health control used, e.g., the presence or absence of cultivar resistance and the use of untreated saved seed.

Of all the viruses surveyed, PMoV was the only virus detected in pepper crops irrespective of the geographic and agronomic parameters. PMoV was found inland and at the coast of Andalusia, both in open field and in greenhouse crops, and in summer and winter crops (Table 4). To date, no pepper cultivars resistant to PMoV are available. PMoV has emerged as a pathogen of Solanaceous crops in the Mediterranean countries of Spain, France, Italy and Greece [14,32–34]. Recent studies suggest that pollen transmitted from surrounding *Parietaria officinalis* plants could be a major source of infection in Solanaceous crops, and several insect species of the orders Hemiptera and Thysanoptera were candidates as vectors. These include biological control agents such mirids that are predators of arthropod pests [15]. These features within the context of integrated pest management in greenhouses could pose one of the greatest challenges for successful control of PMoV among all of the viruses of pepper in Andalusia.

5. Conclusions

The present study compared virus species found in pepper crops that were grown in different locations, in open air fields and in greenhouses, with chemical or with integrated disease control, and as summer or as winter crops. During this survey in southern Spain, significant differences were evident as aphid-transmitted CMV was mostly found in open field crops. Only few cases of infection with tobamoviruses ToMV and PMMoV were registered, probably since most farms use tobamovirus-resistant cultivars. Trips-transmitted TSWV was found in greenhouses located near the coast, where no extreme temperatures and percentages of humidity occur that can harm the vector. In contrast, PMoV, which is transmitted by pollen and to which there are no resistant cultivars, was detected in pepper crops irrespective of the geographic and agronomic parameters, and therefore constitutes an emerging virus that could pose a serious threat to pepper crops in the Mediterranean.

Acknowledgments: This study was financed by project INIA-FEDER: RTA09-00016-00-00. The research contract of L.R. is 80% supported by the Andalusian Operational Programme, European Social Fund (2007–2013). We thank Manuel Arriaza (IFAPA, Centro Alameda del Obispo) for help on the statistical analyses of data, and Vince Reid for advice during the preparation of the manuscript.

Author Contributions: D.J. and C.G. conceived and designed the experiments; A.S. and F.P. performed the experiments; A.S. and D.J. analyzed the data; L.R. contributed analysis tools; D.J. and A.S. wrote the paper.

Conflicts of Interest: The authors declare no conflict of interest.

References

1. Malmstrom, C.M.; Melcher, U.; Bosque-Pérez, N.A. The expanding field of plant virus ecology: Historical foundations, knowledge gaps, and research directions. *Virus Res.* **2011**, *159*, 84–94. [CrossRef] [PubMed]

2. Reitz, S.R.; Yearby, E.L.; Funderburk, J.; Stavisky, J.; Olson, S.M.; Momol, M.T. Integrated management tactics for *Frankliniella thrips* (Thysanoptera: Thripidae) in field-grown pepper. *J. Econ. Entomol.* **2003**, *96*, 1201–1214. [CrossRef] [PubMed]

3. Pitrat, M. Vegetable crops in the Mediterranean Basin with an overview of virus resistance. *Adv. Virus Res.* **2012**, *84*, 1–29. [PubMed]

4. Berlinger, M.J.; Jarvis, W.R.; Jewett, T.J.; Lebiush-Mordechi, S. Managing the greenhouse, crop and crop environment. In *Integrated Pest and Disease Management in Greenhouse Crops*; Albajes, R., Lodovica Gullino, M., van Lenteren, J.C., Eds.; Kluwer Academic Publishers: New York, NY, USA, 2002; pp. 97–123.

5. Soler, S.; Prohens, J.; López, C.; Aramburu, J.; Galipienso, L.; Nuez, F. Viruses infecting tomato in Valencia, Spain: Occurrence, distribution and effect of seed origin. *J. Phytopathol.* **2010**, *158*, 797–805. [CrossRef]

6. Ministerio de Medio Ambiente y Medio Rural y Marino. *Análisis Provincial de Superficie, Rendimiento y Producción*; Ministerio de Medio Ambiente y Medio Rural y Marino: Madrid, Spain, 2009. (In Spanish)

7. Cuadrado, I.M. Situación de las virosis en la horticultura almeriense. *Vida Rural* **1999**, *97*, 38–39. (In Spanish)

8. Alonso, E.; Garcia Luque, I.; Avila Rincón, M.J.; Wicke, B.; Serra, M.T.; Diaz Ruiz, J.R. A tobamovirus causing heavy losses in protected pepper crops in Spain. *J. Phytopathol.* **1989**, *125*, 67–76. [CrossRef]

9.   Velasco, L.; Simón, B.; Janssen, D.; Cenis, J.L. Incidences and progression of tomato chlorosis virus disease
     and tomato yellow leaf curl virus disease in tomato under different greenhouse covers in southeast Spain.
     Ann. Appl. Biol. **2008**, *153*, 335–344. [CrossRef]

10. Luis Arteaga, M.; Saez, E.; Berdiales, B.; Rodriguez Cerezo, E. Detección del virus del moteado verde
     atenuado del tabaco (TMGMV) en cultivos de berenjena y pimiento en Almería. In Proceedings of the Ninth
     Congress, Salamanca, Spain, 19–23 October 1998; Sociedad Española de Fitopatologia: Madrid, Spain, 1998;
     p. 69. (In Spanish)

11. Avilla, C.; Collar, J.L.; Duque, M.; Fereres, A. Yield of bell pepper (*Capsicum annuum*) inoculated with CMV
     and/or PVY at different time intervals. *Z. Pflanzenkrankh. Pflanzenschutz* **1997**, *104*, 1–8.

12. Jorda, C.; Nuez, F.; Lacasa, A. Situación actual de la virosis del bronceado en los cultivos españoles.
     Phytoma Esp. **1995**, *66*, 20–23. (In Spanish)

13. Fortes, I.M.; Moriones, E.; Navas Castillo, J. Tomato chlorosis virus in pepper: Prevalence in commercial
     crops in southeastern Spain and symptomatology under experimental conditions. *Plant Pathol.* **2012**, *61*,
     994–1001. [CrossRef]

14. Janssen, D.; Sáez, E.; Segundo, E.; Martín, G.; Gil, F.; Cuadrado, I.M. *Capsicum annuum*—A new host of
     Parietaria mottle vírus in Spain. *Plant Pathol.* **2005**. [CrossRef]

15. Aramburu, J.; Galipienso, L.; Aparicio, F.; Soler, S.; López, C. Mode of transmission of *Parietaria mottle virus*.
     J. Plant Pathol. **2010**, *92*, 679–684.

16. Aznar-Sánchez, J.A.; Galdeano-Gómez, E.; Pérez-Mesa, J.C. Intensive horticulture in Almería (Spain):
     A counterpoint to current European rural policy strategies. *J. Agrar. Chang.* **2011**, *11*, 241–261. [CrossRef]

17. Leyva, R.; Constán-Aguilar, C.; Blasco, B.; Sánchez-Rodríguez, E.; Romero, L.; Soriano, T.; Ruíz, J.M. Effects
     of climatic control on tomato yield and nutritional quality in Mediterranean screenhouse. *J. Sci. Food Agric.*
     2014, *94*, 63–70. [CrossRef] [PubMed]

18. Glass, R. Biological control in the greenhouses of Almeria and challenges for a sustainable intensive
     production. *Outlooks Pest Manag.* **2012**, *23*, 276–279. [CrossRef]

19. Córdoba-Sellés, C.; Cebrián, M.C.; Sánchez-Navarro, J.A.; Espino, A.; Martín, R.; Jordá, C. First report of
     Tomato torrado virus in tomato in the Canary Islands, Spain. *Plant Dis.* **2007**. [CrossRef]

20. Galdeano-Gómez, E.; Aznar-Sánchez, J.A.; Pérez-Mesa, J.C. Sustainability dimensions related to
     agricultural-based development: The experience of 50 years of intensive farming in Almeria (Spain).
     Int. J. Agric. Sustain. **2013**, *11*, 125–143. [CrossRef]

21. Moury, B.; Verdin, E. Viruses of pepper crops in the Mediterranean basin: A remarkable stasis. *Adv. Virus Res.*
     2012, *84*, 127–162. [PubMed]

22. Marchoux, G.; Ginoux, G.; Morris, C.; Nicot, P. Pepper: The breakthrough of viruses. *PHM Rev. Hortic.* **2000**,
     410, 17–20.

23. Zitter, T.A.; Murphy, J.F. Cucumber mosaic. *Plant Health Instr.* **2009**. [CrossRef]

24. Arli-Sokmen, M.; Mennan, H.; Sevik, M.A.; Ecevit, O. Occurrence of viruses in field-grown pepper crops and
     some of their reservoir weed hosts in Samsun, Turkey. *Phytoparasitica* **2005**, *33*, 347–358. [CrossRef]

25. Hiskias, Y.; Lesemann, D.-E.; Vetten, H.J. Occurrence, distribution and relative importance of viruses infecting
     hot pepper and tomato in the major growing areas of Ethiopia. *J. Phytopathol.* **1999**, *147*, 5–11. [CrossRef]

26. Petrovic, D.; Bulajic, A.; Stankovic, I.; Ignjatov, M.; Vujakovic, M.; Krstic, B. Presence and distribution of
     pepper viruses in Serbia. *Ratar. i Povrt.* **2010**, *47*, 567–576.

27. Contreras, J.; Pedro, A.; Sanchez, J.A.; Lacasa, A. The influence of extreme temperatures on the development
     of *Frankliniella occidentalis* (Pergande) (Thysanoptera: Thripidae). (Influencia de las temperaturas extremas
     en el desarrollo de *Frankliniella occidentalis* (Pergande) (Thysanoptera: *Thripidae*)). *Bol. Sanid. Veg. Plagas* **1998**,
     24, 251–266. (In Spanish)

28. Chyzik, R.; Ucko, O. Seasonal abundance of the western flower thrips *Frankliniella occidentalis* in the arava
     valley of Israel. *Phytoparasitica* **2002**, *30*, 335–346. [CrossRef]

29. Tscharntke, T.; Rand, T.A.; Bianchi, F.J.J.A. The landscape context of trophic interactions: Insect spillover
     across the crop-noncrop interface. *Ann. Zool. Fenn.* **2005**, *42*, 421–432.

30. Fraile, A.; Pagán, I.; Anastasio, G.; Sáez, E.; García-Arenal, F. Rapid genetic diversification and high fitness
     penalties associated with pathogenicity evolution in a plant virus. *Mol. Biol. Evol.* **2011**, *28*, 1425–1437.
     [CrossRef] [PubMed]

31. Power, A.G.; Mitchell, C.E. Pathogen spillover in disease epidemics. *Am. Nat.* **2004**, *164*, S79–S89. [CrossRef] [PubMed]

32. Roggero, P.; Ciuffo, M.; Katis, N.; Alioto, D.; Crescenzi, A.; Parrella, G.; Gallitelli, D. Necrotic disease in tomatoes in Greece and south Italy caused by tomato strain of *Parietaria mottle virus*. *J. Plant Pathol.* **2000**. [CrossRef]

33. Marchoux, G.; Parrella, G.; Gebre-Selassie, K.; Gagnalon, P. Identification de deux ilarvirus sur tomate dans le sud de la France. *Phytoma* **1999**, *522*, 53–55. (In French)

34. Aramburu, J. First report of *Parietaria mottle virus* on tomato in Spain. *Plant Dis.* **2001**. [CrossRef]

14

Thai Orchid Genetic Resources and Their Improvement

Kanchit Thammasiri

Department of Plant Science, Faculty of Science, Mahidol University, Rama VI Road, Phayathai, Bangkok 10400, Thailand; kanchitthammasiri@gmail.com

Abstract: Thailand is the origin of about 1300 species and 180–190 genera of orchids, comprising the major tropical orchids in the world. These wild Thai orchids grow naturally in various habitats and have unique flowers, stems, leaves, and roots. Many genera, including *Vanda, Rhynchostylis, Ascocentrum, Aerides, Phalaenopsis, Doritis, Dendrobium, Bulbophyllum, Cirrhopetalum, Spathoglottis,* and *Paphiopedilum*, contribute significantly to the Thai orchid industry for cut-flowers and potted plants. The improvement of these orchids' horticultural characteristics has been significant through breeding, tissue culture, and cultural practices, as well as by technological applications and extension. Orchids will continue to dominate other ornamental crops in Thailand due to their diversity, better technologies, know-how from research, suitable climatic conditions, and experienced and skillful growers and exporters, as well as their nationwide popularity.

Keywords: Thai orchid species; orchid diversity; breeding

1. Introduction

Thailand is the origin of about 1300 species and 180–190 genera of orchids, comprising the major tropical orchids in the world. Although Thailand is a natural habitat for several diverse species of orchids, interest in growing cultivars for their economic value was first recorded in 1913 with the introduction of some exotic materials by a foreigner working in Thailand. Having a hobby of growing orchids, he brought cattleyas and some other genera to Bangkok, all of which were subsequently sold to a high ranking officer. Several other high rank and file individuals in the country also became interested in orchid growing as a hobby during the same period. Information concerning orchid growing, an expensive hobby suitable only for the rich and elite, was provided by a small group of high-ranking officials and older, wealthy people in Thailand.

Orchids have the highest value among several tropical ornamental crops, especially among cut-flower crops which are important to Thai agriculture and the economy. Although orchid growing started as a hobby about 100 years ago, until 1966 only a small amount of orchid cut-flowers were exported from Thailand to Europe. The introduction of the *Dendrobium* cultivar "Pompadour" proved to be a landmark that also increased the popularity for orchid cultivation in Thailand. This hybrid was easy to grow and propagate by division, and was high yielding with a long vase-life. Thailand attained the status of the world's leading producer and exporter of orchids in a little over a decade, and it has held the top rank since 1979.

Orchid cultivation in Thailand has been confined to the Central Plain, mainly in Bangkok and its nearby provinces, where there are favorable climatic conditions, water, transportation and a marketing system. Orchid production takes place mainly in three provinces, Samut Sakhon, Bangkok, and Nakhon Pathom, followed by the nearby central provinces and those to the north and the south of Thailand. The estimated total area for orchid cultivation in 2012 was 3003 ha. The suitable environment, high orchid genetic diversity, efficient infrastructure, experienced growers, technological applications, extension, training, teaching, and research, as well as business skills, have each contributed to the success of orchid production and trade in Thailand [1]. Many farmers have made orchid cultivation their main occupation, often with higher returns than other crops. Commercial orchid production has been facilitated by over 10 tissue culture laboratories. Marketing has been facilitated for growers by over 50 export organizations engaged exclusively in export of Thai orchids [2].

Thailand has a long history of orchid trade, especially for export. It is estimated that 54% of the orchids produced in Thailand are currently exported, while the remaining 46% are consumed in the domestic markets. Exports of orchid cut-flowers began in 1963 with a few hundred thousand sprays of mostly "Pompadour" sent to European markets. The export value of orchid cut-flowers has increased sharply to about 60 million U.S. $ in 2014. The export of orchid cut-flowers still predominates, but the export of orchid plants has also rapidly increased to about 22 million U.S. $ in 2014 (Figure 1) [2].

The imports of ornamental plants and flowers into Thailand, especially orchid plants and cut-flowers, has been relatively low because there are a large variety of indigenous tropical orchids which are relatively inexpensive and good quality.

The success of Thai orchid cultivation and trade of cut-flowers and potted plants was a result of concerted national efforts with respect to germplasm resource management, research, training, extension, technological applications, and communication networks [3,4].

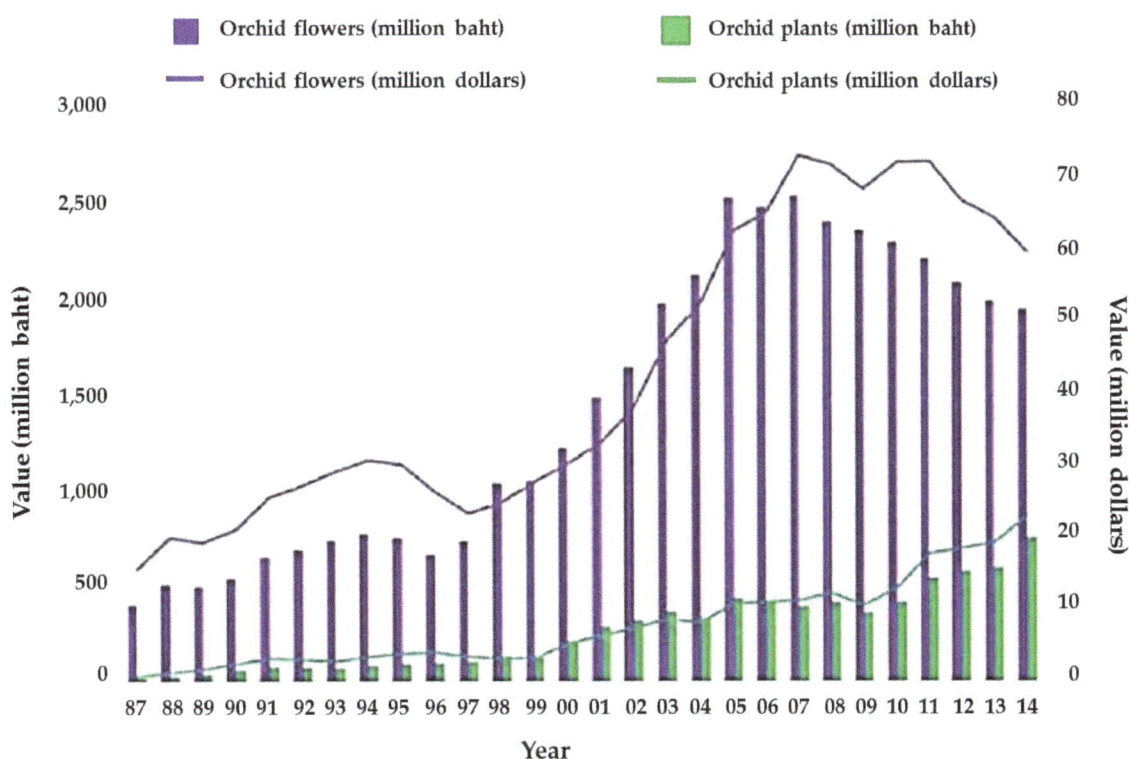

Figure 1. Orchid flower and plant export value from Thailand during 1987–2014.

2. Thai Orchid Genetic Resources

Orchidaceae (Orchid family) is subdivided into six subfamilies [5], with Thailand as the origin of all six subfamilies. They have distinguishing characteristics described briefly [6] as follows:

1. APOSTASIOIDEAE: This subfamily is sometimes separated into a new family, named Apostasiaceae by some authors as it is quite different in size and shape of sepals and petals with two or three fertile stamen and an ovary with three locules. In Thailand, it is represented by two genera and five species, and is mainly confined to moist evergreen lowland forests.

2. CYPRIPEDIOIDEAE: The well-known slipper orchids belong to this subfamily. Their common characteristic feature is a deeply saccate or slipper-shaped lip. The unusual structure includes a column with two lateral stamen, a large shield-like median staminode, and lateral sepals connate along the inner margin forming one sepal mostly hidden behind the lip. The genus *Paphiopedilum* is only found in Thailand with about 14 species.

3. NEOTTIOIDEAE: This orchid group is mainly terrestrial in moist evergreen forests with more or less fleshy rhizomes or tubers. There are about 24 genera and 68 species in Thailand.

4. ORCHIDOIDEAE: This subfamily is very close to the subfamily Neottioideae in general habit, as well as in habitat, but differs in having tubers or root-stem tuberoids, an anther usually basally attached to the viscidium (or visidia), and with the apex of the rostellum often protruding between the thecae. There are about 10 genera and 65 species recorded in Thailand. *Habenaria* is the most well-known one.

5. EPIDENDROIDEAE: This is the largest subfamily, growing in various habitats. Their soft, waxy pollinia, normally without stipe and viscidium, is very characteristic of this group. There are about 63 genera and 692 species found in Thailand. Many of them are very attractive in floral size, color and longevity. Thus, they are very popular among orchid growers.

6. VANDOIDEAE: This subfamily is very close to the Epidendroideae and has recently been included in the latter subfamily. The difference is mainly in the hard or cartilaginous pollinia which usually has a stipe and viscidium. There are about 78 genera and 289 species in Thailand. Many of them are also very popular, as are those in the Epidendroideae.

3. Cultivated Thai Orchid Genera and Their Improvement

Among 180–190 genera and 1300 species of wild Thai orchids, about 11 genera are cultivated as a hobby and for commercial purposes. In addition, they have been improved through breeding and production technologies. The details of these genera are as follows:

3.1. Vanda

The genus *Vanda* is an epiphyte and monopodial orchid with large stems. There are about 40–50 species in this genus. They grow in tropical Asia from East India to Southeast Asia, Indochina, New Guinea, Australia, Solomon Islands, the Philippines, Taiwan and nearby islands. In Thailand, 9 species are found. They are popular for growing, and are used as parents, especially *Vanda coerulea* (Figure 2) because they give large flowers, and a long spray with a good arrangement of flowers. Flower colors of the hybrids are blue to red-purple (Figure 3). *Vanda* can be crossed with other monopodial genera, such as *Rhynchostylis*, *Aerides*, *Ascocentrum*, etc., and produce many outstanding cultivars for cut-flowers and potted plants.

Figure 2. Variation in flower colors of *Vanda coerulea*.

Figure 3. *Vanda* hybrids.

3.2. Rhynchostylis

This genus is epiphytic, monopodial, and has medium to large stems with thick leaves. In Thailand, there are 3 species in this genus, namely *R. gigantea* (Figure 4), *R. retusa* (Figure 5) and *R. coelestis* (Figure 6). The flower colors of *Rhynchostylis gigantea* are white, white with red-purple spots, red-purple and orange. This genus is popularly grown as potted plants and as parents for producing interspecific hybrids (Figures 7 and 8).

Figure 4. *Rhynchostylis gigantea.*

Figure 5. *Rhynchostylis retusa.*

Figure 6. *Rhynchostylis coelestis.*

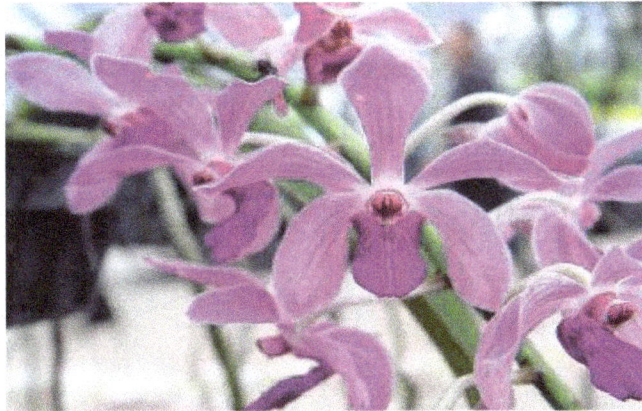

Figure 7. A hybrid of *Rhynchostylis gagantea* and *Vanda coerulea*.

Figure 8. A hybrid of *Rhynchostylis coelestis* and *Seidenfadenia mitrata*.

3.3. Ascocentrum

This genus is a small epiphyte and monopodial orchid with 3 types of flower color, purple-red (*A. ampullaceum*), red (*A. curvifolium*), and orange (*A. miniatum*) (Figure 9).

Figure 9. *Ascocentrum miniatum*.

3.4. Aerides

This genus (Figure 10) is a small epiphyte and monopodial orchid and can grow in strong sunlight. It has beautiful flowers and sprays, so it is popular for hobbyists to grow and use for orchid improvement.

Figure 10. *Aerides odorata.*

3.5. Phalaenopsis

This genus (Figure 11) is a short epiphyte and monopodial orchid with long leaves. It has one to many flowers per spray, and some continuously bloom one at a time. This genus, especially for its hybrids (Figure 12), is economically important in Taiwan.

Figure 11. Variation in flower color of *Phalaenopsis cornu-cervi.*

Figure 12. *Phalaenopsis* hybrids.

3.6. Doritis

This genus (Figure 13) is terrestrial and can grow in sand or rock. Its sprays are straight and hard with 5–15 flowers which continuously bloom for 5–6 months. It grows easily as a potted plant and has been crossed with *Phalaenopsis* to get a medium-sized straight flower spray (Figure 14).

Figure 13. *Doritis pulcherrima.*

Figure 14. *Doritis* hybrids.

3.7. Dendrobium

This genus (Figure 15) is a sympodial orchid with diversity in size and shape of stems, leaves, its habitats and growth habits. They are over 1000 species in the genus around the world. They are popularly grown and used in breeding to get outstanding hybrids (Figure 16).

Figure 15. *Dendrobium cruentum.*

Figure 16. *Dendrobium cruentum* hybrids.

3.8. Bulbophyllum

This genus (Figure 17) is a small sympodial orchid. Pseudobulbs always have angles with thick leaves. They are easy to propagate by division with only one pseudobulb and are popular for growing as potted plants.

Figure 17. *Bulbophyllum putidum.*

3.9. Cirrhopetalum

This genus (Figure 18) was separated from *Bulbophyllum* because its dorsal sepal is smaller than the lateral sepals. The flower spray is umbellate and the pseudobulb is cone shaped.

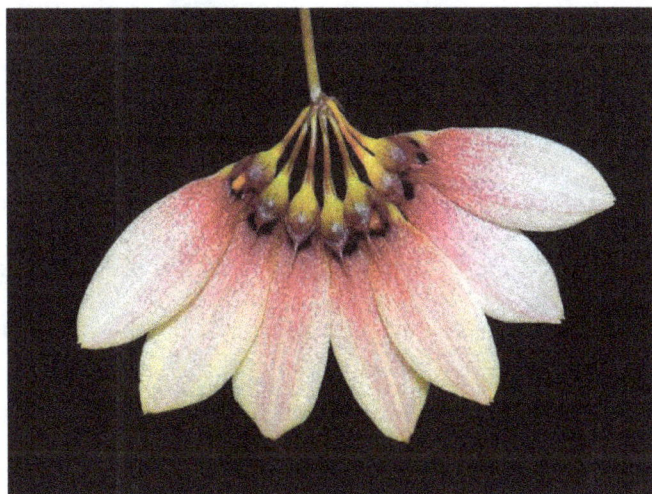

Figure 18. *Cirrhopetalum lepidum.*

3.10. Spathoglottis

This genus (Figure 19) is a terrestrial orchid with a pseudobulb and rhizome. They are used for landscaping since they can grow well in the soil. In Thailand, there are 5 species. *Spathoglottis plicata* is used to make hybrids (Figure 20) with many colorful orchids.

Figure 19. *Spathoglottis plicata.*

Figure 20. *Spathoglottis* hybrids.

3.11. Paphiopedilum

This genus is called "Lady's slipper" because the lip looks like lady's shoes. There are 15 species reported in Thailand. They are popular to grow from seeds as potted plants. The popular species used for trade are *P. bellatulum* (Figure 21a), *P. concolor*, *P. exul* (Figure 21b). *P. godefroyae*, *P. niveum*, *P. callosum*, *P. parishi* and *P. hirsutissimum*.

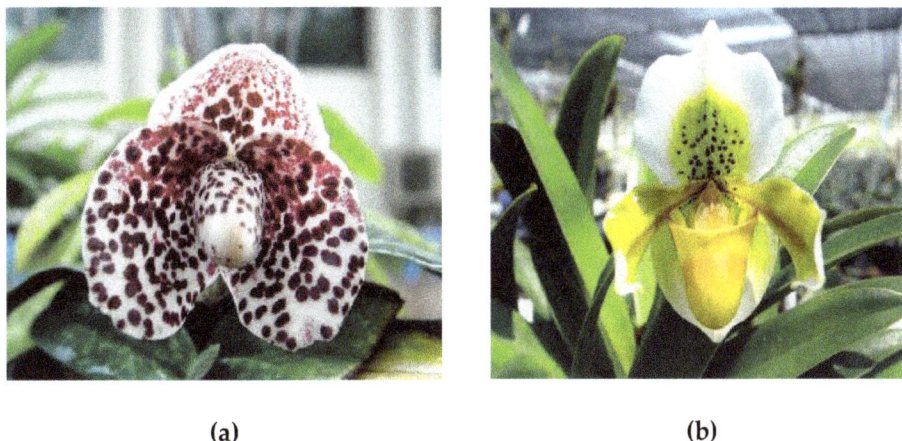

(a)   (b)

Figure 21. *Paphiopedilum bellatulum* (**a**) and *Paphiopedulum exul* (**b**).

In addition to the breeding of cultivated Thai orchid genera, Thai orchidists from both the public and private sectors have improved the orchid production technologies for sustainable use, modifying greenhouses, water resources, planting material and containers, pest control, plant breeding, tissue culture [7] and postharvest management. These improvements have resulted in increased yield and quality and have thereby directly enhanced production and increased quantity of orchids for export [8].

4. Conclusions

Thai orchid production has a bright future where export values are expected to remain high and stable. Orchids will continue to dominate other ornamental crops in Thailand due to high orchid diversity, better technological know-how, suitable climatic conditions, and experienced and skillful growers and exporters, as well as their popularity with the public.

The success story of orchids in Thailand is a good example of the development of an ornamental crop through its diversity in natural habitats to become a major crop. The development of the Thai orchid industry has taken a long time to develop, but it is now a high-income business and an important part of the national economy.

Conflicts of Interest: The author declares no conflict of interest.

References

1. Thammasiri, K. *Orchids in Thailand: A Success Story*; APAARI, FAO, Angkor Publishers Ltd.: New Delhi, India, 1997.
2. Department of Agricultural Extension. *Statistics of Cut-Flowers and Orchid Production*; Ministry of Agriculture and Cooperatives: Bangkok, Thailand, 2015.
3. Thammasiri, K. Sustainable management of Thai orchid species. In Proceedings of the 11th Asia Pacific Orchid Conference, Okinawa, Japan, 2–11 February 2013; pp. 92–96.
4. Thammasiri, K. Ex situ conservation of Thai orchid species. In Proceedings of the 20th World Orchid Conference, Sands Expo and Convention Center, Singapore, 13–20 November 2011; pp. 141–148.
5. Dress, R.L. *The orchids, Natural History and Classification*; Harvard University Press: Cambridge, MA, USA, 1981.

6.   Thaithong, O. *Orchids of Thailand*; Office of Environmental Policy and Planning: Bangkok, Thailand, 1999; p. 239.

7.   Arditti, J.; Ernst, R. *Micropropagation of Orchids*; John Wiley & Sons, Inc.: New York, NY, USA, 1993.

8.   Thammasiri, K. Current status of orchid production in Thailand. *Acta Hortic.* **2015**, *1078*, 25–33. [CrossRef]

15

Regionalization of Maize Responses to Climate Change Scenarios, N Use Efficiency and Adaptation Strategies

Frank Eulenstein [1,2,*], Marcos Alberto Lana [1], Sandro Luis Schlindwein [3],
Askhad Khasrethovich Sheudzhen [2], Marion Tauscke [1], Axel Behrendt [1],
Edgardo Guevara [4] and Santiago Meira [4]

[1] Leibniz-Centre for Agricultuiral Landscape Research (ZALF) Müncheberg, Eberswalder Straße 84, Müncheber 15374, Germany; Marcos.Lana@zalf.de (M.A.L.); mtauschke@zalf.de (M.T.); abehrendt@zalf.de (A.B.)

[2] Department Agro-Chemistry, Kuban State Agrarian University, Krasnodar 350044, Russia; rgpzkrs@mail.kuban.ru

[3] Departamento de Engenharia Rural, Universidade Federal de Santa Catarina, Florianópolis 88034-000, Brazil; sandro.schlindwein@ufsc.br

[4] Department of Crop Production INTA—Instituto Nacional de Tecnologia Agropecuaria, Pergamino 2700, Argentina; guevara.edgardo@inta.gob.ar (E.G.); meira.santiago@inta.gobv.ar (S.M.)

* Correspondence: feulenstein@zalf.de

Abstract: As with any other crop, maize yield is a response to environmental factors such as soil, weather, and management. In a context of climate change, understanding responses is crucial to determine mitigation and adaptation strategies. Crop models are an effective tool to address this. The objective was to present a procedure to assess the impacts of climate scenarios on maize N use efficiency and yield, with the effect of cultivar ($n = 2$) and planting date ($n = 5$) as adaptation strategies. The study region was Santa Catarina, Brazil, where maize is cultivated on more than 800,000 ha (average yield: 4.63 t·ha^{-1}). Surveying and mapping of crop land was done using satellite data, allowing the coupling of weather and 253 complete soil profiles in single polygons ($n = 4135$). A Decision Support System for Agrotechnology Transfer (DSSAT) crop model was calibrated and validated using field data (2004–2010 observations). Weather scenarios generated by Regional Climatic Models (RCMs) were selected according their capability of reproducing observed weather. Simulations for the 2012–2040 period (437 ppm CO_2) showed that without adaptation strategies maize production could be reduced by 12.5%. By only using the best cultivar for each polygon (combination of soil + weather), the total production was increased by 6%; when using both adaptation strategies—cultivar and best planting date—the total production was increase by 15%. The modelling process indicated that the N use efficiency increment ranged from 1%–3% (mostly due to CO_2 increment, but also due to intrinsic soil properties and leaching occurrence). This analysis showed that N use efficiency rises in high CO_2 scenarios, so that crop cultivar and planting date are effective tools to mitigate deleterious effects of climate change, supporting energy crops in the study region.

Keywords: climate change; crop model; efficiency use

1. Introduction

Understanding climate change and impacts on crops is critical to determining anthropogenic responses of mitigation and adaptation. While the soil resource is relatively constant over time, the weather is subject to remarkable variability, making it the most challenging and unpredictable aspect of agriculture [1]. Despite all efforts mobilized to identify the directions of climate change, the high level of uncertainty entrenched in climate scenarios is still a challenge for accurate future predictions. On the other hand, the comprehension on how crops respond to environmental factors and management variables relies in mathematical models that could explain with accuracy key physiological and phenological processes.

2. Experimental Section

Crop Models for Impact Assessment

As crop science represents an integration of the disciplines of biology, physics and chemistry, plant and crop simulation models are mathematical representations of this system [2]. Lobell and Burke [3] stated that process-based models are typically developed and tested using experimental trials, and thus offer the distinct advantage of leveraging knowledge from decades of research in crop physiology and reproduction, agronomy, and soil science, among other disciplines. In order to correctly simulate physiological and phenological processes, these models require the input of specific data of soil, weather, and crop or cultivar, as well as management practices such as planting date, plant population and fertilization. The models need to be calibrated to local conditions in order to achieve high levels of accuracy. Simulations are usually run for a single or few sites, and input from farmers and technicians during the decision making process are important. Nevertheless, the scaling up of model outcomes can bring a more complete picture, allowing a better understanding of regional impacts. Examples of regional applications are food security early warning systems and market decisions [4]. If site-specific estimates are averaged to obtain regional mean yields, the procedure is likely to introduce aggregation errors that depend on the degree of non-linearity of the crop model functions, as well as the heterogeneity of sites in the region. Furthermore, in climate change studies, the selection of individual sites that are representative of present-day conditions may be inappropriate if the projected future climate is likely to shift the suitability of a crop into new regions [5].

3. Methodology

The study region was the state of Santa Catarina, located in Southern Brazil. The state covers an area of 95,703 km^2, with a population estimated at 6.2 million inhabitants [6]. Agricultural production plays an important role in the state, contributing more than 7% of the GDP [7]. The state has a structured agro-industrial complex, where the main part is coarse grains produced for animal feed. In the case of maize, 100% of the local production is directed to local agro-industry demand.

In order to assess the actual land use of Santa Catarina, data from different sources was merged within a single GIS data base. Further environmental information like bodies of water, islands, wetlands, and urban regions were obtained from 1:50,000 and 1:100,000 scale digital maps [8]. In addition, maps of vegetation and categories of land use (like crop land, pastures, crop land + pasture, etc.) were obtained from Environment [9] in a 1:250,000 scale, and a soil map (1:250,000 scale) was also obtained [10].

3.1. Impact Assessment

The Decision Support System for Agrotechnology Transfer—DSSAT v. 4.5 contains the CERES—Maize model [11] and was used to determine best planting dates, fertilization strategies, and N use efficiency, and to examine potential effects of climate change on agriculture. In the embedded model, the development and growth of the crop was simulated on a daily basis from planting until physiological maturity. In order to ensure a correct simulation of the model, field experiments were set up in different locations to obtain genetic coefficients required for each maize cultivar, as proposed by He et al. [12], as well as to proceed with model validation. Four varieties of maize were tested: three open pollinated varieties (MPA01, Ivanir and Fortuna, respectively) and one commercial hybrid (AS 1548). The trials were conducted according to the recommendations of Soler et al. [13], and validation was done using observed data from different field experiments containing all required data needed for model validation [14]. A specific fertilization scheme was use for each soil (to attend the crop N requirements), ranging from 50 kg/N/ha to 120 kg/N/ha. The two best calibrated cultivars were chosen to perform further assessments.

3.2. Weather Scenarios

Regional Climatic Models (RCMs) were provided by CLARIS-LPB Project Data Archive Center and were restructured to match the model input criteria using Weatherman Software version 4.5.0.0 [15]. In order to define which RCM was to be used in the simulation, a comparison was done between yields simulated using RCMs as well as yields simulated using observed weather. The formulated hypothesis stated that the best RCM—or ensemble of RCMs—would be the one providing the best fit of yields generated using RCMs and observations. Yield for different planting dates (01/08 until 01/12, every 15 days) was then simulated using observed weather (1982–2012, 30 cropping seasons) from selected weather stations (those with more than 30 years of observations).

3.3. Regionalization

Results were organized and plotted on GIS maps only on areas identified for agricultural land use. Yield change in the scenarios was calculated based on recorded yields from the last 30 years. Nitrogen use efficiency was calculated according to the recommendations of Eulenstein et al. [16] using simulations with past weather as a base line, followed by comparisons with efficiency obtained using climate scenarios.

4. Results and Discussion

The main findings from the results were the following:

- Simulations run for the 2012–2040 period (437 ppm of CO_2) without adaptation strategies showed reductions of 12.5% in maize total production (Figure 1);
- The modelling process indicated that the N use efficiency increment ranged from −20% to +12% (according to the model, mostly due to CO_2 increment, but also due to soil properties and leaching) (Figure 2);
- By only using the best maize cultivar for each polygon (soil + weather), total production increased by 6%; when using both adaptation strategies—cultivar and best planting date—total production increased by 15% (Figure 3);
- N use efficiency rose in high CO_2 scenarios, but was also influenced by soil and weather in nonlinear relationships;
- Crop cultivar and planting date were effective tools for mitigating deleterious effects of climate change, supporting energy crops in the study region;
- The potential for maize production—and therefore ethanol—will be increased in the South-eastern region, while the Western region will suffer strong reductions in its production potential.

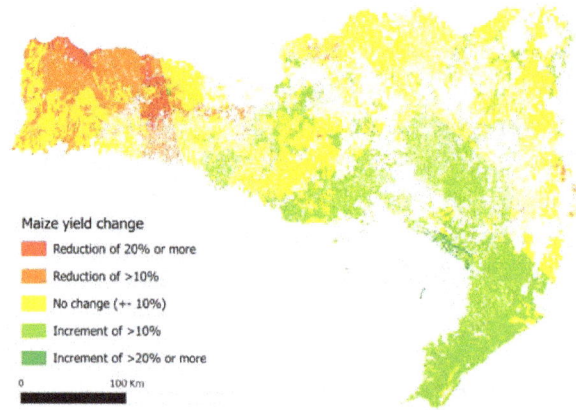

Figure 1. Change in maize yield (%) in the state of Santa Catarina, Brazil, for the 2012–2040 period. The impact was calculated using climate scenarios and without any kind of adaptation strategy (planting date or cultivar).

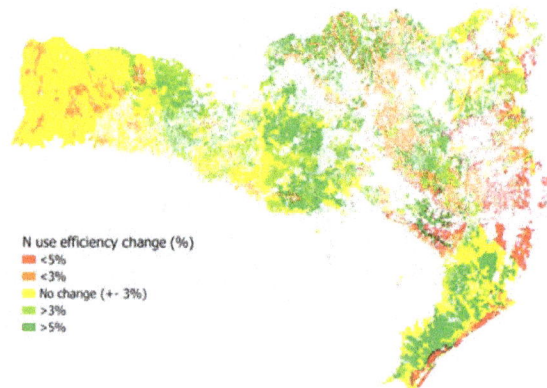

Figure 2. Nitrogen use efficiency change under impact of climatic scenarios for the state of Santa Catarina, Brazil. Changes are relative to the actual values (baseline calculated with climatic data from the last 30 years).

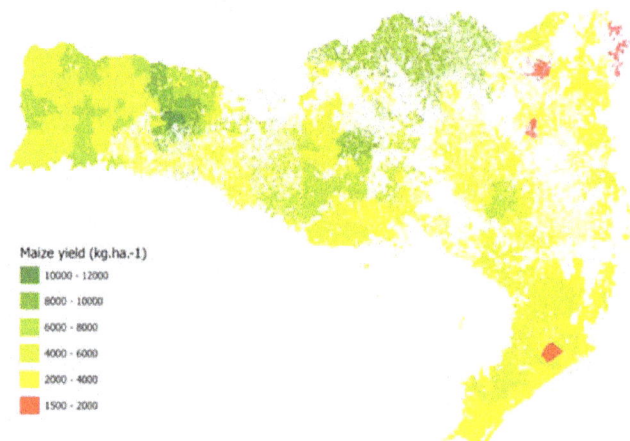

Figure 3. Expected maize yields under climatic scenarios for the 2012–2040 period. When adaptation strategies are employed, yields can potentially raise in many regions, increasing the production up to 15% from actual levels.

5. Conclusions

The effect of the weather scenarios on maize yield and N use efficiency is dependent on local conditions, being influenced mainly by soil parameters. The selected weather scenarios indicate that climate change, despite the increment of CO_2 concentration and increase in the N use efficiency, will have deleterious effects on maize production if no adaptation strategy is employed. Different planting dates and adapted cultivars are an effective tool to overcome this reduction. Overall, despite localized changes, the study region will still present a good potential for maize production.

Acknowledgments: The research leading to these results has received funding from the European Community's Seventh Framework Programme (FP7/2007–2013) under Grant Agreement No. 212492: CLARIS LPB. A Europe-South America Network for Climate Change Assessment and Impact Studies in La Plata Basin.

Author Contributions: Marcos Alberto Lana, Edgardo Guevara and Santiago Meira organized the data, run the calibration and generated the results. Frank Eulenstein, Sandro Luis Schlindwein, Askhad Khasrethovich Sheudzhen, Marion Tauschke and Axel Behrendt interpreted the results and wrote this text. All authors contributed equally to this paper.

Conflicts of Interest: The authors declare no conflict of interest.

References

1.   Nadler, A.; Bullock, P. Long-term changes in heat and moisture related to corn production on the Canadian Prairies. *Clim. Chang.* **2010**, *104*, 339–352. [CrossRef]

2.   Hoogenboom, G. Contribution of agrometeorology to the simulation of crop production and its applications. *Agric. For. Meteorol.* **2000**, *103*, 137–157. [CrossRef]

3.   Lobell, D.B.; Burke, M.B. On the use of statistical models to predict crop yield responses to climate change. *Agric. For. Meteorol.* **2010**, *150*, 1443–1452. [CrossRef]

4.   Parry, M.L.; Canziani, O.F.; Palutikof, J.P.; Van der Linden, P.J.; Hanson, C.E. *Climate Change: Impacts, Adaptation and Vulnerability. Contribution of Working Group II to the Fourth Assessment Report of the Intergovernmental Panel on Climate Change*; Cambridge University Press: Cambridge, UK, 2007.

5.   Saarikko, R.A. Applying a site based crop model to estimate regional yields under current and changed climates. *Ecol. Model.* **2000**, *131*, 191–206. [CrossRef]

6.   Brazilian Institute of Geography and Statistics (IBGE). *SIDRA-IBGE: Aggregated Database*; IBGE: Brasília, Brazil, 2012.

7.   Epagri/Cepa. *Síntese Anual da Agricultura de Santa Catarina*; Epagri/Cepa: Florianópolis, Brazil, 2011. (In Portuguese)

8.   Epagri/IBGE. *Mapoteca Topográfica Digital de Santa Catarina*; Epagri/IBGE: Florianópolis, Brazil, 2012. (In Portuguese)

9.   Brazilian Ministry of Environment. *Mapa de Cobertura Vegetal*; Brasil: Brasília, Brazil, 2012. (In Spanish)

10.   Embrapa Solos. *Sistema Brasileiro de Classificação de Solos*; Embrapa Solos: Rio de Janeiro, Brazil, 1999. (In Portuguese)

11.   Jones, C. *CERES-Maize: A Stimulation Model of Maize Growth and Development*; Texas A&M University Press: College Station, TX, USA, 1986.

12.   He, J.; Dukes, M.D.; Jones, J.W.; Graham, W.D.; Judge, J. Applying glue for estimating Ceres-maize genetic and soil parameters for sweet corn production. *Trans. ASABE* **2009**, *52*, 1907–1921. [CrossRef]

13.   Soler, C.M.T.; Sentelhas, P.C.; Hoogenboom, G. Application of the CSM-CERES-maize model for planting date evaluation and yield forecasting for maize grown off-season in a subtropical environment. *Eur. J. Agron.* **2007**, *27*, 165–177. [CrossRef]

14.   Hunt, L.A.; Boote, K.J. Data for model operation, calibration, and evaluation. In *Understanding Options for Agricultural Production*; Tsuji, G.Y., Hoogenboom, G., Thornton, P.K., Eds.; Kluwer Academic Publishers: Dordrecht, The Netherlands, 1998; pp. 9–40.

15. Wilkens, P.W. *DSSAT v4 Weather Data Editing Program (Weatherman). Data Management and Analysis Tools—Decision Support System for Agrotechnology Transfer Version 4.0: DSSAT v4: Data Management and Analysis Tools*; University of Hawaii: Honolulu, HI, USA, 2004; pp. 92–151.

16. Eulenstein, F.; Tauschke, M.; Lana, M.; Sheudshen, A.; Dannowski, R.; Schindler, R.; Drechsler, H. Nutrient balances in agriculture: A basis for the efficiency survey of agricultural groundwater conservation measures. In *Novel Measurement and Assessment Tools for Monitoring and Management of Land and Water Resources in Agricultural Landscapes of Central Asia*; Mueller, L., Saparov, A., Lischeid, G., Eds.; Springer: Cham, Switzerland, 2014; pp. 263–273.

16

Encapsulation of Shoot Tips and Nodal Segments for in Vitro Storage of "Kober 5BB" Grapevine Rootstock

Carla Benelli

Trees and Timber Institute (IVALSA-Istituto per la Valorizzazione del Legno e delle Specie Arboree), National Research Council, 50019 Florence, Italy; benelli@ivalsa.cnr.it

Abstract: In vitro preservation of the "Kober 5BB" rootstock (*Vitis berlandieri* × *Vitis riparia*) was assessed with the encapsulation technique and slow growth storage. Shoot tips and nodal segments excised from in vitro cultures were encapsulated in calcium-alginate beads. A 30 min ion exchange time proved optimal for forming proper beads. The encapsulated and naked explants were stored at 4 °C in the dark or light. After 9 months of cold storage, the highest regrowth, 83.3%, was recorded for the encapsulated shoot tips maintained in darkness. The development of the encapsulated nodal segments was 55.6% under the same storage conditions. The encapsulated explants had a better regrowth capacity after storage than the naked explants.

Keywords: alginate bead; grapevine rootstock; in vitro culture; slow growth storage

1. Introduction

The encapsulation technique for creating synthetic seeds is an important application for in vitro culture. Shoot tips, axillary buds or nodal segments may be used to develop the synthetic seeds. Synthetic seeds have been defined as artificially encapsulated somatic embryos or non-embryogenic in vitro-derived propagules and are used for sowing under in vitro or ex vitro conditions [1–3]. Synthetic seed technology combines the advantages of clonal propagation with those of seed propagation (i.e., storability, easy to handle and transport, protection against diseases and pests). The most recent application foresees the use of synthetic seeds in medium and long-term storage. The encapsulation technique is used today in advanced procedures of cryopreservation such as encapsulation-dehydration and encapsulation-vitrification methods [4], obtaining very promising results for the long-term preservation of plant germplasm [5–7]. The encapsulation of shoot tips, axillary buds or nodal segments has been reported for various species [8–13], and slow growth storage has been applied as a method for medium-term conservation [14–16]. Slow growth storage involves limiting development by reducing temperature and/or light intensity, adding osmotic compounds such as mannitol or sucrose in the culture medium, and use of growth retardants [17]. Among these, the most commonly used are the reduction of temperature and light intensity. These two parameters have physiological consequences that result in a significant reduction in cell metabolism and, consequently, shoot growth. With in vitro slow growth storage, it is possible to extend the intervals between subcultures, thus reducing the cost of stock plant maintenance as well as the risk of contamination during subculturing [15,18].

There are a large number of grapevine cultivars and rootstocks, many of which are unique to small, remote environments, and are therefore important for preserving biodiversity and reducing genetic erosion. The aim of this study was to optimize an encapsulation and storage protocol for shoot tips and nodal segments of "Kober 5BB" grapevine rootstock (*Vitis berlandieri* × *V. riparia*) as an alternative to traditional conservation methods. The experiments evaluated the effect of storage on the regrowth ability of encapsulated shoot tips and nodal segments following different storage conditions. This report describes the initial results of an investigation aimed at the development of a protocol for encapsulation of "Kober 5BB" explants.

2. Materials and Methods

2.1. Culture Media and Conditions

In vitro stock cultures of grapevine rootstock "Kober 5BB" were proliferated on MS semisolid medium [19] with 30 g/L sucrose at pH 5.7 (Figure 1A). The proliferation medium contained 1.5 mg/L benzyladenine (BA) and 3 g/L Gelrite®. The salts, vitamins and gelling agent used for the culture media were purchased from Sigma-Aldrich (St. Louis, MI, USA). The medium was autoclaved for 20 min at 121 °C. The cultures were maintained at 23 ± 1 °C in a growth chamber under a 16 h photoperiod with light intensity 60 μmol·m^{-2}·s^{-1} (standard culture conditions).

Figure 1. (**A**) In vitro stock culture of "Kober 5BB" rootstock; (**B**) Encapsulated explants in 3% sodium alginate, 100 mM CaCl$_2$·2H$_2$O with ion exchange time of 30 min. Regrowth of explants under standard conditions (21 days) after 3 months of storage at 4 °C in darkness; (**C**) encapsulated shoot tip; (**D**) encapsulated nodal segments and (**E**) naked shoot tips.

2.2. Encapsulation Procedure and Storage

Shoot tips (2–3 mm) and nodal segments (5–6 mm) were excised from in vitro stock cultures of "Kober 5BB". For encapsulation, the explants were plunged into a solution of 3% (w/v) sodium alginate (medium viscosity, Carlo Erba, Cornaredo, Italy) and MS liquid medium. Drops of alginate solution with shoot tip or nodal segment were sucked into a micropipette with sterile plastic tips and dropped into MS liquid medium supplemented with 100 mM calcium chloride (CaCl$_2$·2H$_2$O) for complexation. All operations were performed under sterile conditions. In order to achieve polymerization and prepare beads of an ideal shape and size with a uniform texture (Figure 1B), in the first experiment, different time periods (20, 30 or 40 min) for Na$^+$/Ca^{2+} ion exchange in the calcium chloride solution were tested. After each incubation period in the complexing agent, the encapsulated explants were retrieved and rinsed three times in sterile distilled water in order to remove traces of calcium chloride. Regrowth of the explants was then assessed as described below.

The following experiment was carried out on both encapsulated, with 30 min of ion exchange time, and naked explants, which were placed in 90 cm diameter Petri dishes containing semisolid, hormone-free MS medium and held at 4 °C in a growth chamber in darkness or in light (30 μmol·m^{-2}·s^{-1}, 8 h photoperiod) for 3, 6 or 9 months of storage.

Regrowth of the explants (encapsulated and naked) was evaluated for bead incubation time and for each cold storage period after transferring them onto proliferation medium for 21 days, under standard conditions. Regrowth ability of encapsulated explants was determined as the time required for the shoot to appear and break through the gel [20]. Ten explants of each type were placed in a Petri dish, and dishes were replicated three times for each incubation time or treatment combination by storage period.

2.3. Data Analysis

The average regrowth time was calculated as follows: Σ (NxTx)/number of developed shoots; where Nx is the number of developed shoots within consecutive intervals of time, Tx is the number of days between the beginning of the test and the end of the specific time interval.

Data on regrowth were recorded and presented as means with standard error of the mean (SEM). Significant differences among means were analyzed following analysis of variance using Duncan's Multiple Range Test at $p < 0.05$. Statistical analysis of percentages was carried out by a non-parametric Chi square test ($p \leqslant 0.05$) for pairwise comparisons. All statistical tests were performed with Systat 13 (Systat Software, Inc., San Jose, CA, USA).

3. Results

3.1. Shoot Regrowth from Beads

Based on regrowth, optimal ion exchange between 3% sodium alginate and 100 mM calcium chloride was observed at 30 min (Table 1). This proved to be the most effective time interval for producing reasonably smooth beads around the explants. With this time, more than 96% of "Kober 5BB" beads containing shoot tips and 85% containing nodal segments were regrown in the shortest average regrowth time, 15 and 25 days, respectively. Low percentages of regrowth ability were observed when the complexation period was shorter or longer than 30 min. In addition, at 20 min of complexation time, the beads were too soft and difficult to handle, while they showed a well-defined shape but regrowth was slower at 40 min. Following encapsulation, shoot tips developed more quickly (15 vs. 25 days) and at a higher percentage (96% vs. 85%) than nodal segments.

Table 1. Effect of ion exchange time on regrowth ability and average regrowth time of encapsulated explants of "Kober 5BB" rootstock.

Ion Exchange Time (Min)	Shoot Tips (%)	Average Regrowth Time (Days)	Nodal Segments (%)	Average Regrowth Time (Days)
20	53.3 ± 2.6 b	18	46.7 ± 2.8 b	25
30	96.7 ± 0.3 a	15	85.3 ± 0.6 a	25
40	33.7 ± 2.0 b	19	31.2 ± 2.1 b	26

Means ± standard error mean (SEM). Within each column, different letters indicate significantly different percentages using Duncan's Multiple Range Test, $p < 0.05$.

3.2. Storage in Slow-Growth Conditions

Irrespective of tissue type and storage condition, the explants exhibited some level of regrowth through 9 months of cold storage (Figure 2). No significant differences were observed in the regrowth ability of encapsulated versus naked shoot tip or nodal explants at 3 months, but encapsulated shoot tip and nodal explants exhibited significantly higher regrowth percentages than naked explants in both light and dark storage at 6 and 9 months (Figure 2B,C).

Regrowth percentages of all explants in both growing conditions decreased over storage time (Figures 2 and 3), and the decrease was most evident between 3 and 6 months of storage. At 6 and 9 months, shoot tip explants showed higher regrowth ability than nodal explants (Figure 3). Furthermore, the highest regrowth ability at 6 and 9 months was recorded for the encapsulated shoot tips kept in the dark (Figure 2).

Overall, darkness had a positive influence on the regrowth of the explants at each storage period (Figure 4). The preservation conditions assessed proved to be more suitable for the encapsulated shoot tips than for the encapsulated nodal segments. No morphological changes were observed for the encapsulated explants that developed into shoots.

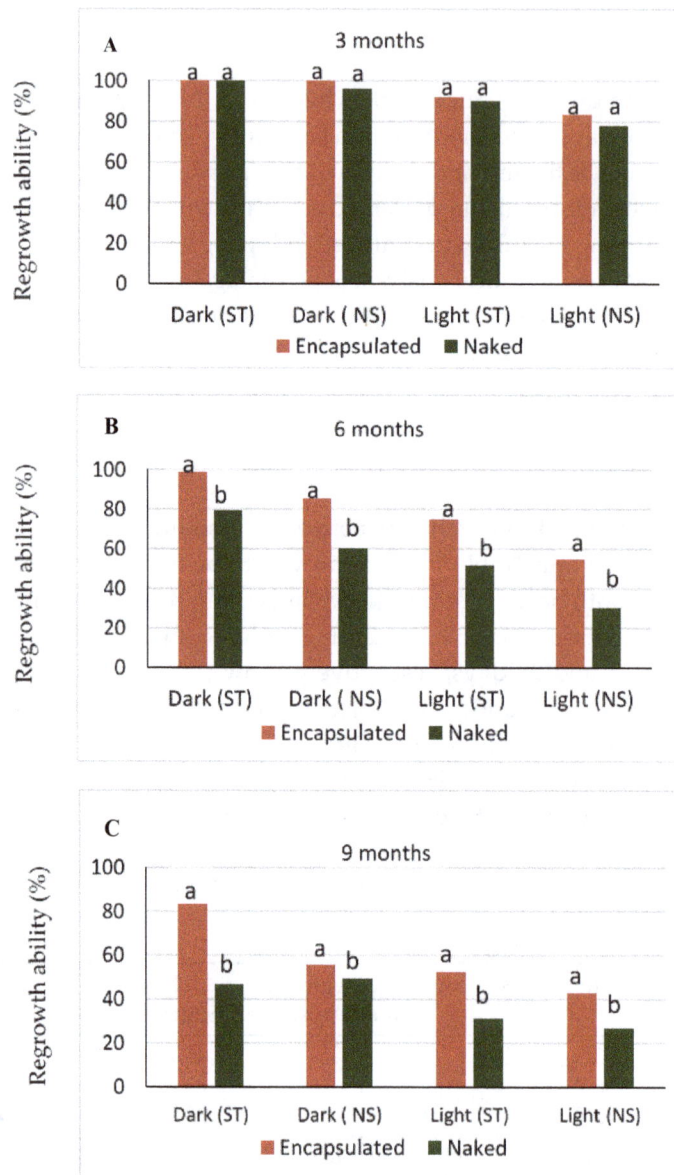

Figure 2. Regrowth ability (%) of encapsulated and naked shoot tips (ST) or nodal segments (NS) of "Kober 5BB" rootstock stored at 4 °C in darkness and light conditions for each storage period. (**A**) three months; (**B**) six months; (**C**) nine months. Different letters indicate percentages of encapsulated versus naked means within tissue type and light versus dark conditions that significantly differed (χ^2 test, $p \leqslant 0.05$) at each storage period.

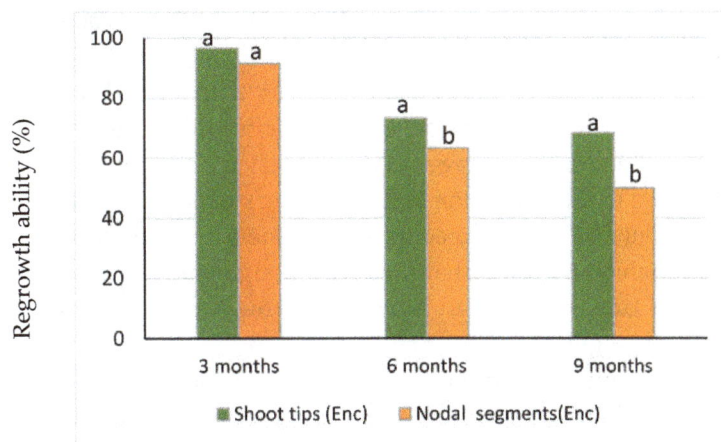

Figure 3. Regrowth ability of encapsulated shoot tips versus nodal segments after 3, 6, and 9 months of 4 °C storage. Different letters indicate percentages of shoot tip versus nodal segments that significantly differed (χ^2 test, $p \leqslant 0.05$) at each storage period.

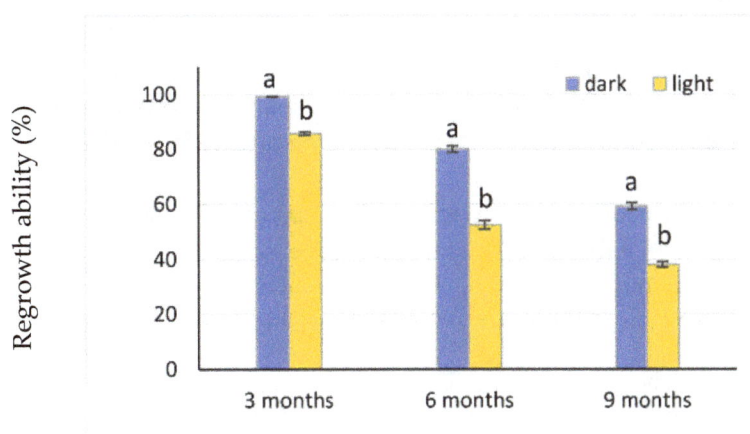

Figure 4. Effect of dark versus light on explant regrowth after 3, 6, and 9 months of 4 °C storage. Different letters indicate percentages for dark versus light that significantly differed (χ^2 test, $p \leqslant 0.05$) at each storage period.

4. Discussion

Encapsulation of explants is affected by sodium alginate and calcium chloride concentrations, and by ion exchange time, which together can give beads their optimum characteristics [11]. Sodium alginate and calcium salt have proven to be the best combination for encapsulation since they are low cost, non-damaging, easy to use and give high propagule-to-plant conversion [21]. The hardness of beads mainly depends on the number of sodium ions exchanged with calcium ions, and complexing time should be optimized in order to form uniformly spherical and firm beads. The exchange time generally has ranged between 20–40 min [22]. For "Kober 5BB" beads, the best results were obtained with the 30 min hardening process; this ion exchange time showed the best regrowth for shoot tips (96.7%) and nodal segments (85.3%). Most studies on fruit, ornamental and vegetable species suggest that a 30 min exchange time is best for obtaining a uniform spherical shape and a good level of sprouting from the beads [9,23–29]. The shortest average regrowth time of encapsulated "Kober 5BB" shoot tips was 15 d when the encapsulated explants were hardened for 30 min, while it was 25–26 days for nodal segments, for the ion exchange time tested. Regrowth time has varied with species; for example, when applying the same protocol for encapsulation used in this study, regrowth in *Photinia fraserii* occurred in 10 days [30], while for *Nerium oleander* and *Kalanchoe* spp., it took 18 and 19 days, respectively [8].

The percentage of encapsulated and naked explants that formed shoots depended on the duration and conditions of storage. Under slow growth in vitro culture, the encapsulated explants of "Kober 5BB" were well suited for the storage conditions. Their better regrowth compared to naked explants could be attributed to their inclusion in the gel matrix which acts as an "artificial endosperm" for the explants. The present results confirmed the effectiveness of the protective coating. This agreed with previous literature which reported that the sodium alginate matrix provided protection and supplied nutrients to explants resulting in a higher regrowth percentage of encapsulated nodal segments than naked nodal segments [31].

The encapsulation technology could represent a useful technique for plant material conservation when used for medium or long-term preservation. In cryopreservation, this technique has been used for different species [32–34]. Moreover, various studies have described encapsulation for short and medium conservation (from 1 month to 1 year) without losing explant viability [35–38] and for increasing the interval between subcultures by reducing growth.

5. Conclusions

Implementation of the encapsulation technique may be used for clonal propagation and for effective storage at low temperatures. This study has shown that in vitro explant encapsulation could be used for storing "Kober 5BB" rootstock tissues up to 9 months at 4 °C without subcultures. The results indicated that the method could have great potential for preserving desirable elite genotypes of grapevine for medium periods, and were a useful prerequisite for further studies extending the storage period in this species.

In some plant species, poor regrowth of encapsulated propagules into plants has been a limitation for commercial application of the encapsulation technique, so further evaluation of the technique is needed. In addition, plant nurseries may be interested in achieving the development and growth of plants from synthetic seeds in soil without the aseptic conditions required for in vitro culture.

Conflicts of Interest: The author declares no conflict of interest.

References

1. Murashige, T. Plant cell and organ cultures as horticultural practices. *Acta Hortic.* **1977**, *78*, 17–30.
2. Aitkens-Christie, J.; Kozai, T.; Takayama, S. Automation in plant tissue culture: General introduction and overview. In *Automation and Environmental Control in Plant Tissue Culture*; Aitken-Christie, J., Kozai, T., Smith, M.A.L., Eds.; Kluwer Academic Publication: Dordrecht, The Netherlands, 1995; pp. 1–18.
3. Standardi, A.; Piccioni, E. Recent perspectives on the synthetic seed technology using non-embryogenic vitro-derived explants. *Int. J. Plant Sci.* **1998**, *159*, 968–978.
4. Dereuddre, J.; Blandin, S.; Hassen, N. Resistance of alginate-coated somatic embryos of carrot (*Daucus carota* L.) to desiccation and freezing in liquid nitrogen: 1. Effects of preculture. *CryoLetters* **1991**, *12*, 125–134.
5. Sakai, A.; Engelmann, F. Vitrification, encapsulation-vitrification and droplet-vitrification: A review. *CryoLetters* **2007**, *28*, 151–172. [PubMed]
6. Fabbri, A.; Ganino, T.; Lambardi, M.; Nisi, R. Crioconservazione di gemme di portinnesto 'Kober' 5BB (*Vitis berlandieri* × *Vitis riparia*): Aspetti anatomici. *Ital. Hortus* **2007**, *14*, 82–86.
7. Benelli, C.; de Carlo, A.; Engelmann, F. Recent advances in the cryopreservation of shoot-derived germplasm of economically important fruit trees of *Actinidia, Diospyros, Malus, Olea, Prunus, Pyrus* and *Vitis*. *Biotechnol. Adv.* **2013**, *31*, 175–185. [PubMed]
8. Lambardi, M.; Benelli, C.; Ozudogru, E.A.; Ozden-Tokatli, Y. Synthetic seed technology in ornamental plants. *Floric. Ornam. Plant Biotechnol. Adv. Top. Issues* **2006**, *2*, 347–354.
9. Piccioni, E.; Standardi, A. Encapsulation of micropropagated buds of six woody species. *Plant Cell Tissue Org. Cult.* **1995**, *42*, 221–226. [CrossRef]

10. Micheli, M.; Hafiz, I.A.; Standardi, A. Encapsulation of in vitro-derived explants of olive (*Olea europaea* L. cv. Moraiolo) II. Effects of storage on capsule and derived shoots performance. *Sci. Hortic.* **2007**, *113*, 286–292. [CrossRef]

11. Rai, M.K.; Asthana, P.; Singh, S.K.; Jaiswal, V.S.; Jaiswal, U. The encapsulation technology in fruit plants—A review. *Biotechnol. Adv.* **2009**, *27*, 671–679. [CrossRef] [PubMed]

12. Reddy, C.M.; Murthy, K.S.R.; Pullaiah, T. Synthetic seeds: A review in agriculture and forestry. *Afr. J. Biotechnol.* **2012**, *11*, 14254–14275.

13. Kulus, D.; Zalewska, M. In vitro plant recovery from alginate-encapsulated *Chrysanthemum × grandiflorum* (Ramat.) Kitam shoot tips. *Propag. Ornam. Plants* **2014**, *14*, 3–12.

14. Nower, A.A.; Ali, E.A.; Rizkalla, A. Synthetic seeds of pear (*Pyrus communis* L.) rootstock storage in vitro. *Aust. J. Basic Appl. Sci.* **2007**, *1*, 262–270.

15. Benelli, C.; Ozudogru, E.A.; Lambardi, M.; Dradi, G. In vitro conservation ornamental plants by slow growth storage. *Acta Hortic.* **2012**, *961*, 89–93. [CrossRef]

16. Cordeiro, S.Z.; Simas, N.K.; Henriques, A.B.; Sato, A. In vitro conservation of *Mandevilla moricandiana* (Apocynaceae): Short-term storage and encapsulation–dehydration of nodal segments. *In Vitro Cell Dev. Biol. Plant* **2014**, *50*, 326–336. [CrossRef]

17. Grout, B.W.W. In vitro conservation of germplasm. In *Plant Tissue Culture: Application and Limitations*; Bhojwani, S.S., Ed.; Elsevier: Amsterdam, The Netherlands, 1990; pp. 394–411.

18. Ikhlaq, M.M.; Hafiz, I.A.; Micheli, M.; Ahmad, T.; Abbasi, N.A.; Standardi, A. In vitro storage of synthetic seeds: Effect of different storage conditions and intervals on their conversion ability. *Afr. J. Biotechnol.* **2010**, *9*, 5712–5721.

19. Murashige, T.; Skoog, F. A revised medium for rapid growth and bio assays with tobacco tissue cultures. *Physiol. Plantarum.* **1962**, *15*, 473–497. [CrossRef]

20. Standardi, A.; Micheli, M. Encapsulation of in vitro-derived explants: An innovative tool for nurseries. In *Protocols for Micropropagation of Selected Economically-Important Horticultural Plants*; Lambardi, M., Ozudogru, E.A., Jain, S.M., Eds.; Humana Press: New York, NY, USA, 2013; pp. 397–418.

21. Sharma, S.; Shahzad, A.; Teixeira da Silva, J.A. Synseed technology—A complete synthesis. *Biotechnol. Adv.* **2013**, *31*, 186–207. [CrossRef] [PubMed]

22. Standardi, A. Encapsulation: Promising technology for nurseries and plant tissue laboratories. *AgroLife Sci. J.* **2012**, *1*, 48–54.

23. Sicurani, M.; Piccioni, E.; Standardi, A. Micropropagation and preparation of synthetic seed in M.26 apple rootstock I: Attempts towards saving labor in the production of adventitious shoot tips suitable for encapsulation. *Plant Cell Tissue Org. Cult.* **2001**, *66*, 207–216.

24. Lisek, A.; Orlikowska, T. In vitro storage of strawberry and raspberry in calcium-alginate beads at 4 °C. *Plant Cell Tissue Org. Cult.* **2004**, *78*, 167–172. [CrossRef]

25. Naik, S.K.; Chand, P.K. Nutrient-alginate encapsulation of in vitro nodal segments of pomegranate (*Punica granatum* L.) for germplasm distribution and exchange. *Sci. Hortic.* **2006**, *108*, 247–252. [CrossRef]

26. Germanà, M.A.; Micheli, M.; Pulcini, L.; Standardi, A. Perspective of the encapsulation technology in the nursery activity of *Citrus*. *Caryologia* **2007**, *60*, 192–195.

27. Daud, N.; Taha, R.M.; Hasbullah, N.A. Artificial seed production from encapsulated micro shoots of *Saintpaulia ionantha* Wendl. (African Violet). *J. Appl. Sci.* **2008**, *8*, 4662–4667.

28. Sundararaj, S.G.; Agrawal, A.; Tyagi, R.K. Encapsulation for in vitro short-term storage and exchange of ginger (*Zingiber officinale* Rosc.) germplasm. *Sci. Hortic.* **2010**, *125*, 761–766.

29. Nor Asmah, H.; Nor Hasnida, H.; Nashatul Zaimah, N.A.; Noraliza, A.; Nadiah Salmi, N. Synthetic seed technology for encapsulation and regrowth of in vitro-derived Acacia hybrid shoot and axillary buds. *Afr. J. Biotechnol.* **2011**, *10*, 7820–7824.

30. Ozden-Tokatli, Y.; de Carlo, A.; Gumusel, F.; Pignattelli, S.; Lambardi, M. Development of encapsulation techniques for the production and conservation of synthetic seeds in ornamental plants. *Propag. Ornam. Plants* **2008**, *8*, 17–22.

31. Singh, S.K.; Rai, M.K.; Asthana, P.; Sahoo, L. Alginate-encapsulation of nodal segments for propagation, short-term conservation and germplasm exchange and distribution of *Eclipta alba* (L.). *Acta Physiol. Plant* **2010**, *32*, 607–610. [CrossRef]

32. Gonzalez-Arnao, M.T.; Engelmann, F. Cryopreservation of plant germplasm using the encapsulation-dehydration technique: Review and case study on sugarcane. *CryoLetters* **2006**, *27*, 155–168. [PubMed]

33. Engelmann, F.; Arnao, M.T.G.; Wu, Y.; Escobar, R. Development of encapsulation dehydration. In *Plant Cryopreservation: A Practical Guide*; Reed, B., Ed.; Springer: New York, NY, USA, 2008; pp. 59–75.

34. Padrò, M.D.A.; Frattarelli, A.; Sgueglia, A.; Condello, E.; Damiano, C.; Caboni, E. Cryopreservation of white mulberry (*Morus alba* L.) by encapsulation-dehydration and vitrification. *Plant Cell Tissue Org. Cult.* **2012**, *108*, 167–172.

35. Hung, C.D.; Trueman, S.J. Encapsulation technology for short-term preservation and germplasm distribution of the African mahogany *Khaya senegalensis*. *Plant Cell Tissue Org. Cult.* **2011**, *107*, 397–405. [CrossRef]

36. Hung, C.D.; Trueman, S.J. Alginate encapsulation of shoot tips and nodal segments for short-term storage and distribution of the eucalypt *Corymbia torelliana* × *C. citriodora*. *Acta Physiol. Plant* **2012**, *34*, 117–128.

37. Sakhanokho, H.F.; Pounders, C.T.; Blythe, E.K. Alginate encapsulation of Begonia microshoots for short-term storage and distribution. *Sci. World J.* **2013**. [CrossRef]

38. Srivastava, V.; Khan, S.A.; Banerjee, S. An evaluation of genetic fidelity of encapsulated microshoots of the medicinal plant: *Cineraria maritima* following six months of storage. *Plant Cell Tissue Org. Cult.* **2009**, *99*, 193–198. [CrossRef]

17

Effect of Methyl Jasmonate on Physical and Chemical Properties of Mango Fruit cv. Nam Dok Mai

Panida Boonyaritthongchai *, Chalida Chimvaree, Mantana Buanong, Apiradee Uthairatanakij and Pongphen Jitareerat

Division of Postharvest Technology, King Mongkut's University of Technology Thonburi, Bangkok 10150, Thailand; chalida_tai@hotmail.com (C.C.); mantana.bua@kmutt.ac.th (M.B.); apiradee.uth@kmutt.ac.th (A.U.); pongphen.jit@kmutt.ac.th (P.J.)
* Correspondence: panida.boo@kmutt.ac.th

Abstract: The effect of methyl jasmonate (MeJA) on anthracnose severity and physical and chemical properties of mango fruit cv. Nam Dok Mai was investigated. The mango fruit were harvested at the mature-green stage and the fruit surface was disinfected with 100 ppm sodium hypochlorite solution. The fruit samples were then fumigated with 30 ppm MeJA in an enclosed container at 25 °C for 6 h, and subsequently stored at 13 °C for 18 days. Non-treated fruit were used as the control. The results showed that MeJA had no effect on anthracnose severity, stem end rot disease and color change. MeJA treatment induced ethylene production and enhanced the accumulation of β-carotene content throughout the storage period compared with non-treated fruit. This result indicated that MeJA treatment may be used to increase the accumulation of β-carotene content of mango fruit during storage at low temperature.

Keywords: fumigation; mango; methyl jasmonate

1. Introduction

Mango (*Mangifera indica* L.) cv. Nam Dok Mai No. 4 is the most popular cultivar in Thailand with lusciously fragrant, sweet juicy flesh and a golden yellow skin color that attracts consumers. In Thailand, anthracnose and stem-end rot are considered the most serious and destructive diseases of mango. It is considered a major problem limiting the storage and shelf life of mango fruit. Methyl jasmonate (MeJA) is a plant hormone and endogenous signal molecule that has important roles in responses to environmental stresses [1–4]. Postharvest application of methyl jasmonate is a new approach for maintenance of quality of fruit, including induction of plant defense systems to protect from disease. Methyl jasmonate can cause changes in physical appearance, mechanical properties, and the composition of bioactive compounds of fruit [5]. It is an important cellular regulator that is involved in diverse developmental stages including fruit ripening, accumulation of pigments, phenolic compounds, antioxidants, and sugars [6–9], and has increased β-carotene accumulation in ripening tomato fruit [10]. MeJA treatments have reduced deterioration and the development of chilling injury symptoms in three cultivars of mango, Tommy Atkins, Kent, and Zill [11,12], and other fruit species such as guava (*Psidium guajava*), loquat (*Eriobotrya japonica* L.), peach (*Prunus persica* L.) and pomegranate (*Punica granatum*) [13,14]. Jasmonates also play an integral role in intracellular signal transduction cascades which occur in the inducible defense mechanisms that plants have developed against pathogens and other stresses. The objective of this study was to determine the effects of MeJA on physical and biochemical properties, antioxidant systems, and bioactive compounds and quality changes of Nam Dok Mai No. 4 mango fruit during cold storage.

2. Experimental Section

2.1. Mango Fruit Preparation

Mango fruit cv. Nam Dok Mai No. 4 were harvested at commercial maturity from Good Agricultural Practices (GAP) farms. They were transported to the Division of Postharvest and Technology, School of Bioresources Technology, King Mongkut's University of Technology Thonburi, Thailand. Fruit were cleaned with tap water, separated by size and color, the fruit surface disinfected with 100 ppm sodium hypochlorite solution, and dried in air at ambient temperature. The fruits were then randomly divided into two groups. One hundred forty-four mangoes were placed in glass jars containing 1.35 mL 30 ppm methyl jasmonate (MeJA) or water (for the control) spotted onto filter paper for 6 h at 25 °C in darkness. After treatment, fruit were transferred from the treatment jars into plastic baskets covered with perforated polyethylene film and stored at 13 °C for 18 days. Three replicates (9 fruit per one replicate) of each treatment were used for all analyses. Fruit decay and physical and chemical properties were evaluated every 3 days during storage.

2.2. Disease Incidence of Anthracnose and Stem End Rot

The severity of anthracnose disease was assessed by the extent of total decayed area on each fruit surface using a 5-point scale [15], where 0 = no disease symptoms, 1 = 1%–10% of the fruit surface, 2 = 11%–20% of the surface, 3 = 21%–30% of the surface, 4 = 31%–40% of the surface, and 5 = >40% of the surface affected by anthracnose disease spots. The severity of stem end rot disease was assessed by the extent of total decayed area on each fruit surface using a 5-point scale, where 0 = no disease symptoms, 1 = 1%–10% of the fruit surface, 2 = 11%–20% of the surface, 3 = 21%–30% of the surface, 4 = 31%–40% of the surface, and 5 = >40% of the fruit surface exhibited brown rot from stem end rot disease.

2.3. Colour Changes

Skin and pulp color were assessed with a tristimulus color different meter (Minolta CR300) and expressed as L, a, and b values. These values were used to calculate hue values. Negative a values indicated green and the higher positive a values indicated red color. Higher positive b values indicated a more yellow color.

2.4. Extraction, Saponification and HPLC Analysis of β-Carotene

All extraction and saponification steps were carried out under yellow or dim lights. A fruit sample was ground in an Oster blender until pulverized. 5 g of sample was added to a 50 mL round-bottomed Nalgene centrifuge tube. 10 mL of cold ethanol was added, and the sample was homogenized with a Polytron homogenizer at medium speed for 3 min. 8 mL of hexane was added, and the sample was homogenized for an additional 2 min. The mixture was then centrifuged for 4 min at $7000 \times g$. The carotenoid-bearing hexane layer was transferred with a Pasteur pipet to a 125 mL screw-cap Ehrlenmeyer flask. 5 mL saturated sodium chloride solution was added to the contents of the centrifuge tube, and the contents were stirred gently to homogenize the mixture. An additional 8 mL of hexane was added, and the mixture was homogenized at low speed with the Polytron for 1 min. The mixture was centrifuged as before, and the hexane layer transferred to the Ehrlenmeyer flask with the first extract.

15 mL of 10% methanolic KOH was added to the contents of the Ehrlenmeyer flask. The flask was flushed with nitrogen, sealed, and wrapped in aluminum foil to exclude light. The flask was left at room temperature for 16 h with gentle shaking. The mixture was then transferred to a separatory funnel and washed to remove the KOH, first with 50 mL of 10% NaCl solution and then with deionized water until the rinse had a neutral pH. The KOH solution was further extracted with an additional 10 mL hexane. The combined hexane extracts were evaporated under nitrogen just to dryness, then redissolved in methylene chloride and brought to volume with mobile phase (see below). The sample was immediately filtered through a 0.45 mm filter into an amber analysis vial and sealed.

The HPLC system consisted of a Hewlett Packard Series 1050 autosampler, Series 1050 pump, and a Series 1040 M diode array detector, operated by HP Chemstation software. A 250 × 4.6 mm, 5 mm internal diameter Vydac 201TP54 reversed phase C18 column fitted with biocompatible titanium frits was used for separation, together with a Vydac 4.6 × 25 mm high performance guard column filled with the same packing material. The mobile phase consisted of acetonitrile, methanol and methylene chloride 75:20:5 *v/v/v*, containing 0.1% butylated hydroxytoluene and 0.05% triethylamine. The methanol contained 0.05 M ammonium acetate. All reagents were HPLC grade. The flow rate was 1.5 mL/min. Detection was at 450 nm. Identification and purity of peaks was confirmed by comparing spectra using the computer software.

2.5. Ethylene Production

Ethylene production was measured during cold storage at 13 °C. Three fruit from each replication were randomly selected to measure respiration rate. Individual fruit were kept in airtight glass jars fitted with a rubber septum for collecting the gases. After 1 h incubation of the fruit, the head space gas was withdrawn from container using a syringe and injected into a gas chromatograph (Shimadzu, Kyoto, Japan). N_2 was used as the carrier gas. Ethylene was quantified using a flame ionization detector. Ethylene was estimated and expressed as $\mu L \ C_2H_4/kg \cdot h$.

2.6. Statistical Analysis

The experiment was arranged in completely randomized design. All data presented were means of three replicates along with standard errors of means, and mean were compared using t-test. Differences at $p \leq 0.05$ and $p \leq 0.01$ were considered significant.

3. Results and Discussion

3.1. Disease Severity

The symptoms of anthracnose disease on MeJA-treated and control mango fruit were observed during storage at 13 °C by 3 days. MeJA-treated mango and control fruit showed no significant difference in anthracnose or stem end rot disease severity through the end of storage at 18 days (Figure 1A,B). Treatments with MeJA have been reported to suppress gray mold rot (*Botrytis cinerea*) on strawberries (*Fragaria* × *ananassa* Duch.) [10], to reduce microbial contamination of fresh cut celery (*Apium graveolens*) and peppers (*Capsicum annum*) [16], and to suppress fungal decay in grapefruit (*Citrus* × *paradisi*) [17], in contrast to the present results. These differences may be due to the wide range in genetic makeup of different horticultural crops and different treatment methods with MeJA among studies [18]. MeJA has been shown to prevent postharvest diseases in a number of horticultural crops, but the mode of action in reducing disease has not been well-characterized, although MeJA effects could be because of a direct inhibitory effect on pathogen growth and/or because of the induction of natural disease resistance [17].

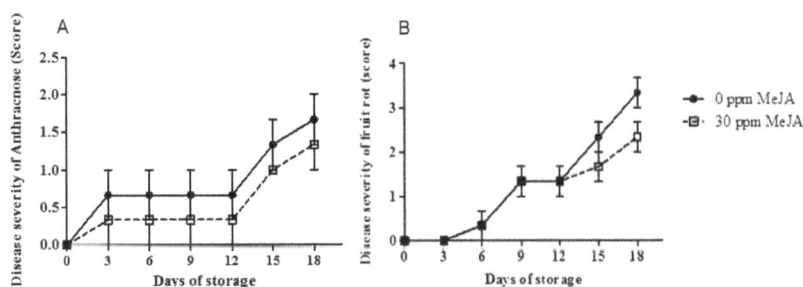

Figure 1. Disease severity of anthracnose (**A**) and stem end rot (**B**) in the mango fruit cv. Nam Dok Mai No.4 fumigated with 30 ppm methyl jasmonate (MeJA) versus untreated controls during storage at 13 °C for 18 days. There was no difference between treated and control fruit on any day.

3.2. Internal Pulp Color

The L value, or the lightness of the yellow pulp, and b value, expressing blue (−) to yellow (+) color did not differ between the MeJA treatment and controls on any day (Figure 2A,B). The a value, expressing green (−) to red (+) color, and hue angle were only higher for MeJA treatment at 18 days (Figure 2C,D). Colour change of mango pulp as L, a, and b values, and hue angle, was constant through 12 days of storage, but changed between 12 and 18 days. The L value and hue angle value of both treatments decreased, while a and b values of both treatments increased. Pulp color of both treatments changed from light yellow to dark yellow after 12 days.

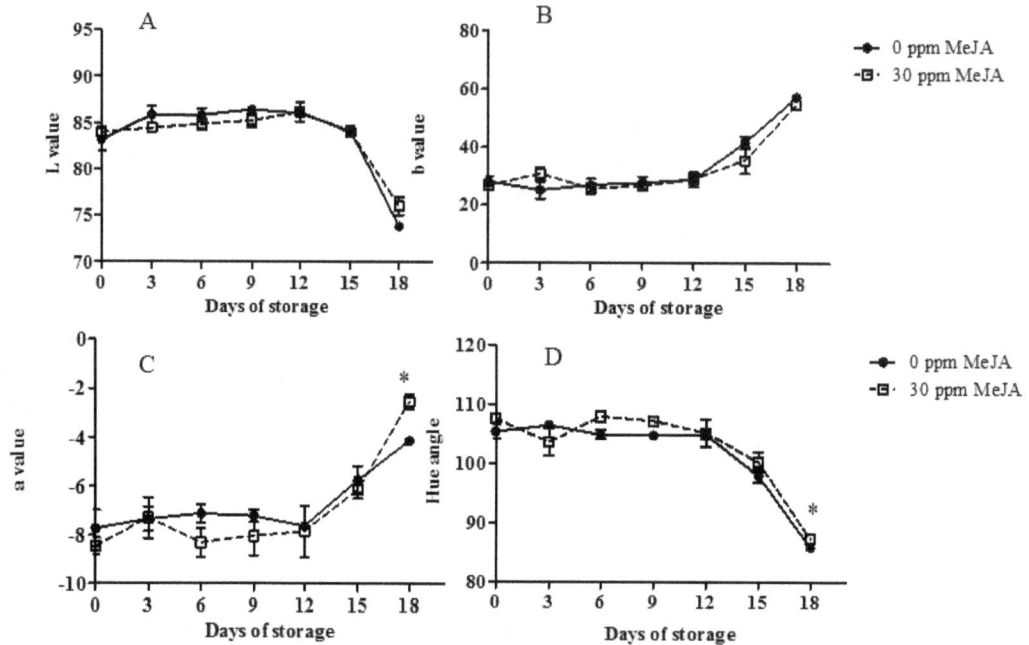

Figure 2. Colour change expressed as L (**A**), b (**B**), a (**C**), and hue angle (**D**) values of the fruit pulp of cv. Nam Dok Mai No.4 fumigated with 30 ppm methyl jasmonate (MeJA) versus untreated (control) during storage at 13 °C for 18 days. An asterisk indicates a statistically significant difference at $p \leq 0.05$; otherwise, treatments did not differ.

3.3. Peel Color

MeJA treatment showed a higher L value than controls at 3 and 15 days of storage, but a lower value than controls at 18 days (Figure 3A). The a and b values and hue angle of both treatments was maintained for 12 days of storage, but a and b values declined from 12 to 18 days while hue angle increased (Figure 3B–D). However, MeJA values did not differ from controls in these traits. The response of mango to MeJA has depended on the maturity stage, cultivar and subsequent storage temperature [19]. Gonzalez-Aguilar et al. [19] reported that use of 10^{-5} M MeJA enhanced yellow and red colour development of mango cv. Kent stored at 20 °C, but in the present research the fruit were kept at a lower temperature (13 °C), so it is possible that the colder storage suppressed color development.

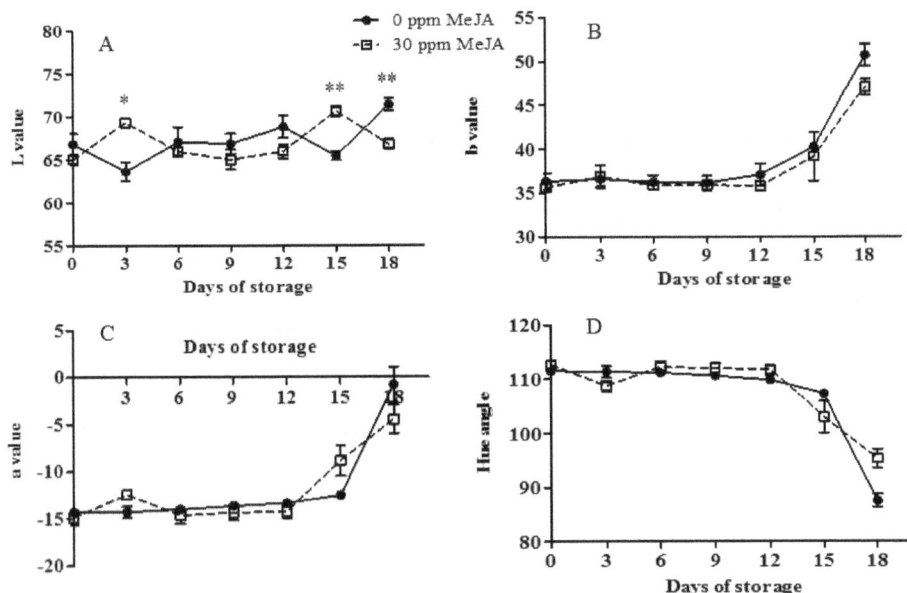

Figure 3. Colour change expressed as L (**A**), b (**B**), a (**C**), and hue angle (**D**) values of the fruit peel of cv. Nam Dok Mai No.4 fumigated with 30 ppm methyl jasmonate (MeJA) versus untreated (control) during storage at 13 °C for 18 days. Asterisks indicate a statistically significant difference at $p \leq 0.05$ (*) or $p \leq 0.01$ (**); otherwise, treatments did not differ.

3.4. Ethylene Production

MeJA-treated fruit showed a significantly higher ethylene production than control fruit at day 9 of storage, while control fruit showed its highest production at 12 days although it was not higher than MeJA-treated fruit (Figure 4). The ethylene production of MeJA treatment decreased after 9 days until the end of storage. The ethylene production of control fruit increased gradually until reaching a peak at day 12 of storage, and then decreased until the end of storage. MeJA enhanced softening in apple, promoting apple ripening as indicated by increased ethylene biosynthesis, acceleration of yellowing and softening [20]. The present results were similar to the effect of MeJA on ethylene production in tomato fruit [21] and increased ethylene-forming enzyme activity in preclimacteric apple fruit [22].

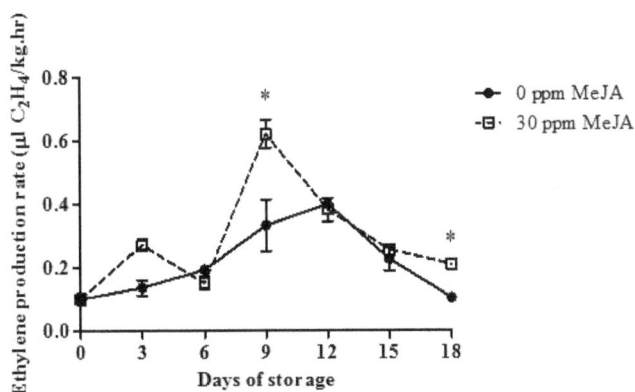

Figure 4. Ethylene production of mango fruit, cv. Nam Dok Mai No.4, fumigated with 30 ppm methyl jasmonate (MeJA) versus untreated (control) during storage at 13 °C for 18 days. An asterisk indicates a statistically significant difference at $p \leq 0.05$; otherwise, treatments did not differ.

3.5. β-Carotene Content

β-Carotene content of mango pulp of MeJA-treated and control fruit increased during storage (Figure 5). MeJA-treated mango showed a significantly higher β-carotene content than control fruit at 15 and 18 days of cold storage. Application of MeJA vapour improved skin color by promoting β-carotene synthesis and chlorophyll degradation in Golden Delicious apple [20]. The response of mango fruit to MeJA depends up on the maturity stage, cultivar and subsequent storage temperature [19]. MeJA has been shown to increase β-carotene accumulation in ripening tomatoes [23].

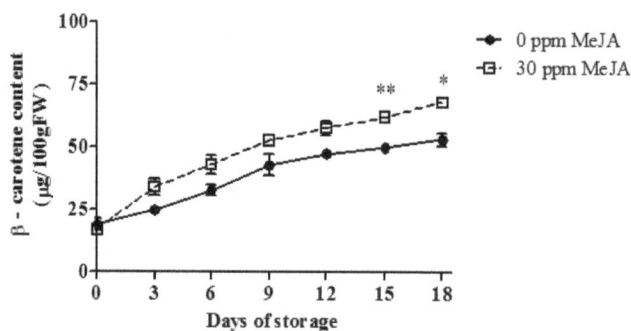

Figure 5. β-carotene content of mango fruit, cv. Nam Dok Mai No.4, fumigated with 30 ppm methyl jasmonate (MeJA) versus untreated (control) fruit during storage at 13 °C for 18 days. Asterisks indicate statistically significant differences at $p \leq 0.05$ (*) and $p \leq 0.01$ (**); otherwise, treatments did not differ.

4. Conclusions

A 30 ppm MeJA fumigation treatment did not reduce disease severity from anthracnose or stem end rot in mango cv. Nam Dok Mai No. 4 during storage at 13 °C. Nor did the MeJA fumigation treatment have an effect on color change of mango peel, although there was a small effect on pulp color. MeJA treatment increased maximum ethylene production and induced greater accumulation of β-carotene in mango pulp.

Acknowledgments: We are grateful to the National Research University Project for financial support of this research.

Author Contributions: P.B., M.B., A.U. and P.J. conceived and designed the experiments. M.B., P.B. and C.C. performed the experiments. P.J. and A.U. analyzed the data. P.B. and P.J. drafted and revised it critically for important intellectual content.

Conflicts of Interest: The authors declare no conflict of interest.

References

1. Raskin, I. Salicylic acid, a new plant hormone. *Plant Physiol.* **1992**, *99*, 799–803. [CrossRef] [PubMed]

2. Creelman, R.A.; Mullet, J.E. Biosynthesis and action of jasmonate in plants. *Ann. Rev. Plant Physiol. Plant Mol. Biol.* **1997**, *48*, 355–381. [CrossRef] [PubMed]

3. Ananieva, E.A.; Christov, K.N.; Popova, L.P. Exogenous treatment with salicylic acid leads to increased antioxidant capacity in leaves of barley plant sexpose to paraquat. *J. Plant Physiol.* **2004**, *161*, 319–328. [CrossRef] [PubMed]

4. Wang, L.; Chen, S.; Kong, W.; Li, S.; Archbold, D.D. Salicylic acid pretreatment alleviates chilling injury and affects the antioxidant system and heat shock protein of peaches during cold storage. *Postharvest Biol. Technol.* **2006**, *41*, 244–251. [CrossRef]

5. Fan, X.; Mattheis, J.P.; Fellman, J.K.; Patterson, M.E. Effect of methyl jasmonate on ethylene and volatile production by Summered apples depends on fruit developmental stage. *J. Agric. Food Chem.* **1997**, *45*, 208–211. [CrossRef]

6. Cheong, J.J.; Choi, Y.D. Methyl jasmonate as a vital substance in plants. *Trends Genet.* **2003**, *19*, 409–413. [CrossRef]

7. Burhan, O.; Ebubekir, A.; Kenan, Y.; Yakup, O.; Onur, S. Effect of methyl jasmonate treatments on the bioactive compounds and physicochemical quality of "Fuji" apples: Ciencia e InvestigaciÓn Agraria. *Crop Prod.* **2013**, *40*, 201–211.

8. Rohwer, C.L.; Erwin, J.E. Horticultural applications of jasmonates: A review. *J. Hortic. Sci. Biotechnol.* **2008**, *83*, 283–304. [CrossRef]

9. Heridia, J.B.; Zevallos, C. The effects of exogenous ethylene and methyl jasmonate on the accumulation of phenolic antioxidants in selected whole and wounded fresh produce. *Food Chem.* **2009**, *115*, 1500–1508. [CrossRef]

10. Saniewski, M.; Czapski, J. The effect of methyl jasmonate on lycopene and β-carotene accumulation in ripening red tomatoes. *Experientia* **1983**, *39*, 1373–1374. [CrossRef]

11. González-Aguilar, G.A.; Fortiz, J.; Cruz, R.; Baez, R.; Wang, C.Y. Methyl jasmonate reduces chilling injury and maintains postharvest quality of mango fruit. *J. Agric. Food Chem.* **2000**, *48*, 515–519. [CrossRef] [PubMed]

12. González-Aguilar, G.A.; Buta, J.G.; Wang, C.Y. Methyl jasmonate and modified atmosphere packaging (MAP) reduce decay and maintain postharvest quality of papaya "Sunrise". *Postharvest Biol. Technol.* **2003**, *28*, 361–370. [CrossRef]

13. González-Aguilar, G.A.; Tiznado-Hernández, M.E.; Zavaleta-Gatica, R.; Martínez-Téllez, M.A. Methyl jasmonate treatments reduce chilling injury and activate the defense response of guava fruits. *Biochem. Biophs. Res. Commun.* **2004**, *313*, 694–701. [CrossRef]

14. Meng, X.; Han, J.; Wang, Q.; Tian, S. Changes in physiology and quality of peach fruits treated by methyl jasmonate under low temperature stress. *Food Chem.* **2009**, *114*, 1028–1035. [CrossRef]

15. Chantrasri, P. Induction of Resistance to Anthracnose Disease of Postharvest Mango Fruit by Antagonist Yeasts. Master's thesis, Chiangmai University, Chiang Mai, Thailand, 2006.

16. Buta, J.G.; Moline, H.E. Methyl jasmonate shelf life and reduces microbial contamination of fresh cut celery and peppers. *J. Agric. Food. Chem.* **1998**, *46*, 1253–1256. [CrossRef]

17. Droby, S.; Porat, R.; Cohen, L.; Weiss, B.; Shapira, B.; Philosoph-Hasas, S.; Meir, S. Suppressing green mold decay in grapefruit with postharvest jasmonate application. *J. Am. Soc. Hortic. Sci.* **1999**, *124*, 184–188.

18. Wang, K.; Dickinson, R.E.; Liang, S. Clear sky visibility has decreased over land globally from 1973 to 2007. *Science* **2009**, *323*, 1468–1470. [CrossRef] [PubMed]

19. Gonzalez Aguilar, G.A.; Buta, J.G.; Wang, C.Y. Methyl jasmonate reduces chilling injury symptoms and enhance color development of "Kent" mangoes. *J. Sci. Food Agric.* **2001**, *81*, 1244–1249. [CrossRef]

20. Perez, A.G.; Sanz, C.; Richardson, D.G.; Olias, M. Methyl jasmonate promotes β-carotene synthesis and chlorophyll degradation in Golden Delicious apple peel. *Plant Growth Regul.* **1993**, *12*, 163–167. [CrossRef]

21. Saniewski, M.; Czapski, J. Stimulatory effect of methyl jasmonate on ethylene production in tomato fruits. *Experientia* **1985**, *41*, 256–257. [CrossRef]

22. Saniewski, M.; Nowacki, J.; Lange, E.; Czapski, J. The effect of methyl jasmonate on anthocyanin accumulation, ethylene production and ethylene forming enzyme activities in apple. *Fruit Sci. Rep.* **1988**, *15*, 97–102.

23. Olias, J.M.; Rios, J.J.; Valle, M.; Zamora, R.; Sanz, L.C.; Axelroad, B.A. Fatty acid hydroperoxide lyase in germinating soybean seedlings. *J. Agric. Food Chem.* **1990**, *38*, 624–630. [CrossRef]

18

New Approaches to Irrigation Scheduling of Vegetables

Michael D. Cahn [1,*] and Lee F. Johnson [2]

[1]   University of California, Cooperative Extension, Monterey County, 1432 Abbott St., Salinas, CA 93901, USA
[2]   NASA ARC-CREST/CSUMB, MS 232-21, Moffett Field, CA 94035, USA; Lee.F.Johnson@nasa.gov
*   Correspondence: mdcahn@ucdavis.edu

Abstract: Using evapotranspiration (ET) data for scheduling irrigations on vegetable farms is challenging due to imprecise crop coefficients, time consuming computations, and the need to simultaneously manage many fields. Meanwhile, the adoption of soil moisture monitoring in vegetables has historically been limited by sensor accuracy and cost, as well as labor required for installation, removal, and collection of readings. With recent improvements in sensor technology, public weather-station networks, satellite and aerial imaging, wireless communications, and cloud computing, many of the difficulties in using ET data and soil moisture sensors for irrigation scheduling of vegetables can now be addressed. Web and smartphone applications have been developed that automate many of the calculations involved in ET-based irrigation scheduling. Soil moisture sensor data can be collected through wireless networks and accessed using web browser or smartphone apps. Energy balance methods of crop ET estimation, such as eddy covariance and Bowen ratio, provide research options for further developing and evaluating crop coefficient guidelines of vegetables, while recent advancements in surface renewal instrumentation have led to a relatively low-cost tool for monitoring crop water requirement in commercial farms. Remote sensing of crops using satellite, manned aircraft, and UAV platforms may also provide useful tools for vegetable growers to evaluate crop development, plant stress, water consumption, and irrigation system performance.

Keywords: evapotranspiration; crop coefficients; soil moisture sensors; NDVI; web application; decision support tools; UAV; remote sensing

1. Introduction

1.1. Water Scarcity and Commercial Vegetable Production

Most economically important vegetable production regions of the world have Mediterranean, semi-arid, or desert climates in which supplemental irrigation is required to maximize yield and quality. Efficient irrigation management has become a major concern for vegetable farmers in these areas as water supplies have become more restricted and environmental impairments to ground and surface water from agricultural run-off and drainage has received more attention from regulatory agencies. Agricultural water resources have become increasingly stressed in the vegetable producing regions of Australia, Southern Europe, Chile, the Middle East, North Africa, China, and the western United States. California has been under severe drought conditions for four years since 2012 [1,2], and ground water supplies have been depleted to historically low levels on the central coast, where most of the salad vegetables are grown for the US and export markets. Lowering of the ground water table below sea-level through agricultural pumping has caused salt water to intrude into coastal aquifers and threatened the sustainability of thousands of hectares of prime farmland used for vegetable production [3]. Australia also experienced a severe decade-long drought during the early 2000's that impacted vegetable farmers in the Murray-Darling Basin in New South Wales and Queensland [4]. The vegetable production regions of Chile [5] and Spain [6] have also experienced recurring multiyear droughts.

Since many vegetables require high rates of nitrogen fertilizer and have shallow root systems, leaching of nitrate during the irrigation season and during periods of heavy precipitation has resulted in nitrate contamination of aquifers in many key vegetable production regions [7–9]. Irrigation run-off from vegetable fields also contaminates surface water bodies such as rivers, creeks, estuaries, and lakes with nutrients and pesticides [10–12]. Growers in California and Europe must report on the amount of fertilizer nitrogen that they apply to their crops to comply with environmental regulations [13,14]. They are also required to implement best practices to minimize nitrate losses, which will require improving irrigation management.

1.2. Challenges for Irrigation Scheduling in Modern Vegetable Operations

Vegetable production poses several unique challenges in managing irrigation water efficiently. One of the major challenges for growers is the number of fields that must be concurrently managed in a medium to large size vegetable operation. Fields sizes for vegetable crops tend to be small (<5 ha) relative to agronomic crops, especially fields planted with leafy green, crucifer, and other salad vegetables. Large vegetable growing operations in the Salinas valley of California manage 1–2 thousand ha of vegetables, with an average field size of just 4 ha. Small fields permit intensive management, allowing plantings to be staggered so that a steady supply of vegetables can be delivered to buyers and shippers. However, smaller field sizes means that farming decisions must be coordinated for many fields that may vary with respect to maturity, soil texture, cultural practices, microclimate, or other site-specific conditions.

In addition to field size, the diversity of vegetables and number of crop rotations per season increases management complexity. Most diversified operations will produce more than 30 types of vegetables, each with unique nutrient and water requirements. Many vegetables are grown over short cropping cycles. Leafy green salad mixes like leaf and crisphead lettuce, for instance, reach maturity over just 30–65 day intervals during the summer. As a result, farmers in Mediterranean climates often grow two to three rotations of vegetable crops per year. Fields often are intensively tilled between crops to remove compacted zones caused by traffic from harvest machinery and to prepare raised beds for seeding or transplanting the next crop. These activities require that crews retrieve drip tape or sprinkler pipe after each crop is harvested.

Another irrigation scheduling challenge is the number of field operations that must be coordinated during a crop cycle. Fields are typically sprayed to protect against insect and disease pests several times per crop cycle, and must be periodically cultivated to control weeds. Fertilizer applications are made several times during the growth cycle to satisfy the nutrient requirements of vegetables. Irrigations must be timed to accommodate access for tractor and hoeing crews so that activities can be completed on schedule. There may be several extended periods when a vegetable crop cannot be irrigated to permit the soil to dry sufficiently to allow tractor access. Irrigation equipment, such as sprinkler and mainline pipes, may need to be moved before each pass of a tractor through a field.

Irrigation scheduling is also complicated by the numerous methods that vegetable growers use to supply water to their crops. Depending on the crop type, plant density, development stage, water source, and field and soil characteristics, growers may choose to irrigate using overhead sprinklers, drip, furrow, or a combination of methods. In California, leafy green vegetables harvested for salad mixes are seeded at high densities (8.5 million plants per hectare) [15] on 2 m wide raised beds and are irrigated almost exclusively with sprinklers. Small seeded vegetables such as crisphead and romaine lettuce are typically germinated with sprinklers, and after several weeks, irrigated with surface drip until harvest. Celery transplants may be established using overhead sprinklers, then drip irrigated, and occasionally flood irrigated to rewet the shoulders of raised beds.

Since vegetable production is frequently characterized by short crop cycles, intensive crop rotations, and numerous small fields at different stages of maturity, farm managers are challenged to schedule and coordinate all the field activities while taking time to carefully schedule irrigations to optimize water use. Consequently, many farm managers may follow predetermined irrigation schedules to simplify water management and make small adjustments during the cropping season depending on their observations of the crop, soil, and weather conditions. Usually the amount of water applied to high value vegetables exceeds crop evapotranspiration (ET) to avoid water stress. In the Salinas Valley of California, for example, applied water amounts on broccoli, cauliflower, and cabbage averaged >200% of the estimated crop evapotranspiration (ETc) requirement [16]. While these applications rates may avoid water stress so that yield and quality are maximized, over-irrigating may strain water supplies and result in the leaching of nitrate and/or generate run-off that degrades the quality of receiving surface water bodies.

Considering the numerous challenges and limitations of managing vegetables, growers, farm managers, and irrigators need convenient methods to schedule irrigations so that they can better determine how much water to apply to match the requirements of their crops. During the past few decades, significant improvements and lower costs for wireless communications, computing power, sensor technology, and aerial and satellite imagery have increased the potential to develop accurate and intuitive approaches for scheduling irrigations in vegetables.

2. Advances in Soil Moisture Sensor Technology

2.1. Recent Developments in Soil Moisture Sensing

The monitoring of soil moisture has long been a standard way to determine when crops need to be irrigated. Growers and farm managers typically evaluate soil moisture by probing with a shovel or auger or monitoring with sensors. Sensors for volumetric moisture content and soil water tension have been commercially available for more than 40 years but were originally used more in research than for commercial crop production. Volumetric soil moisture sensors provide readings in units of $m^3 \cdot m^{-3}$, and tension-based sensors readings are typically in units of kPa, where a greater absolute value corresponds to drier soil conditions [17]. In the past, the primary barriers to more widespread use of soil moisture sensors in irrigation management have included both cost as well as labor required for the installation, removal, and collection of readings [18]. In recent years, there has been a proliferation of commercially available soil moisture monitoring systems for agriculture. Many sensors interface with dataloggers and wireless communication systems to provide near real-time status of soil moisture from several depths and locations within a field. Data are automatically uploaded by radio or cell phone communications to cloud-based computer servers and are accessible through apps on smartphones and tablet computers. These communication advancements greatly improve the convenience of accessing data and can be configured to provide timely alerts when crops require irrigation. Many of these wireless communication systems for soil moisture sensors also support on-farm weather stations, digital flow meters, and control valves, which facilitates the monitoring of irrigation system operations.

Soil moisture sensors have also evolved during the last few decades in terms of size, cost, and accuracy. Electromagnetic (EM) soil moisture sensors, used to determine volumetric soil moisture content, include time domain reflectometry (TDR) and capacitance sensors. These were once bulky and expensive instruments. With improvements in electronic manufacturing and better designs, current EM sensors (Figure 1A,B) are generally much smaller, cheaper, and more accurate than earlier models. Some versions are integrated with soil temperature and salinity sensors (e.g., Model 5TE, Decagon Devices Inc., Pullman, WA, USA). Others are integrated with dataloggers and radio communications (e.g., Model gStake, gThrive Inc., Santa Clara, CA, USA) to facilitate field installation and removal.

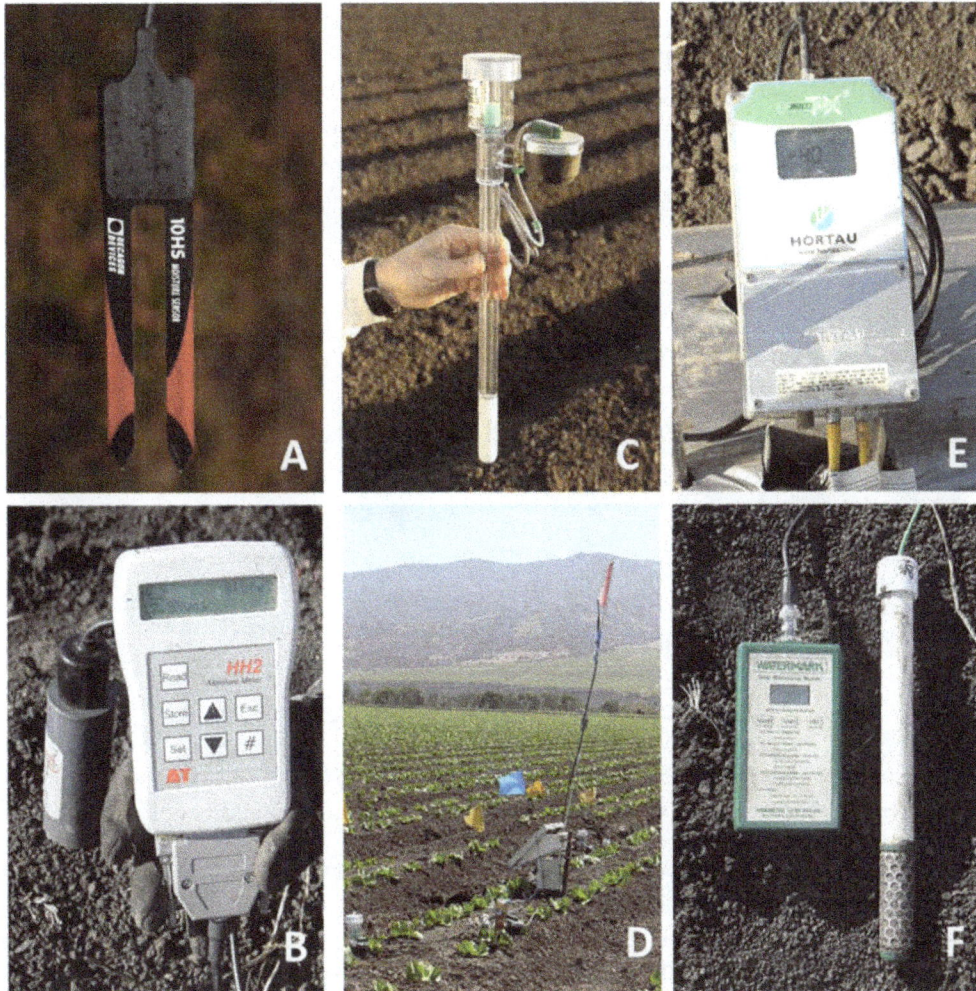

Figure 1. Examples of various soil moisture sensors: (**A**,**B**) capacitance sensors; (**C**) tensiometer with an electronic gauge; (**D**) tensiometers with electronic gauges installed in a lettuce field and interfaced with a datalogger and radio communications; (**E**) tensiometer integrated with pressure transducer, datalogger, and radio communications; (**F**) granular matrix sensor and reader.

Tension-based soil moisture sensors have been considered the best method to determine if a crop needs water since most vegetables experience reductions in yield and quality under prolonged periods of high soil water tensions. Tension thresholds that optimize production have been determined for many vegetable species [17,19], including broccoli [20], cabbage [21], cauliflower [22], lettuce [23,24], potato [25], and tomato [26]. An advantage of tension thresholds is that they are less influenced by soil texture than volumetric moisture thresholds. Tensiometers can accurately measure soil water tension in a range of 0–85 kPa using a mechanical vacuum gauge attached to a water filled tube with a porous ceramic cup [17] (Figure 1C). Tensiometers must be installed without air gaps between the ceramic cup and soil to function properly. To reduce labor for collecting readings, several recent models have sensitive electronic pressure transducers that measure vacuum pressure and can be interfaced to dataloggers and wireless communication services (Figure 1D,E) so that data may be monitored remotely (Figure 2) (e.g., Hortau Inc., Lévis, QC, Canada). Though tensiometers do not need calibration, they usually require periodic maintenance to assure that they are functioning properly. Entrapped air that develops at high tensions must be replaced with de-aired water. Adding a weak solution of algaecide or bleach can prevent biological growth inside the tensiometer tube, which may potentially clog the ceramic cup [26].

Granular matrix sensors (GMS) (Figure 1F) indirectly measure soil water tension using electrical resistance and are often used as an alternative to tensiometers (e.g., Model watermark 200SS, Irrometer

Company Inc., Riverside, CA, USA). Most GMS can be interfaced to dataloggers and wireless communications [26]. An advantage of GMS compared to tensiometers is that they do not require regular servicing. These sensors retain sensitivity at higher soil moisture tensions (up to 200 kPa) but are less accurate at low soil moisture tensions (0–15 kPa) than tensiometers [17,26]. GMS readings are also affected by fluctuations in soil temperature [17]. Another limitation of GMS is that the response time to wetting and drying cycles is slower compared to tensiometers [26,27].

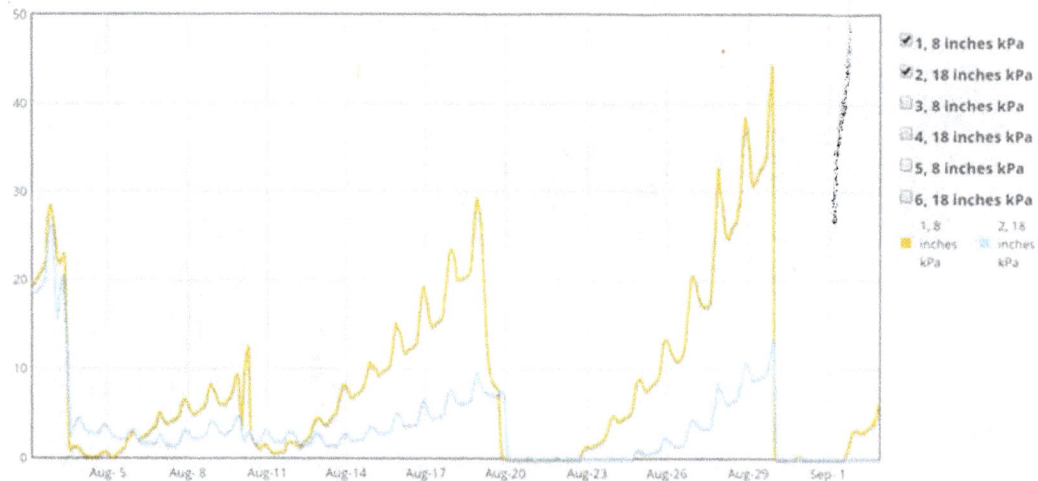

Figure 2. Fifteen minute readings (kPa) from tensiometers installed in a lettuce field at 8 (20 cm) and 18 inch (46 cm) depths are displayed in an online irrigation scheduling application.

A number of studies have evaluated automated irrigation scheduling based on soil moisture readings for improving the water use efficiency of vegetables irrigated with drip tape. Automatic controllers used in these studies irrigated crops for short durations several times per day when soils dried below a predetermined threshold [28,29]. Some control systems have relied on tension based sensors [26,28], while others have used volumetric soil moisture sensors [28,30–34] to determine when to irrigate. The use of TDR soil moisture sensors for triggering irrigations in sub-surface drip irrigated sweet corn resulted in an 11% savings in water use with similar yields compared to a standard sprinkler irrigated treatment [30]. Small plot trials have demonstrated water savings, reduced nitrate leaching, and improvements in yield in drip irrigated fresh market tomatoes, bell peppers, and zucchini squash using capacitance soil moisture sensors to trigger irrigations [31–35]. Using tensiometers, Mũnoz-Carpena et al. [26] were able to reduce water use in fresh market tomato by 67% compared to the grower standard practice without significant reductions in marketable yield. Most of the water savings were during the early crop stages when evapotranspiration rates were low. Capacitance soil moisture sensors were generally found to be more reliable for automated irrigation scheduling than tensiometers and GMS. As discussed earlier, tensiometers require regular maintenance, and the GMS response to changes in soil moisture was too slow to use these sensors for triggering irrigations. While these studies have demonstrated the potential for water savings in small plots, automated irrigation scheduling may have several limitations in commercial vegetable fields. The optimal volumetric soil moisture threshold for triggering an irrigation would need to be empirically determined for different soil types. Pressurized water would need to be continuously available to accommodate frequent short irrigations. In large fields, short irrigation cycles could lead to significant drainage at the lower end of a field when drip lines depressurize.

2.2. Limitations of Soil Moisture Sensors for Irrigation Scheduling in Vegetables

While much progress has been made in improving the accuracy and utility of soil moisture sensors, several factors still limit their use for irrigation scheduling of vegetables. Though costs for individual sensors may be less than in the past, the addition of dataloggers, cell phone modems, and radio communications, which facilitate real-time monitoring, has added to the total costs. Labor for installation and removal of sensors remains a significant cost, especially in vegetable crops with short production cycles. Considering labor and capital costs, many growers elect to use soil moisture monitoring equipment in a small percentage of their vegetable fields.

Though soil moisture sensors are useful for determining when to irrigate vegetables, they are less useful for estimating how much water to apply. Soil moisture tension readings must be converted to volumetric moisture to estimate soil water depletion since a previous irrigation or rain event. Volumetric soil moisture sensors may also need calibration. Most capacitance sensors use the manufacturer's calibration equations to convert readings to volumetric water content. Clay content, organic matter, salinity, bulk density, and temperature can affect the accuracy of capacitance sensors [36–40]. The effect of these factors on water content readings can vary in different soil types. For example, Kargas and Soulis [38] evaluated the accuracy of the 10HS capacitance sensor, and found that temperature had a larger effect on soil water content readings in clay than course textured soils. They concluded that calibration for specific soil types was needed to achieve accurate readings.

Even with accurate sensors, spatial variability can limit the reliability of soil moisture estimates if readings are collected from only a few locations, especially if soil hydraulic properties vary within a field or the irrigation system applies water unevenly. Soil maps can be useful for guiding the placement of soil moisture sensors in locations that represent the dominant soil properties of fields. In drip-irrigated fields, determining the optimal location to accurately monitor soil moisture depletion can be particularly challenging. Soil moisture is typically higher under drip tape than adjacent to the plants, where root activity is concentrated. Placing sensors too close to a drip line may lead to under-irrigating a crop, and placing sensors too far away may result in over-irrigating [41]. Through computer modeling experiments, Soulis et al. [42] concluded that the optimal positioning of soil moisture sensors in drip-irrigated crops was influenced by soil hydraulic properties, crop evapotranspiration rate, and the configuration of the irrigation system. An additional consideration is the ideal depth at which to place sensors relative to the crop root zone. Broccoli roots can reach depths greater than 1.2 m at maturity [16], while roots of many leafy greens such as spinach may reach less than 0.5 m by harvest [43].

3. ET-Based Approaches to Scheduling Irrigations in Vegetables

Water requirements of vegetable crops can also be determined from estimates of crop evapotranspiration (ETc). Using Penman-Monteith [44,45], or similar equations [46], evapotranspiration (ET_0) of a well-watered reference crop, such as grass, or alfalfa, is commonly estimated from measurements of air temperature, relative humidity, wind speed, and solar radiation [47]. The evapotranspiration of a well-watered vegetable crop can be estimated relative to a reference crop by multiplying ET_0 by a crop coefficient (Kc). Crop coefficients for most major vegetable crops were summarized by Allen et al. [48], Guerra et al. [49], and Grattan et al. [50].

Networks of public weather stations to monitor reference ET have been established in many vegetable production regions of the world where irrigation is commonly used. The California Irrigation and Information System (CIMIS) is operated by the Department of Water Resources in California and consists of more than 145 weather stations sited on reference crops throughout the state. ET_0 and other meteorological data, including precipitation, air and soil temperature, wind speed, solar radiation, and relative humidity, are available for users to download from the state operated website. Similar networks have been developed for other states including Arizona (AZMET), Colorado (CoAgMet), Florida (FAWN), Nevada (NICENET), Oklahoma (MesoNet), Oregon (AgriMet), and Washington (AgWeatherNet). Most European countries also have weather station networks that provide daily ET_0 data through public websites (e.g., Spain [51], Italy [52]). In addition, CIMIS offers a satellite based product that estimates ET_0 at a 2 km resolution based on data from the Geostationary Operational Environmental Satellite [53].

Many commercial companies also offer affordable weather stations (e.g., ET107, Campbell Sci. Inc., Logan, UT, USA; WatchDog 2900ET Weather Station, Spectrum Technologies, Aurora, IL, USA; HOBO U30-NRC Weather Station, Onset Computer Corp. Bourne, MA, USA) that can be used to estimate ET_0 on a farm. These systems offer a way to monitor ET_0 in regions not covered by a public network, or for farms that are not located reasonably near an established station. However, in practice, few growers site or maintain stations in accordance with World Meteorological Organization standards for reference evapotranspiration measurements. Many growers site weather stations near trees, buildings, or parking lots, which limits the fetch and can influence the micro-climate, and potentially bias estimates of ET_0. Stations are infrequently sited on a suitable well-watered reference crop. Most farmers do not allocate time to regularly maintain and calibrate meteorological instrumentation on their stations. Commercially available atmometers (ETgage Co., Loveland, CO, USA) can be used to monitor ET_0 but must also be sited over a reference crop.

Although ET_0 data have been useful for water agencies to estimate the seasonal water use of crops at a regional scale, growers have generally considered this approach to be impractical for scheduling irrigations in vegetable systems. A major difficulty is that the Kc value can change daily as vegetables grow and leaf area increases. During the establishment phase, when crop foliage covers a small percentage of the soil surface, ET is mostly from soil evaporation rather than from crop transpiration. For many vegetable crops, such as leafy greens and brassicas, the canopy cover is less than 10% until halfway through the crop cycle. The Kc during this early stage will depend on factors influencing evaporative losses from the soil, such as method and frequency of irrigations, and soil physical properties [48]. Later, in the rapid growth phase, Kc values increase daily as the canopy cover develops and covers the soil.

Solutions for calculating a daily Kc for lettuce were proposed by Gallardo et al. [54] in which soil evaporation from unshaded soil (E) is estimated separately from the water lost by transpiration and soil evaporation under the crop canopy (T). Using this approach, Kc is proportional to the fractional cover (Fc) of the leaf canopy shading the soil, and E is related to the hydraulic properties of the soil and the time since the crop was irrigated. The general approach is consistent with Allen et al. [48,55], who also proposed separating the Kc of crops that have a significant period of minimal canopy cover into dual components consisting of a basal crop coefficient (Kcb) that represents water loss by transpiration and an evaporation coefficient (Ke), representing loss from soil evaporation. Direct relationships between Fc and Kcb have been reported for several crops such as broccoli [56], wheat [57], cotton [58], sugar beets [58], and grapes [59].

Both Gallardo et al. [54] and Allen et al. [48] describe calculations for estimating soil evaporation, which divide Ke into stage 1 and stage 2 periods. Stage 1 occurs immediately following an irrigation or rain event that saturates the soil to field capacity. During stage 1, water loss is limited by the energy available for evaporation. Stage 2 occurs after the soil is visibly dry and water loss is limited by soil hydraulic properties. The Stage 2 rate of evaporation diminishes with time as the soil water content declines.

4. Software for ET Based Scheduling of Vegetables

4.1. Overview of Software Tools

Even with publicly available ET_0 data and accurate Kc models, the implementation of ET-based irrigation scheduling at the scale of a commercial vegetable farm may be difficult for growers. Daily ET_0 values need to be retrieved from a location representative of the field of interest. An estimate of canopy cover is needed to determine the average Kc for each irrigation event. Soil evaporation calculations require knowledge of the soil properties of the field, irrigation method, and interval between irrigations or rainfall events. To convert crop ET into an irrigation recommendation, the application rate and distribution uniformity of the irrigation system, and in some cases the leaching fraction, also needs to be integrated into the calculations.

Recognizing that most growers have limited time to dedicate to making decisions on water management, several universities and public institutions have developed computer programs to facilitate ET based scheduling. Spreadsheet and Windows-based irrigation scheduling programs, such as CROPWAT [60], KanSched 2.0, Basic Irrigation Scheduling (BIS) [61], and consumptive use program (CUP) [62] automate the irrigation scheduling calculations but require users to retrieve and enter daily ET_0 values. Since separate spreadsheet files maybe needed for each crop, these programs may be difficult to implement in large growing operations that manage many fields. Washington Irrigation Scheduling Expert (WISE) [63,64] is a downloadable JavaScript application that runs on personal computers. More recently developed web-based applications such as Irrigation Scheduler mobile (Washington State University) [65], Irrigation Management Online (Oregon State University) [66], Wateright (Fresno State University) [67,68], and CropManage (University of California, Division of Agriculture and Natural Resources) [69,70] were developed to automatically retrieve ET_0 data from weather station networks such as AgWeatherNet and CIMIS and support irrigation scheduling of multiple fields. These applications are accessed through web browsers, some of which automatically resize the user interface to smartphone screens. Colorado State University WISE Irrigation Scheduler [71] and SmartIrrigation (University of Florida, University of Georgia) [72–74] have similar capabilities but can operate exclusively on smartphones through downloadable applications. Outside of the United States, several public agencies have developed online irrigation scheduling services including IRRINET in Italy [75], ISS-ITAP in Spain [76], IRRISA in France [77], and IrriSatSMS [78] in Australia. In addition, a growing number of ET-based software tools are commercially available. Two examples are Probe Schedule (IRRinet LLC, Dalles, OR, USA) and Irrigation Advisor (PowWow Energy Inc., San Mateo, CA, USA). Further details of these commercial products are not provided here since they are proprietary and usually not documented in the scientific literature.

Most online irrigation scheduling tools have cloud-based databases that retain information associated with each planting, such as soil type, planting and harvest dates, weather station name, and irrigation system application rate, so that the user does not need to reenter critical information for each irrigation event. After entering the information required to initiate a new planting, users can quickly look up how long to irrigate their crops. Some scheduling software provides weekly or daily summaries of how much water to apply (e.g., IrrigationScheduler, Wateright, SmartIrrigation, etc.), while others such as CropManage enable the user to input specific dates for each irrigation [70]. The online format facilitates automated retrieval of current and forecasted weather data in advance of planned irrigation events, so that growers can be alerted when a field will need water. In addition to linking with weather station networks, decision support tools can also incorporate other online services such as UC Davis SoilWeb for identifying the soil physical properties of a field using the United States Department of Agriculture (USDA) Soil Survey Geographic (SSURGO) database [79] or Google Maps, which can be customized for determining the longitude and latitude of a field or viewing locations of nearby weather stations. CropManage can also be customized to automatically retrieve flowmeter and soil moisture data from internet-accessible dataloggers.

While many of these irrigation-scheduling programs will provide recommendations for multiple commodities, few have been developed or tested specifically for vegetables. Most irrigation scheduling applications use single Kc values for the four crop growth stages described by Doorenbos and Pruitt [80] to simplify calculations of ETc. CropManage was initially developed for vegetable irrigation and employs a dual crop coefficient approach for estimating crop ET similar to Gallardo et al. (1996), as described in Johnson et al. [81] and Smith et al. [16]. The vegetable crops currently supported include broccoli, cabbage, cauliflower, lettuce, and spinach. Since the user enters the date of each irrigation event, CropManage adjusts watering recommendations for the frequency and method of irrigation. Empirical models of fractional cover are included for each supported vegetable crop so that the user can customize the Kcb curves for a specific season, bed width, and planting configuration. Replicated field trials demonstrated that the CropManage irrigation recommendations using the ET based algorithm with a dual crop coefficient optimized water use and yield in crisp head lettuce and broccoli [81].

4.2. Achieving Widescale Adoption of Irrigation Software

Despite significant progress in developing ET-based irrigation scheduling software for commercial vegetable farms, many challenges exist that will need to be addressed to achieve large-scale adoption by growers. Although public ET weather station networks have been established in many vegetable producing regions of the US and Europe, they are less common in other regions such as India, China, and Latin America. In areas where ET_0 data are available, the density or siting of weather stations may not provide sufficient resolution to fully capture regional variation in microclimates. In these instances, growers may choose to install and operate their own weather stations, as described above.

Another challenge is to deliver accurate irrigation recommendations without requiring users to provide excessive details about their crop, irrigation system, and field conditions. Morrison [82] found that time constraints and concerns about data entry errors were factors that deterred grower adoption of the irrigation scheduling software. Irrigation scheduling apps also need to be intuitive for irrigators and farm managers to use in the field. The user-interface displayed on smartphones and tablet computers must be easy to read and to understand, while providing sufficient detail for the user to verify that the values for ET_0, Kc, and other variables used in the scheduling calculations are accurate. Also, since the principal language of many irrigators in the western US is Spanish, irrigation scheduling applications need to support multiple languages.

The cost of computer programming services is also a major barrier to developing and maintaining irrigation scheduling software. As software has become more sophisticated, and needs to be compatible with personal and tablet computers and smartphones, the complexity and consequently costs of development have increased. Software products will incur annual costs associated with updating to new technology, introducing new features, and troubleshooting identified errors. Most publicly available irrigation scheduling applications were initially developed under research grants, and need either continued grant funding or income generated from user subscriptions to support updating and maintenance costs. Several online irrigation scheduling services that were initially funded by the European Union became inactive after grants ended [83]. It may be cost-effective for institutions to collaborate on the development of software products that are sufficiently flexible to be customized for different growing regions and commodities. SmartIrrigation is an example of such collaboration between the University of Florida and University of Georgia. There also may be opportunities for public agencies, commodity boards, and commercial companies to partner on the development of commercial irrigation scheduling software. For instance, SureHarvest Inc. (Santa Cruz, CA, USA) recently developed an online irrigation-scheduling tool for almonds in collaboration with the University of California and the Almond Board of California.

5. Field Measurements of Crop ET

Field measurements of crop ET are needed to develop and verify crop coefficients as well as to investigate the interaction of water stress on crop yield and quality. Several reviews detail the advantages and disadvantages of various methods of measuring crop ET [84,85]. Weighing lysimeters have often been used to measure crop water use [85,86]. While considered the most accurate ET measurement method, the expense of these instruments limits their use to research stations and constrains the range of sites, crops, and management practices that can reasonably be evaluated. Energy balance methods such as Bowen ratio [85], eddy covariance [85], and surface renewal [87] are more affordable alternatives to lysimetry for evaluating crop ET and can be deployed in commercial fields for evaluating crop water use under a wide range of growing conditions. Instrumentation costs for these methods can exceed $10,000 USD per station, which has limited their use primarily to research studies. Energy balance methods involve monitoring of net radiation, soil heat flux, and sensible heat flux to estimate the latent heat flux (Wm^{-2}), which can be converted to water flux or evapotranspiration rate through the latent heat of vaporization. Generally, the sensors used for energy balance measurements have become more reliable, cheaper, and more accurate in recent years. The capabilities of dataloggers used for operating the instrumentation have also improved, with more memory and faster computation speeds, while post-processing software modules have become more accessible. Energy balance methods have been used for evaluating the water use of vegetable crops in commercial fields, including artichokes, broccoli, beans, lettuce, onions, and processing tomatoes [50,88–92]. These methods have been useful for reevaluating existing crop coefficients, as production practices and varieties have improved during recent years [93]. However, considering the diversity of vegetables and production methods, relatively few ET studies have been published using these techniques during the last decade. Surface renewal, which uses thermocouples to measure rapid changes in air temperature, has greatly reduced the costs of estimating sensible heat flux and may be well suited to vegetable fields due to a smaller fetch requirement compared to eddy covariance. Advances in surface renewal methodology [94] have culminated in a commercial service that provides growers with ET estimates and irrigation recommendations via a web application (Tule Technology Inc., Oakland, CA, USA). However, this tool has primarily been used in perennial crops such as trees, grapes, and strawberries.

6. Satellite-Based Crop ET Determination

6.1. Energy Balance

Researchers have developed satellite-based energy balance models to estimate ET at various spatial and temporal scales (e.g., reviews of Courault et al. [95]; Kalma et al. [96]; Gonzalez-Dugo et al. [97]). The Surface Energy Balance Algorithm for Land (SEBAL) and the Mapping EvapoTranspiration at high Resolution with Internalized Calibration (METRIC) are two widely used approaches [98,99]. Optical and thermal band data from Landsat or other satellites is used to help parameterize the energy balance equation [100]. At the moment of satellite overpass, latent heat flux (LE) is retrieved through the calculation of surface energy balance that is based on radiative, aerodynamic, and energy balance physics. LE is converted to ET for the corresponding hour and subsequently to reference ET fraction (ETrF) by comparison with the measured reference ET for the hour. Finally, daily (2) ET is derived as the product of ETrF and reference ET for the day, under the assumption that ETrF approximates the average daylight evaporative fraction. The energy balance approach accounts for the effects of ET reduction in water-stressed crops, due to stomatal regulation and increased ET due to bare-soil evaporation. Linear or spline interpolation can be used to estimate ETrF (hence daily ET) between satellite overpasses, which occur every eight days in the case of Landsat, assuming clear-sky conditions. Manual intervention is used to calibrate the sensible heat flux computation by identifying portions of the image (hot and cold 'anchor' pixels) that represent extreme ET conditions, where ET can be estimated and assigned a priori. In a typical agricultural situation, cold pixels are associated with well-irrigated agricultural fields with high Fc and hot pixels with bare dry fields. Overall the ET retrieval error for the satellite-based energy balance is typically 5–15%, when estimates are produced by an experienced operator, and rises to 30–40% when operated by non-specialists or novices to the fields of hydrologic science, environmental physics, remote sensing, or agricultural systems [85]. An automated calibration method has been recently developed to facilitate and improve operation by non-specialists [101].

6.2. Vegetation Index

Hybrid approaches estimate fractional cover (Fc) from remote sensing and use ground-based ET_0 data. Fc is a good indicator of light interception, which is a strong driver of ETc [48]. Weighing lysimeter observations by Bryla et al. [102] revealed a strong relationship between Fc and ETc for vegetable crops in California's San Joaquin Valley, and the potential for using satellite-based Fc to estimate ETc in vegetable crops was demonstrated by Johnson and Trout [103]. Fc can be estimated from various spectral visible-region bands, where chlorophyll absorption dominates, combined with near-infrared (NIR), where vegetation is highly reflective. A common formulation is the normalized difference vegetation index, or NDVI, which is derived as (NIR − red)/(NIR + red) [104]. Trout et al. [105] and Johnson and Trout [103] found a strong relationship between NDVI and Fc for vegetables and other crop types in California's San Joaquin Valley. NDVI began to lose sensitivity above 80 percent cover, a point generally regarded as effective full cover for water management applications [106]. As above, interpolation can be applied to estimate NDVI, and hence ETc, for days between satellite overpasses as needed.

7. Satellite Based Irrigation Management Services

7.1. Prototype Systems

By furnishing field scale estimates of Fc, a key factor for estimating Kcb and basal ETc (ETcb), optical remote sensing may potentially improve the accuracy of ET estimates in vegetables. Several satellite-based models have been developed for irrigation management. An early proof-of-capability satellite demonstration on wheat and corn was undertaken by the DEMETER project in Europe, which involved timely delivery of Landsat-based crop coefficients to agricultural end-users, with available online visualization and analysis capabilities [107].

More recently, the fully-automated Satellite Irrigation Management Support (SIMS) [108] uses atmospherically corrected Landsat data to map NDVI, Fc, and Kcb for multiple crop types, including vegetables across about eight million acres of California irrigated farmland [109]. These variables are updated every eight days at 30 m spatial resolution (cloud cover and data availability permitting) from 2010 to the present. The SIMS uses daily 2 km ET_0 statewide grids produced by the California Irrigation Management Information System [53] to generate basal crop ET. Web data services allow users to display annual time-series graphs and download data for any given location (Figure 3). An application programming interface (API) enables on-demand transfer of SIMS data products to support external irrigation advisory services such as CropManage and also allows the user and other software tools to specify the crop type via the API and retrieve crop-specific Kcb data from SIMS. Where information is available on applied water, an FAO56 based soil water balance model can be used to adjust for soil evaporation and crop stress and to retrospectively derive agricultural water use fractions at the field scale for the evaluation of irrigation management and system performance [110].

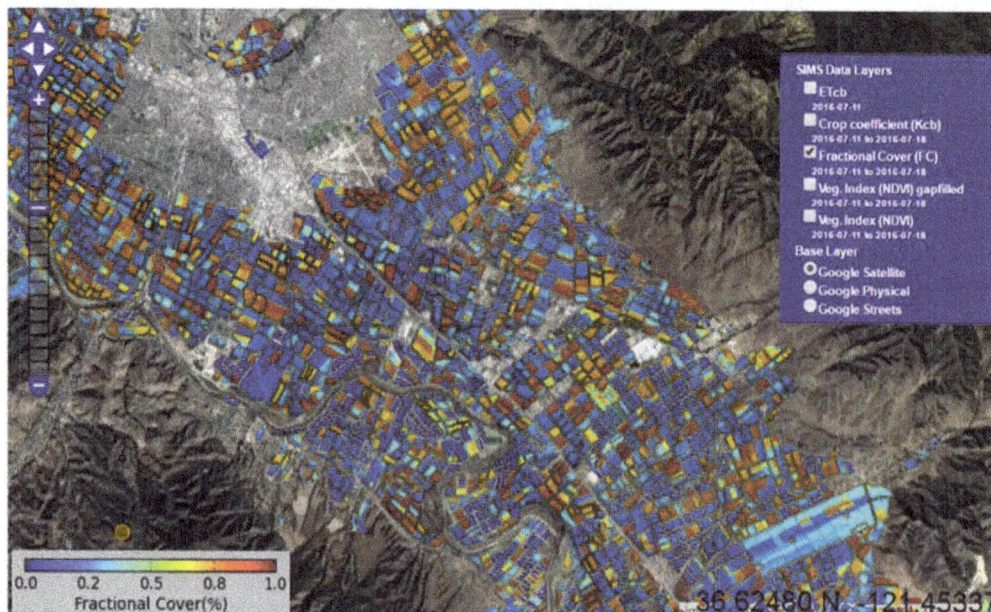

Figure 3. SIMS map of fractional cover of vegetable crops near Salinas, California as of mid-July 2016.

IrriSAT [111] is a weather based irrigation management and benchmarking technology that uses satellite remote sensing to generate crop water management information in Australia, primarily for grape and cotton growers at present [112,113]. As with SIMS, Landsat is used to estimate Kc at a 30 m resolution. Unlike SIMS, Kc is derived directly from a single linear relationship with satellite NDVI. Daily crop water use is determined as the product of Kc and ET_0 observations from nearby weather stations, and a seven-day ET_0 forecast is also produced. The delivery platform, which is built on the Google App Engine helps irrigators track soil moisture and better manage irrigation schedules. In addition, IrriSAT facilitates the calculation of seasonal agronomic performance metrics, including the irrigation water use index and gross production water use index.

Additional satellite-based services have been implemented Europe and Australia [114]. These include IRRISAT-Italy (as distinct from IrriSAT-Australia described above), EO4Water (Austria), and IrriEye (southern Australia). These services generally involve the use of optical DEIMOS 20 m satellite imagery to generate and deliver leaf area index with derived Kc and related map overlays within a browsing and querying toolbox [115]. Personalized irrigation guidance regarding crop water requirement is dispensed through a secure website, SMS, and email. End-users are able to provide evaluative feedback. A variety of crops are being tested, including vegetables, sugar beet, corn, alfalfa, and orchards.

Earth Engine Evapotranspiration Flux (EEFlux) [116] is a METRIC implementation on the Google Earth Engine cloud computing platform [117]. Landsat scenes from 1984 to the present are combined with gridded weather data to allow the analysis of most land areas worldwide. The products include maps of ET, associated energy balance flux components, and land cover. A time-series can be constructed by processing several scenes and interpolating between image dates. Automated image calibration for anchor pixels is offered in a Level 1 version to accommodate operation by non-experts, with some sacrifice in accuracy. A Level 2 version, which allows custom calibration by the operator, requires an annual license fee. EEFlux is a general tool that operates on both agricultural and non-agricultural land cover types.

7.2. Satellite Remote Sensing Considerations for Irrigation Scheduling

Satellite observation is recognized as a useful tool that will continue to support crop coefficient development and ET monitoring for new types and varieties of crops, as facilitated by the wider availability of gridded weather data and continued advancement in convenient and user-friendly mapping technologies [118]. The use of public domain satellites with open-data policies, i.e., the data source is available at no charge, is an advantage of this approach, as is the capability to collect data for many fields simultaneously. Vegetable crops, however, do present some unique challenges for satellite-based evaluation. The frequency of satellite overpasses, presence of cloud cover, and added time needed to perform custom atmospheric correction (e.g., Vermote and Saleous [119]) may limit the use of remote sensing for real time irrigation scheduling of fast-growing vegetable crops. As more satellite platforms become available, such as the European Space Agency Sentinel-2 system, the observation interval between images may soon decrease to four to five days for NDVI-based ET models. Also, upon further testing, it may be shown that use of a simplified atmospheric correction approach (e.g., Tasumi et al. [120]) may be adequate for application in some or most regions. Where real-time irrigation scheduling operations are unsupported, remote sensing may still prove useful for the retrospective evaluation of irrigation system performance or compilation of water use metrics. Another challenge is the typically small field size for vegetables as noted above (4–5 ha), which is near the minimum land area recommended for evaluation by Landsat or other public Earth observation satellites. An additional challenge is posed by leaf color, especially for crops such as red lettuce that depart widely from the green-leaf norm. A recent study has also shown that while NDVI was strongly related to Fc in both leafy greens ($r^2 = 0.88$) and cole crops ($r^2 = 0.93$), the relationships were different [121]. Thus, the development of customized relationships between NDVI and Fc by crop category is recommended to increase the accuracy of Fc estimates.

8. Remote Sensing Using Manned Aircraft and Unmanned Aerial Vehicles (UAV)

The use of manned aircraft and UAVs for monitoring vegetables has become a viable alternative or complement to satellite observation. Image resolution is generally much finer than that of Landsat or other open-data satellites, and passes over fields can be scheduled on a more frequent and flexible basis. However, data calibration to assure image consistency in support of time-series observations may be lacking. Several commercial companies offer NDVI and thermal spectrum images at <1 m resolution taken at weekly intervals using small planes in California and elsewhere (e.g., CERESimaging Inc., Oakland, CA, USA; TerrAvion Inc., San Leandro, CA, USA). Images can be accessed from the web generally within 24 h of collection. The high resolution of these scenes can provide enhanced information about within-field spatial variability in crop growth and water stress and for determining Fc used to parameterize ET equations. Patterns can show when a crop is under-irrigated such as in Figure 4, where the NDVI values for a romaine lettuce crop are lowest midway between sprinkler lines, presumably caused by poor irrigation distribution uniformity.

Figure 4. NDVI image of a romaine lettuce field irrigated with sprinklers on lateral pipes (**A**) spaced 12.2 m apart; Plants were frequently smaller in areas midway between pipes (**B**) where the NDVI values were low. Photo credit: CERESimaging Inc., Oakland, CA, USA.

The acquisition costs of UAVs and the training needed for mission planning, flight operation, and image processing have diminished significantly during recent years. A number of studies describe the use of UAVs for monitoring water stress [122,123] and vegetative cover [124,125] of crops, though few have specifically focused on vegetables. Due to the ease of operation, arranging flights at specific times and locations with UAVs may be simpler than with manned aircraft, which usually need to image a minimal number of fields to justify flight costs. This flexibility may be useful in vegetables when timely field scouting is needed to make decisions or diagnose a problem with a fast-growing crop. UAVs can fly at very low altitudes and carry sensor packages that can provide high-resolution imagery in the order of a few centimeters. A disadvantage of UAV operation, in the case of an owned system, would be a higher up-front cost than the engagement of a manned aircraft provider and additional time required for data retrieval and image processing.

9. Conclusions

Water scarcity will most likely continue to be a significant problem in many of the important vegetable production regions of the world [126]. Improving water use efficiency through more accurate scheduling of irrigation can conserve water and address water quality impacts associated with commercial vegetable operations. However, irrigation scheduling in vegetables presents some unique challenges due to the diversity of crop types, intensive rotations, and number of fields that must be managed, as well as competing cultural operations that are involved in growing marketable crops. Advances in soil moisture sensors, wireless communications, ET measurements, remote sensing, computer technology, and cloud computing offer many potential opportunities to develop robust irrigation advisory tools to help farmers accurately determine and meet crop water needs. While much progress has already been made in soil moisture sensing and ET-based irrigation scheduling, achieving wide-scale adoption in the vegetable industry will require continued innovation. Continued collaboration between public research institutions, universities, commodity boards, and commercial firms will likely be needed to develop simple-to-use tools that will be broadly accepted by vegetable growers.

Acknowledgments: The authors wish to thank Forrest Melton, Richard Smith, and Rick Snyder for their careful review and edits in the preparation of this manuscript.

Author Contributions: Michael D. Cahn was the primary author of Sections 1–5, 8 and 9. Lee F. Johnson was the primary author of Sections 6 and 7 and substantially contributed to editing of the entire manuscript.

Conflicts of Interest: The authors declare no conflict of interest. Mention of companies, proprietary products, or trade names does not imply endorsement by the authors or their associated institutions and does not imply approval to the exclusion of other products or vendors that may also be suitable.

Abbreviations

The following abbreviations are used in this manuscript:

API	application programming interface
BIS	basic irrigation scheduling
CIMIS	California irrigation management and information system
CUP	consumptive use program
EEFlux	earth engine evapotranspiration flux
EM	electromagnetic
ET	evapotranspiration
ET_0	reference evapotranspiration
ETc	crop evapotranspiration
Fc	fractional cover
GMS	granular matrix sensor
Kc	crop coefficient
LE	latent heat flux
METRIC	mapping evapotranspiration at high resolution with internalized calibration
NDVI	normalized difference vegetation index
NIR	near infra-red
SEBAL	surface energy balance algorithm for land
SIMS	satellite irrigation management system
SSURGO	soil survey geographic database
UAV	unmanned aerial vehicle
UC	University of California
USDA	United States Department of Agriculture
WISE	Washington irrigation scheduling expert

References

1. Mount, J.; Chappelle, C.; Gray, B.; Hanak, E.; Howitt, R.; Lund, J.; Frank, R.; Gartrell, G.; Grantham, T.; Medellín-Azuara, J.; et al. California's Water: Managing Droughts. Public Policy Institute of California, Water Policy Center, 2016; p. 4. Available online: http://www.ppic.org/content/pubs/report/R_1016JM2R.pdf (accessed on 18 February 2017).

2. Hanak, E.; Mount, J.; Chappelle, C.; Lund, J.; Medellín-Azuara, J.; Moyle, P.; Thompson, B.; Viers, J.; Seavy, N. What If California's Drought Continues? Public Policy Institute of California, Water Policy Center, 2015; p. 20. Available online: http://www.ppic.org/content/pubs/report/R_815EHR.pdf (accessed on 18 February 2017).

3. Pedrero, F.; Alarcon, J.J.; Nicolas, E.; Asano, T. Monterey Regional Water Pollution Control Agency (MRWPCA): Transforming wastewater into a profitable resource. *Agric. Vergel Frutic. Hortic. Floric. Citric. Vid Arroz* **2009**, *28*, 464–469.

4. Khan, S. Managing climate risks in Australia: Options for water policy and irrigation management. *Aust. J. Exp. Agric.* **2008**, *48*, 265–273. [CrossRef]

5. Arumí, J.L.; Rivera, D.; Holzapfel, E.; Muñoz, E. Effect of drought on groundwater in a Chilean irrigated valley. *Proc. Inst. Civ. Eng. Water Manag.* **2013**, *166*, 231–241. [CrossRef]

6. Ortega-Reig, M.; Palau-Salvador, G.; Cascant i Sempere, M.J.; Benitez-Buelga, J.; Badiella, D.; Trawick, P. The integrated use of surface, ground and recycled waste water in adapting to drought in the traditional irrigation system of Valencia. *Agric. Water Manag.* **2014**, *133*, 55–64. [CrossRef]

7. Harter, T.; Lund, J.R. *Addressing Nitrate in California's Drinking Water with a Focus on Tulare Lake Basin and Salinas Valley Groundwater. Report for the State Water Resources Control Board Report to the Legislature*; Center for Watershed Sciences, University of California: Davis, CA, USA, 2012; p. 78.

8. Hallberg, G.R. *Nitrate in Ground Water in the United States*; Follett, R.F., Ed.; Elsevier Science Publishers: Amsterdam, The Netherlands, 1989.

9. Zhang, L.; Ju, X.; Liu, C.; Kou, C. A study on nitrate contamination of ground water sources in areas of protected vegetables-growing fields—A case study in Huimin County, Shandong Province. *Sci. Agric. Sin.* **2010**, *43*, 4427–4436.

10. Howarth, R.W.; Sharpley, A.; Walker, D. Sources of nutrient pollution to coastal waters in the United States: Implications for achieving coastal water quality goals. *Estuaries* **2002**, *25*, 656–676. [CrossRef]

11. Ng, C.M.; Weston, D.P.; Lydy, M.J. Pyrethroid insecticide transport into Monterey Bay through riverine suspended solids. *Arch. Environ. Contam. Toxicol.* **2012**, *63*, 461–470. [CrossRef] [PubMed]

12. Anderson, B.S.; Hunt, J.W.; Phillips, B.M.; Nicely, P.A.; Gilbert, K.D.; Vlaming, V.D.; Connor, V.; Richard, N.; Tjeerdema, R.S. Ecotoxicologic impacts of agricultural drain water in the Salinas River, California, USA. *Environm. Toxicol. Chem.* **2003**, *22*, 2375–2384. [CrossRef]

13. Agricultural Order No. R3-2012-0011 Conditional Waiver of Waste Discharge Requirements for Discharges from Irrigated Lands. 2012; p. 94. Available online: http://www.waterboards.ca.gov/centralcoast/water_issues/programs/ag_waivers/docs/ag_order/agorder_final_011014.pdf (accessed on 20 January 2017).

14. Council Directive 91/676/EEC of 12 December 1991 Concerning the Protection of Waters against Pollution Caused by Nitrates from Agricultural Sources. 1991. Available online: http://eur-lex.europa.eu/legal-content/EN/TXT/?uri=CELEX%3A31991L0676 (accessed on 20 January 2017).

15. Koike, S.; Cahn, M.D.; Cantwell, M.; Fennimore, S.; LeStrange, M.; Natwick, E.; Smith, R.; Takale, E. Spinach production in California. In *Vegetable Production Series*; Vol. ANR Publication 7212; UCANR: Richmond, CA, USA, 2011.

16. Smith, R.; Cahn, M.; Hartz, T.; Love, P.; Farrara, B. Nitrogen dynamics of cole crop production: Implications for fetility management and environmental protection. *HortScience* **2016**, *51*, 1586–1591. [CrossRef]

17. Shock, C.C.; Wang, F. Soil water tension, a powerful measurement for productivity and stewardship. *HortScience* **2011**, *46*, 178–185.

18. Pardossi, A.; Incrocci, L. Traditional and New Approaches to Irrigation Scheduling in Vegetable Crops. *HortTechnology* **2011**, *21*, 309–313.

19. Shock, C.; Pereira, A.; Hanson, B.; Cahn, M. Vegetable irrigation. In *Irrigation of Agricultural Crops*, 2nd ed.; Lanscano, R., Sojka, R., Eds.; ASA-CSSA-SSSA (American Society of Agronomy-Crop Science Society of America-Soil Science Society of America): Madison, WI, USA, 2007; pp. 535–606.

20. Thompson, T.L.; Doerge, T.A.; Godin, R.E. Subsurface drip irrigation and fertigation of broccoli: I. Yield, quality, and nitrogen uptake. *Soil Sci. Soc. Am. J.* **2002**, *66*, 186–192. [CrossRef]

21. Smittle, D.A.; Dickens, W.L.; Stansell, J.R. Irrigation regimes affect cabbage water use and yield. *J. Am. Soc. Hortic. Sci.* **1994**, *119*, 20–23.

22. Thompson, T.L.; Doerge, T.A.; Godin, R.E. Nitrogen and water interactions in subsurface drip-irrigated cauliflower: I. Plant response. *Soil Sci. Soc. Am. J.* **2000**, *64*, 406–411. [CrossRef]

23. Gallardo, M.; Jackson, L.E.; Schulbach, K.; Snyder, R.L.; Thompson, R.B.; Wyland, L.J. Production and water use in lettuces under variable water supply. *Irrig. Sci.* **1996**, *16*, 125–137. [CrossRef]

24. Thompson, T.L.; Doerge, T.A. Nitrogen and water rates for subsurface trickle-irrigated romaine lettuce. *HortScience* **1995**, *30*, 1233–1237.

25. Wang, F.-X.; Wu, X.-X.; Shock, C.C.; Chu, L.-Y.; Gu, X.-X.; Xue, X. Effects of drip irrigation regimes on potato tuber yield and quality under plastic mulch in arid Northwestern China. *Field Crops Res.* **2011**, *122*, 78–84. [CrossRef]

26. Munoz-Carpena, R.; Dukes, M.D.; Li, Y.C.; Klassen, W. Field comparison of tensiometer and granular matrix sensor automatic drip irrigation on tomato. *HortTechnology* **2005**, *15*, 584–590.

27. Markovic, M.P.; Josipovic, M.A.; Sostaric, J.I.; Zebec, V.B.; Rapcan, I.A. Effectiveness of granular matrix sensors in different irrigation treatments and installation depths. *J. Agric. Sci. Belgrad.* **2016**, *61*, 257–269. [CrossRef]

28. Munoz-Carpena, R.; Dukes, M.D.; Li, Y.; Klassen, W. Design and Field Evaluation of a New Controller for Soil-Water Based Irrigation. *Appl. Eng. Agric.* **2008**, *24*, 183–191. [CrossRef]

29. Phene, C.J.; Howell, T.A. Soil sensor control of high-frequency irrigation systems. *Trans. Am. Soc. Agric. Eng.* **1984**, *27*, 392–396. [CrossRef]

30. Dukes, M.D.; Scholberg, J.M. Soil moisture controlled subsurface drip irrigation on sandy soils. *Appl. Eng. Agric.* **2005**, *21*, 89–101. [CrossRef]

31. Zotarelli, L.; Dukes, M.D.; Barreto, T.P. Use of Soil Moisture Sensor-based Irrigation on Vegetable Crops. *Proc. Fla. State Hort. Soc.* **2009**, *122*, 218–220.

32. Zotarelli, L.; Dukes, M.D.; Scholberg, J.M.S.; Femminella, K.; Munoz-Carpena, R. Irrigation scheduling for green bell peppers using capacitance soil moisture sensors. *J. Irrig. Drain. Eng.* **2011**, *137*, 73–81. [CrossRef]

33. Zotarelli, L.; Dukes, M.D.; Scholberg, J.M.S.; Munoz-Carpena, R.; Icerman, J. Tomato nitrogen accumulation and fertilizer use efficiency on a sandy soil, as affected by nitrogen rate and irrigation scheduling. *Agric. Water Manag.* **2009**, *96*, 1247–1258. [CrossRef]

34. Zotarelli, L.; Scholberg, J.M.; Dukes, M.D.; Munoz-Carpena, R.; Icerman, J. Tomato yield, biomass accumulation, root distribution and irrigation water use efficiency on a sandy soil, as affected by nitrogen rate and irrigation scheduling. *Agric. Water Manag.* **2009**, *96*, 23–34. [CrossRef]

35. Zotarelli, L.; Dukes, M.D.; Scholberg, J.M.; Hanselman, T.; Le Femminella, K.; Muñoz-Carpena, R. Nitrogen and water use efficiency of zucchini squash for a plastic mulch bed system on a sandy soil. *Sci. Hortic.* **2008**, *116*, 8–16. [CrossRef]

36. Thompson, R.B.; Gallardo, M.; FerncLndez, M.D.; Valdez, L.C.; Martcnez-GaitcLn, C. Salinity Effects on Soil Moisture Measurement Made with a Capacitance Sensor. *Soil Sci. Soc. Am. J.* **2007**, *71*, 1647–1657. [CrossRef]

37. Visconti, F.; de Paz, J.M.; Martcnez, D.; Molina, M.J. Laboratory and field assessment of the capacitance sensors Decagon 10HS and 5TE for estimating the water content of irrigated soils. *Agric. Water Manag.* **2014**, *132*, 111–119. [CrossRef]

38. Kargas, G.; Soulis, K.X. Performance analysis and calibration of a new low-cost capacitance soil moisture sensor. *J. Irrig. Drain. Eng.* **2012**, *138*, 632–641. [CrossRef]

39. Kelleners, T.J.; Soppe, R.W.O.; Ayars, J.E.; Skaggs, T.H. Calibration of capacitance probe sensors in a saline silty clay soil. *Soil Sci. Soc. Am. J.* **2004**, *68*, 770–778. [CrossRef]

40. Parvin, N.; Degré, A. Soil-specific calibration of capacitance sensors considering clay content and bulk density. *Soil Res.* **2016**, *54*, 111–119. [CrossRef]

41. Soulis, K.X.; Elmaloglou, S. Optimum soil water content sensors placement in drip irrigation scheduling systems: Concept of time stable representative positions. *J. Irrig. Drain. Eng.* **2016**, *142*, 04016054. [CrossRef]

42. Soulis, K.X.; Elmaloglou, S.; Dercas, N. Investigating the effects of soil moisture sensors positioning and accuracy on soil moisture based drip irrigation scheduling systems. *Agric. Water Manag.* **2015**, *148*, 258–268. [CrossRef]

43. Heinrich, A.; Smith, R.; Cahn, M. Nutrient and water use of fresh market spinach. *HortTechnology* **2013**, *23*, 325–333.

44. Penman, H.L. Natural evaporation from open water, bare soil and grass. *Proc. R. Soc.* **1948**, 120–145. [CrossRef]

45. Monteith, J.L. Evaporation and the environment. *Symp. Soc. for Exp. Biol.* **1965**, *19*, 205–234.

46. Allen, R.G.; Walter, I.A.; Elliot, R.; Itenfisu, D.; Brown, P.; Jensen, M.E.; Mecham, B.; Howell, T.A.; Snyder, R.; Eching, S.; et al. The ASCE standardized reference evapotranspiration equation. In *Water for a Sustainable World, Limited Supplies and Expanding Demand*, Proceedings of the Second International Conference on Irrigation and Drainage, Phoenix, AZ, USA, 12–15 May 2003.

47. Allen, R.G.; Pruitt, W.O.; Wright, J.L.; Howell, T.A.; Ventura, F.; Snyder, R.; Itenfisu, D.; Steduto, P.; Berengena, J.; Baselga Yrisarry, J.; et al. A recommendation on standardized surface resistance for hourly calculation of reference ETo by the FAO56 Penman-Monteith method. *Agric. Water Manag.* **2006**, *81*, 1–22. [CrossRef]

48. Allen, R.G.; Pereira, L.S.; Raes, D.; Smith, M. Crop evapotranspiration: Guidelines for computing crop water requirements. In *FAO Irrigation and Drainage Paper*; Food and Agriculture Organization (FAO): Rome, Italy, 1998.

49. Guerra, E.; Ventura, F.; Snyder, R.L. Crop coefficients: A literature review. *J. Irrig. Drain. Eng.* **2016**, *142*, 06015006. [CrossRef]

50. Grattan, S.R.; Bowers, W.; Dong, A.; Snyder, R.L.; Carroll, J.J.; George, W. New crop coefficients estimate water use of vegetables, row crops. *Calif. Agric.* **1998**, *52*, 16–21. [CrossRef]

51. Estaciones Agroclimáticas, Consejería de Agricultura, Pesca y Desarrollo Rural. Available online: http://www.juntadeandalucia.es/agriculturaypesca/ifapa/ria/servlet/FrontController?action=Init (accessed on 27 January 2017).

52. Rete Agrometeorologica Nazionale, Ministero delle Politiche Agricole Alimentari e Forestali. Available online: https://www.politicheagricole.it/flex/FixedPages/Common/miepfy200_reteAgrometeorologica.php/L/IT (accessed on 27 January 2017).

53. Hart, Q.; Brugnach, M.; Temesgen, B.; Rueda, C.; Ustin, S.; Frame, K. Daily reference evapotranspiration for California using satellite imagery and weather station measurement interpolation. *Civ. Eng. Environ. Syst.* **2009**, *26*, 19–33. [CrossRef]

54. Gallardo, M.; Snyder, R.L.; Schulbach, K.; Jackson, L.E. Crop growth and water use model for lettuce. *J. Irrig. Drain. Eng.* **1996**, *122*, 354–359. [CrossRef]

55. Allen, R.G.; Pereira, L.S.; Smith, M.; Raes, D.; Wright, J.L. FAO-56 dual crop coefficient method for estimating evaporation from soil and application extensions. (Special Issue: Evapotranspiration prediction in irrigated agriculture). *J. Irrig. Drain. Eng.* **2005**, *131*, 2–13. [CrossRef]

56. El-Shikha, D.W.P.; Hunsaker, D.; Clarke, T.; Barnes, E. Ground-based remote sensing for assessing water and nitrogen status of broccoli. *Agric. Water Manag.* **2007**, *92*, 183–193. [CrossRef]

57. Hunsaker, D.F.G.; French, A.; Clarke, T.; Ottman, M.; Pinter, P. Wheat irrigation management using multispectral crop coefficients. *Trans. ASABE* **2007**, *50*, 2017–2033. [CrossRef]

58. Gonzalez-Dugo, M.M.L. Spectral vegetation indices for benchmarking water productivity of irrigated cotton and sugarbeet crops. *Agric. Water Manag.* **2008**, *95*, 48–58. [CrossRef]

59. Campos, I.N.C.; Calera, A.; Balbontin, C.; Gonzalez-Piqueras, J. Assessing satellite-based basal crop coefficients for irrigated grapes. *Agric. Water Manag.* **2010**, *98*, 45–54. [CrossRef]

60. Smith, M. CROPWAT. A computer program for irrigation planning and management. In *FAO Irrigation and Drainage Paper*; FAO Land and Water Development Division: Rome, Italy, 1992.

61. Snyder, R.L.; Orang, M.N.; Bali, K.M.; Eching, S. Basic Irrigation Scheduling (BIS), 2000. p. 10. Available online: https://www.researchgate.net/publication/255580842_BASIC_IRRIGATION_SCHEDULING_BIS (accessed on 20 January 2017).

62. Orang, M.N.; Matyac, J.S.; Snyder, R.L. Consumptive use program (CUP) model. *Acta Hortic.* **2004**, *664*, 461–468. [CrossRef]

63. Lieb, B. Washington Irrigation Scheduling Expert. Available online: https://sourceforge.net/projects/wsuwise/ (accessed on 20 January 2017).

64. Leib, B.G.; Elliott, T.V. Washington Irrigation Scheduling Expert (WISE) Software. In *National Irrigation Symposium*, Proceedings of the 4th Decennial Symposium, Phoenix, AZ, USA, 14–16 November 2000; pp. 540–548.

65. Washington State University IrrigationScheduler Mobile. Available online: http://weather.wsu.edu/is/ (accessed on 20 January 2017).

66. Hillyer, C.; English, M. Irrigation Management Online. Available online: http://oiso.bioe.orst.edu/ RealtimeIrrigationSchedule/index.aspx (accessed on 20 January 2017).

67. Wateright. Available online: http://www.wateright.org (accessed on 20 January 2017).

68. Thompson, W. Irrigation scheduling made simple(r): Wateright website does the hard work for you. *Calif. Grow. Avocados Citrus Subtrop.* **1998**, *22*, 22–23.

69. CropManage. Available online: https://cropmanage.ucanr.edu (accessed on 20 January 2017).

70. Cahn, M.; Smith, R.; Farrara, B.; Hartz, T.; Johnson, L.; Melton, F.; Post, K. Irrigation and nitrogen management decision support tool for vegetables and berries. In Proceedings of the U.S Committee on Irrigation and Drainage Conference: Groundwater Issues and Water Management—Strategies Addressing the Challenges of Sustainability USCID, Sacramento, CA, USA, 4–7 March 2014; pp. 53–64.

71. Bartlett, A.C.; Andales, A.A.; Arabi, M.; Bauder, T.A. A smartphone app to extend use of a cloud-based irrigation scheduling tool. *Comput. Electron. Agric.* **2015**, *111*, 127–130. [CrossRef]

72. SmartIrrigation Apps. Available online: http://smartirrigationapps.org/ (accessed on 20 January 2017).

73. Migliaccio, K.W.; Morgan, K.T.; Vellidis, G.; Zotarelli, L.; Fraisse, C.; Zurweller, B.A.; Andreis, J.H.; Crane, J.H.; Rowland, D.L. Smartphone apps for irrigation scheduling. *Trans. ASABE* **2016**, *59*, 291–301.

74. Vellidis, G.; Liakos, V.; Andreis, J.H.; Perry, C.D.; Porter, W.M.; Barnes, E.M.; Morgan, K.T.; Fraisse, C.; Migliaccio, K.W. Development and assessment of a smartphone application for irrigation scheduling in cotton. *Comput. Electron. Agric.* **2016**, *127*, 249–259. [CrossRef]

75. Mannini, P.; Genovesi, R.; Letterio, T. IRRINET: Large Scale DSS Application for On-farm Irrigation Scheduling. *Procedia Environ. Sci.* **2013**, *19*, 823–829. [CrossRef]

76. Montoro, A.; Lopez-Fuster, P.; Fereres, E. Improving on-farm water management through an irrigation scheduling service. *Irrig. Sci.* **2011**, *29*, 311–319. [CrossRef]

77. Deumier, J.M.; Leroy, P.; Peyremorte, P. Tools for improving management of irrigated agricultural crop systems. *Water Rep.* **1996**, *8*, 39–49.

78. Car, N.J.; Christen, E.W.; Hornbuckle, J.W.; Moore, G.A. Using a mobile phone Short Messaging Service (SMS) for irrigation scheduling in Australia Farmers participation and utility evaluation. *Comput. Electron. Agric.* **2012**, *84*, 132–143. [CrossRef]

79. USDA SSURGO. Available online: https://sdmdataaccess.nrcs.usda.gov/ (accessed on 20 January 2017).

80. Doorenbos, J.; Pruitt, W.O. *Guidelines for Predicting Crop-Water Requirements*, 2nd ed.; Food and Agriculture Organization of the United Nations: Rome, Italy, 1977; p. 156.

81. Johnson, L.F.; Cahn, M.; Martin, F.; Melton, F.; Benzen, S.; Farrara, B.; Post, K. Evapotranspiration-based irrigation scheduling of head lettuce and broccoli. *HortScience* **2016**, *51*, 935–940.

82. Morrison, M. Encouraging The Adoption of Decision Support Systems by Irrigators. *Rural Soc.* **2009**, *19*, 17–31. [CrossRef]

83. Giannakis, E.; Bruggeman, A.; Djuma, H.; Kozyra, J.; Hammer, J. Water pricing and irrigation across Europe: Opportunities and constraints for adopting irrigation scheduling decision support systems. *Water Sci. Technol. Water Supply* **2016**, *16*, 245–252. [CrossRef]

84. Rana, G.; Katerji, N. Measurement and estimation of actual evapotranspiration in the field under Mediterranean climate: A review. *Eur. J. Agron.* **2000**, *13*, 125–153. [CrossRef]

85. Allen, R.G.; Pereira, L.S.; Howell, T.A.; Jensen, M.E. Evapotranspiration information reporting: I. Factors governing measurement accuracy. *Agric. Water Manag.* **2011**, *98*, 899–920. [CrossRef]

86. Allen, R.G. *Lysimeters for Evapotranspiration and Environmental Measurements*; American Society of Civil Engineers: New York, NY, USA, 1991.

87. Snyder, R.L.; Spano, D. Using surface renewal measurements to estimate crop evapotranspiration. In Proceedings of the Third Congress of the European Society for Agronomy, Abano Padova, Italy, 18–22 September 1994.

88. Hanson, B.R.; May, D.M. Crop coefficients for drip-irrigated processing tomato. *Agric. Water Manag.* **2006**, *81*, 381–399. [CrossRef]

89. Cahn, M.; Snyder, R.; Hanson, B. Estimating evapotranspiration in processing tomato. *Acta Hortic.* **1999**, *487*, 493–497. [CrossRef]

90. Hanson, B.R.; May, D.M. Crop evapotranspiration of processing tomato in the San Joaquin Valley of California, USA. *Irrig. Sci.* **2006**, *24*, 211–221. [CrossRef]

91. Rosa, R.; Dicken, U.; Tanny, J. Estimating evapotranspiration from processing tomato using the surface renewal technique. (Special Issue: Sensing technologies for sustainable agriculture.). *Biosyst. Eng.* **2013**, *114*, 406–413. [CrossRef]

92. Lopez Avendano, J.E.; Diaz Valdes, T.; Watts Thorp, C.; Rodriguez, J.C.; Castellanos Villegas, A.E.; Partida Ruvalcaba, L.; Velazquez Alcaraz, T.D.J. Evapotranspiration and crop coefficient of bell pepper in Culiacan Valley, Mexico. *Terra Latinoam.* **2015**, *33*, 209–219.

93. Hanson, B.R.; May, D.M. New crop coefficients developed for high-yield processing tomatoes. *Calif. Agric.* **2006**, *60*, 95–99. [CrossRef]

94. Shapland, T.M.; McElrone, A.J.; Khaw Tha Paw, U.; Snyder, R.L. A turnkey data logger program for field-scale energy flux density measurements using eddy covariance and surface renewal. *Ital. J. Agrometeorol.* **2013**, *18*, 5–16.

95. Courault, D.; Seguin, B.; Olioso, A. Review on estimation of evapotranspiration from remote sensing data: From empirical to numerical modeling approaches. *Irrig. Drain. Syst.* **2005**, *19*, 223–249. [CrossRef]

96. Kalma, J.; McVicar, T.; McCabe, M. Estimating land surface evaporation: A review of methods using remotely sensed surface temperature data. *Surv. Geophys.* **2008**, *29*, 421–469. [CrossRef]

97. Gonzalez-Dugo, M.; Neale, C.; Mateos, L.; Kustas, W.; Prueger, J.; Anderson, M.; Li, F. A comparison of operational remote sensing-based methods for estimating crop evapotranspiration. *Agric. For. Meteorol.* **2009**, *149*, 1843–1853. [CrossRef]

98. Bastiaanssen, W.; Meneti, M.; Feddes, R.; Holtslag, A. A remote sensing surface energy balance algorithm for land (SEBAL). *J. Hydrol.* **1998**, *212*, 198–212. [CrossRef]

99. Allen, R.G.; Tasumi, M.; Trezza, R. Satellite-based energy balance for mapping evapotranspiration with internalized calibration (METRIC) Model. *J. Irrig. Drain. Eng.* **2007**, *133*, 380–394. [CrossRef]

100. Teixeira, A.; Bastiaanssen, W.; Ahmad, M.; Bos, M. Reviewing SEBAL input parameters for assessing evapotranspiration and water productivity for the Lower-Middle Sao Francisco river basin, Brazil, Part A: Calibration and validation. *Agric. For. Meteorol.* **2008**, *149*, 462–476. [CrossRef]

101. Allen, R.; Burnett, B.; Kramber, W.; Huntington, J.; Kjaersgaard, J.; Kilic, A.; Kelly, C.; Trezza, R. Automated calibration of the METRIC-Landsat evapotranspiration process. *J. Am. Water Resour. Assoc.* **2013**, *49*, 563–576. [CrossRef]

102. Bryla, D.R.; Trout, T.J.; Ayars, J.E. Weighing Lysimeters for Developing Crop Coefficients and Efficient Irrigation Practices for Vegetable Crops. *HortScience* **2010**, *45*, 1597–1604.

103. Johnson, L.F.; Trout, T. Satellite NDVI assisted monitoring of vegetable crop evapotranspiration in California's San Joaquin Valley. *Remote Sens.* **2012**, *4*, 439–455. [CrossRef]

104. Tucker, C. Red and photographic infrared linear combinations for monitoring vegetation. *Remote Sens. Environ.* **1979**, *8*, 127–150. [CrossRef]

105. Trout, T.; Johnson, L.; Gartung, J. Remote sensing of canopy cover in horticultural crops. *HortScience* **2008**, *43*, 333–337.

106. Neale, C.; Jayanthi, H.; Wright, J. Irrigation water management using high resolution airborne remote sensing. *Irrig. Drain. Syst.* **2005**, *19*, 321–326. [CrossRef]

107. Calera-Belmonte, A.; Jochum, A. Operational Tools for Irrigation Water Management Based on Earth Observation: The DEMETER Project. In *Remote Sensing for Agriculture, Ecosystems, and Hydrology VIII*; SPIE: Stockholm, Sweden, 2006.

108. Melton, F.; Johnson, L.F. Satellite Irrigation Management System. Available online: https://ecocast.arc.nasa.gov/dgw/sims (accessed on 27 January 2017).

109. Melton, F.; Johnson, L.; Lund, C.; Pierce, L.; Michaelis, A.; Hiatt, S.; Guzman, A.; Adhikari, D.; Purdy, A.; Rosevelt, C.; et al. Satellite Irrigation Management Support with the Terrestrial Observation and Prediction System. *IEEE J. Sel. Top. Appl. Earth Obs. Remote Sens.* **2012**, *5*, 1709–1721. [CrossRef]

110. Johnson, L.; Cassel-Sharma, F.; Goorahoo, D.; Melton, F. Landsat-based calculation of agricultural water use fractions in California. In Proceedings of the 19th William T. Pecora Memorial Remote Sensing Symposium, Denver, CO, USA, 17–20 November 2014.

111. IrriSat. Available online: https://irrisat-cloud.appspot.com/ (accessed on 27 January 2017).

112. Montgomery, J.; Hornbuckle, J.; Hume, I.; Vleeshouwer, J. IrriSAT—Weather based scheduling and benchmarking technology. In Proceedings of the 17th ASA Conference, Hobart, Australia, 20–24 September 2015.
113. Hornbuckle, J.; Vleeshouwer, J.; Ballester, C.; Montgomery, J.; Hoogers, R.; Bridgart, R. *IrriSAT Technical Reference*; Deakin University: Victoria, Australia, 2016.
114. Vuolo, F.; D'Urso, G.; De Michele, C.; Bianchi, B.; Cutting, M. Satellite-based irrigation advisory services: A common tool for different experiences from Europe to Australia. *Agric. Water Manag.* **2015**, *147*, 82–95. [CrossRef]
115. D'Urso, G. Current status and perspectives for the estimation of crop water requirements from earth observation. *Ital. J. Agron.* **2010**, *5*, 107–120. [CrossRef]
116. EEflux. Available online: http://eeflux-level1.appspot.com/ (accessed on 27 January 2017).
117. Allen, R.; Morton, C.; Kamble, B.; Kilic, A.; Huntington, J.; Thau, D.; Ratcliffe, I. EEFlux: A Landsat-based evapotranspiration mapping tool on the Google Earth Engine. In Proceedings of the ASABE/IA Irrigation Symposium, Long Beach, CA, USA, 10–12 November 2015.
118. Pereira, L.; Allen, R. Crop evapotranspiration estimation with FAO56: Past and future. *Agric. Water Manag.* **2015**, *147*, 4–20. [CrossRef]
119. Vermote, E.; Saleous, N. *LEDAPS Surface Reflectance Product Description*; University of Maryland: College Park, MD, USA, 2007.
120. Tasumi, M.; Allen, R.; Trezza, R. At-surface albedo from Landsat and MODIS satellites for use in energy balance studies of evapotranspiration. *J. Hydrol. Eng.* **2008**, *13*, 51–63. [CrossRef]
121. Johnson, L.; Cahn, M.; Rosevelt, C.; Guzman, A.; Melton, F.; Lockhart, T.; Farrara, B.; Melton, F. *Satellite Estimation of Fractional Cover in Several California Specialty Crops*; American Geophysical Union: San Francisco, CA, USA, 2016.
122. Gago, J.; Douthe, C.; Coopman, R.E.; Gallego, P.P.; Ribas-Carbo, M.; Flexas, J.; Escalona, J.; Medrano, H. UAVs challenge to assess water stress for sustainable agriculture. *Agric. Water Manag.* **2015**, *153*, 9–19. [CrossRef]
123. Gonzalez-Dugo, V.; Zarco-Tejada, P.; Nicolas, E.; Nortes, P.A.; Alarcon, J.J.; Intrigliolo, D.S.; Fereres, E. Using high resolution UAV thermal imagery to assess the variability in the water status of five fruit tree species within a commercial orchard. *Precis. Agric.* **2013**, *14*, 660–678. [CrossRef]
124. Torres-Sanchez, J.; Pena, J.M.; de Castro, A.I.; Lopez-Granados, F. Multi-temporal mapping of the vegetation fraction in early-season wheat fields using images from UAV. *Comput. Electron. Agric.* **2014**, *103*, 104–113. [CrossRef]
125. Torres-Sanchez, J.; Lopez-Granados, F.; Pena, J.M. An automatic object-based method for optimal thresholding in UAV images: Application for vegetation detection in herbaceous crops. *Comput. Electron. Agric.* **2015**, *114*, 43–52. [CrossRef]
126. Food and Agriculture Organization of the United Nations; The World Water Council. Toward a Water and Food Secure Future: Critical Perspectives for Policy-Makers. 2015, p. 61. Available online: http://www.fao.org/nr/water/docs/FAO_WWC_white_paper_web.pdf (accessed on 20 January 2017).

19

Occurrence and Distribution of Entomopathogenic Nematodes in Sweet Potato Fields in the Philippines and Their Implication in the Biological Control of Sweet Potato Weevil

Ruben Madayag Gapasin [1,*], **Jesusito Laborina Lim** [1], **Elvira Lopez Oclarit** [1], **Leslie Toralba Ubaub** [2] **and Mannylen Coles Alde** [1]

[1] Department of Pest Management, College of Agriculture and Food Science, VisayasState University, Visca, Baybay City 6521-A, Philippines; jesslim24@yahoo.com (J.L.M.); elvie_oclart@yahoo.com (E.L.O.) aldemannylen@yahoo.com (M.C.A.)

[2] University of Southeastern Philippines (USeP Tagum Campus), Davao City 6521-A, Philippines; leslieubaub@gmail.com

* Correspondence: rmgapasin1952@yahoo.com or ruben.gapasin@vsu.ph.com

Abstract: The sweet potato weevil (*Cyclas formicarius* Fabr.) remains a serious threat to sweet potato (*Ipomoea batatas* Poir.) production and is considered the most destructive pest of sweet potatoes in the field and storagein the Philippines. Chemical control of the weevil is seldom practiced by farmers because they find it too costly, it may increase the chance for pesticide resistance, and because of public concern of its effectson non-target organisms. The use of biological controls such as entomopathogenic nematodes (EPN) could offer an effective, economical, and environmentally-friendly alternative management of the weevil. This study determined the occurrence and distribution of entomopathogenic nematodes in selected sweet potato growing areas in the Philippines. Using soil from 13 sweet potato growing areas, EPNs were recovered using the insect baiting method. Morbid insect larvae were suspended in sterile water for 48 h, and the suspension was examined under a stereomicroscope for the presence of EPN. Out of 47 samples collected from the 13 sweet potato production areas, 39 (82%) were positive for the presence of EPNs. Preliminary identification of the EPNs through morphological characters showed that they belonged to Rhaditida: Heterorhabditidae and Steinernematidae. This is the first report on the occurrence of EPNs in sweet potato fields in the Philippines, and their distribution strongly supports the possibility of utilizing them in an IPM management approach as biological agents against the sweet potato weevil. Morphometric and molecular-based identification and pathogenicity studies are underway.

Keywords: *Ipomoea batatas*; *Cylasformicarius*; biological control agent; integrated pest management; IPM; Rhabditida; Heterorhabditidae; Steinernematidae

1. Introduction

The sweet potato weevil (*Cylasformicarius* Fabr.) is a serious threat to sweet potato production, not only locally in the Philippines but also globally. It is the most destructive pest of sweet potato in the field and in storage. Reports indicated that losses in sweet potato production due to this pest ranged from 5% to 97%, especially in areas where the weevil occurs [1]. The insect attacks both the fleshy roots and the stems by tunneling into them and tainting them with a disagreeable odor and a bitter taste which renders them unfit for human and animal consumption.

The use of chemicals to control the weevil is seldom practiced, especially by small farmers because they find them too costly. The chemicals are also considered hazardous to the environment and non-target organisms. Thus, there is a need for cheaper and safer control methods against the sweet potato weevil. The use of a biological control using entomopathogenic nematodes (EPNs) is one alternative. Nematodes are non-polluting and, thus, environmentally-safe and acceptable. Infective juveniles can be applied with conventional equipment, and they are compatible with most pesticides [2]. They find their host either actively or passively, and, in cryptic habitats and sometimes in soil, they have proven superior to chemicals in controlling the target insects [3].

An excellent example of a situation in which a nematode may replace chemicals for control of an insect is the black vine weevil (*Otiorhynchussulcatus*) in cranberries. The use of chemical insecticides is restricted or has not provided adequate control of black vine weevil larvae. When *Heterorhabditis bacteriophora* "NC" strain was applied, it provided 70% control of the black vine weevil soon after treatment and was still providing the same level of control a year later [4].

Entomopathogenic nematodes seem to be organisms with the greatest potential for practical biological suppression of sweet potato weevil [1]. In the United States, studies have shown the superiority of some genera of EPNs for suppressing or reducing weevil populations. To date, no studies have been conducted utilizing EPNs against the sweet potato weevil in the Philippines. Thus, this study determined the occurrence and distribution of EPNs in selected sweet potato growing areas in the Philippines.

2. Experimental Section

2.1. Collection of Soil Samples

A survey was conducted in selected sweet potato growing areas of Visayas, Mindanao and Luzon, Philippines (Table 1). In each of the provinces/municipalities selected, specific areas/sites where sweet potato were grown were identified and sampling sites were established.

Soil samples from around the roots of sweet potato plants at each site were collected. From each sampling site, 10 core samples were taken, and 10 kg composite soil samples were derived. The soil samples were placed in plastic bags and transported to the laboratory. An aliquot of 300 g from each 10 kg sample was processed in the Nematology Laboratory of Visayas State University (VSU), Visca, Baybay City, and Leyte.

2.2. Rearing of Galleria mellonella

Larvae of *Galleria mellonella* were collected from honeybee hives infested with *G. mellonella*. Larvae were reared in the laboratory by feeding a nutrient diet [5]. The insects were allowed to develop into adults in a rearing box, and then were transferred to an ovipositor box to mate and oviposit the eggs. Eggs were collected and placed ina new, clean rearing box, wherein they were reared and maintained to serve as a source of insects for EPN baiting. The larvae were supplied regularly with fresh nutrient diet.

2.3. Nematode Extraction

Two methods of EPN extraction from the soil samples were employed. In the first method, a soil sample was placed in a large pail and mixed thoroughly. Then, water was added and it was mixed again. The mixture was sieved through a 120 μm mesh sieve, and then transferred to a 45 μm mesh sieve. The nematodes were collected from the 45 μm sieve in a beaker and transferred to an improvised Baerman funnel. The EPN suspension was kept in a test tube. In the second method, soil samples were placed in small plastic containers and healthy *G. mellonella* larvae were added to act as bait for the EPNs. After 48 h, dead larvae were recovered and examined under the microscope for the presence of EPN.

2.4. Killing, Staining and Mounting of Nematodes

Nematodes were teased out, killed and stained following standard procedures and techniques used for nematodes [6].   Representative of each EPN were then mounted on slides to facilitate identification.

2.5. Identification of Nematodes

Nematodes were identified up to genera using available keys and other references.

3. Results and Discussion

Results from 47 areas in 13 provinces showed the presence of EPNs, with most found in sweet potato growing areas in Negros Oriental, Visayas (Table 1). Their abundance in these sweet potato areas could be due to the soil type, moisture and temperature in the area at the time of sampling which could have favored the reproduction of EPNs in the soil. Soil from these areas were sandy or sandy loams. According to Molyneux [7], survival of juveniles was high in these types of soils. The unlimited host range and niche conditions may also have contributed to the high frequency of the nematodes [8].

Table 1. A list of the sweet potato growing areas in the Philippines that were sampled, and occurrence of entomopathogenic nematodes (EPNs) from the soil at each site.

Major Islands	Province	City/Barangay (Brgy)/Municipality	Occurrence of EPN [z]
Luzon	Benguet	Benguet State University (BSU)	+
	Laguna	IPB, UP Los Banos	+
	Nueva Ecija	Central Luzon State University (CLSU)	-
	Tarlac	Tarlac College Of Agriculture	-
Visayas	Biliran	DA, Naval	Steinernema
		Provincial scion grove and nursery, Sitio San Roque, Brgy. Larrazabal	-
		Sitio San Roque, Larrazabal,	+
	Cebu	Dakit, Barili	Steinernema
		Cebu Technological University (CTU), Barili Campus	-
		Cambinocot	+
	Eastern Samar	Lawaan	+
		Brgy. Dos, Giporlos	-
		Brgy. Cabong, Borongan	Steinernema
		Balangiga	+
		Brgy. San Pablo, Taft	+
	Leyte	Brgy. San Esteban, Burauen	Steinernema
		Brgy. Maliwaliw, Dagami	-
		Brgy. Tinocdugan, Leyte	Steinernema
		DA, Babatngon (SP)	-
		DA, Babatngon (Okra)	-
		Del Pilar, Dulag	-
		Fatima, Dulag	-
		Gakat, Baybay	-
		San Vicente, Dulag	Heterorhabditis
		VSU, Agromet	-
		Zone 11, Mayorga	Steinernema
		Sitio Madocao, Brgy. Damos, Leyte	+

Table 1. *Cont.*

Major Islands	Province	City/Barangay (Brgy)/Municipality	Occurrence of EPN [z]
Visayas	Negros Oriental	Brgy. Apolong, Valencia	Heterorhabditis
		Brgy. North Poblacion, Valencia	Heterorhabditis
		Brgy. Lepayo, Dauin	+
		Bacong	Heterorhabditis
		Brgy. MaayongTubig, Dauin	-
		Sito May-abo, Brgy. Maluwag, Zamboangita	-
		Zamboangita	+
	Western Samar	Brgy. Bachao, Basey	-
		Brgy. Mabuhay, Marabut	-
		Brgy. Osmena, Hinabangan	-
Mindanao	Agusan del Sur	Sta. Josefa	+
		Trento	-
	Compostela Valley	Mabini	Steinernema
		New Bataan	-
		Kapatagan, Laak	-
	South Cotabato	Sulit, Polomolok	Steinernema
		Sarabia, Koronadal	-
		Sitio Cabuling, Sarabia, Koronadal	-
		Crossing, Polkan, Polomolok (Peanut)	+
		Crossing, Polkan, Polomolok (SP)	-

[z] Genera, or - = absent, and + = possible EPNs/to be determined.

There were two genera of EPNs extracted from the samples, *Steinernema* (Figure 1) and *Heterorhabditis* (Figure 2). Based on morphology of males, females, and juveniles, the most prevalent EPN was *Steinernema*, in agreement with Molyneux that Heterorhabditidae do not survive as well as do Steinernematidae [7]. All members of the order Rhabditidae are bacteriophagous, with many of them forming phoretic association with insects. Through time, some of these nematodes may have evolved into insect pathogens [9]. This may explain the occurrence of EPNs in the different areas surveyed. In Visayas, 18 out of 33 areas were positive for EPNs with 6 identified as *Steinernema*, and 3 as *Heterorhabditis*. Nine are yet to be identified. In the 10 areas of Mindanao sampled, 2 were positive for *Steinernema* while 2 were not identified. Out of 4 areas in Luzon, 2 were positive for the occurrence of EPNs. However, the nematodes in these samples have yet to be identified.

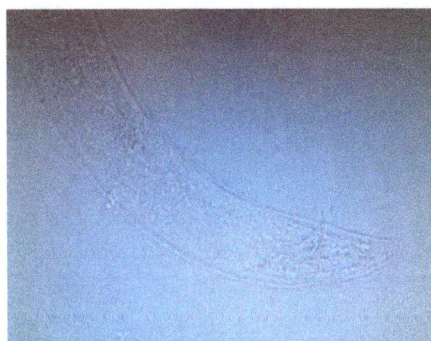

Figure 1. *Steinernema* sp. isolated from Sulit, Polomolok, and South Cotabato showing thespicule withoutthe bursa.

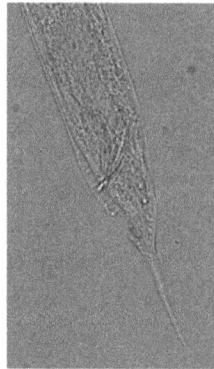

Figure 2. *Heterorhabditis* sp. isolated from Apolong, Valencia, Negros Oriental showing the spicule with bursa.

4. Conclusions

The results established the occurrence of *Heterorhabditis* and *Steinernema* EPNs in some sweet potato growing areas in the Philippines. This implies their abundance which is encouraging for their possible use as biological control agents against sweet potato weevil.

Acknowledgments: The authors would like to thank Department of Agriculture—Bureau of Agriculture Research (DA-BAR) for the financial support.

Author Contributions: Ruben Madayag Gapasin, Jesusito Laborina Lim and Leslie Toralba Ubaub conceived and designed the experiments and wrote the paper; Ruben Madayag Gapasin, Jesusito Laborina Lim, Leslie Toralba Ubaub, Elvira Lopez Oclarit and Mannylen Coles Alde performed the experiments.

Conflicts of Interest: The authors declare no conflict of interest.

References

1. Capinera, J.L. *Sweet Potato Weevil, Cylas formicarius (Fabricius)*; Institute of Food and Agricultural Sciences, University of Florida: Gainesville, FL, USA, 1998.

2. Rovesti, L.; Desco, K.V. Compatibility of chemical pesticides with the entomopathogenic nematodes, *Steinernema carpocapsae* Weiser and *S. feltiae* Feilipjev (Nematoda: Steinernematidae). *Nematologica* **1990**, *36*, 237–245. [CrossRef]

3. Gaugler, R. Biological control potential of neoaplectanid nematodes. *J. Nematol.* **1981**, *13*, 241–249. [PubMed]

4. Shanks, C.H.; Agudelo-Silva, F. Field pathogenicity and persistence of heterorhabditid and steinernematid nematodes (Nematoda) infecting black vine weevil larvae (Coleoptera: Curculionidae) in cranberry bogs. *J. Econ. Entomol.* **1990**, *83*, 107–110. [CrossRef]

5. Wiesner, A. *Die Indduction der Immunabwehr eines Inaeks (Galleria mellonella, Lepidoptera) Durxh Arteigene Haemolymph Faltoren*; Springer: Berlin, Germany, 1993.

6. Goodey, J.B. *Laboratory Methods for Work with Plant and Soil Nematodes*; Ministry of Agriculture, Fisheries and Food: London, UK, 1957.

7. Molyneux, A.S. Survival of infective juveniles of *Heterorhabditis* spp. and *Steinernema* spp. (Nematoda: Rhabditida) at various temperatures and their subsequent infectivity for insects. *Rev. Nematol.* **1985**, *8*, 165–170.

8. Eivazian, K.N.; Niknam, G.; Griffin, C.T.; Mohammadi, S.A.; Moghaddam, S.A. A survey of entomopathogenic nematodes of the families Steinernematidae and Heterorhabditidae (Nematoda: Rhabditida) from north-west of Iran. *Nematology* **2009**, *11*, 107–116.

9. Smart, G.C. Entomopathogenic nematode for the biological control of insects. *Suppl. J. Nematol.* **1995**, *27*, 529–534.

20

Relationship Marketing: A Qualitative Case Study of New-Media Marketing Use by Kansas Garden Centers

Scott Stebner [1], **Cheryl R. Boyer** [2,*], **Lauri M. Baker** [3] **and Hikaru H. Peterson** [4]

[1]   Former Graduate Research Assistant, Department of Communications and Agricultural Education, Kansas State University, 1612 Claflin Rd., Manhattan, KS 66506, USA; scottstebner@icloud.com

[2]   Department of Horticulture and Natural Resources, Kansas State University, 1712 Claflin Rd., Manhattan, KS 66506, USA

[3]   Department of Communications and Agricultural Education, Kansas State University, 1612 Claflin Rd., Manhattan, KS 66506, USA; lmbaker@ksu.edu

[4]   Department of Applied Economics, University of Minnesota, 1994 Buford Ave., St. Paul, MN 55108, USA; hhp@umn.edu

*   Correspondence: crboyer@ksu.edu

Abstract: A primary factor limiting the expansion of many Kansas garden centers is marketing. Most of these businesses spend the majority of advertising dollars on traditional media (newspaper, radio, etc.). However, new-media tools such as social-media can be an effective method for developing profitable relationships with customers. The purpose of this qualitative study was to explore the perceptions and experiences of garden center stakeholders as they use new-media to market their businesses. Grunig's Excellency Theory served as the theoretical framework for this study. Results indicate garden center operators prefer to use traditional media channels to market to their customers and asynchronously communicate with their target audiences. Stakeholders often have inaccurate or conflicting views of traditional media and new-media in regard to advertising and tend to approach new-media marketing from a public information or asynchronous viewpoint.

Keywords: marketing; relationship marketing; social-media marketing; new-media marketing; green industry; qualitative; garden center; nursery; landscape

1. Introduction

The green industry (garden centers, nurseries, landscaping companies, etc.) generates over $200 billion in annual revenue [1] and employs over 450,000 workers [2]. However, the retail garden center industry is highly seasonal and competes with many outside influences that can negatively affect sales, such as poor weather and competition from mass merchandisers [3]. According to Hodges et al. [4], mass merchants have acquired almost half the market share from smaller, local garden centers. Although mass merchants can offer prices that local garden centers cannot match, consumers are sometimes willing to pay higher prices for the increased selection, higher quality plants, and expert knowledge offered by small garden centers [5].

One factor limiting the expansion of garden centers and nurseries within the Great Plains region is marketing [6]. Insufficient funds for marketing is a common problem with smaller retailers who must try to find ways to generate maximum income potential with limited marketing and advertising budgets [7]. Small, family farms that have a yearly revenue not exceeding $50,000 rely heavily on marketing directly to the consumer [8]. Family-owned garden centers are no exception and have traditionally invested the majority of advertising dollars on the Yellow Pages, print media, and direct mail [6]. Such print material most often includes newsletters and direct-mail promotional pieces that seek to educate consumers about sales or offer coupons for seasonal goods.

Although direct marketing of agricultural goods to the public has proven profitable with an association of increased sales [9], a limited marketing budget can prove detrimental to direct-mail marketing because the potential to reach the desired target audience is limited by the resource capital the business is able to allocate to the campaign [10]. Even though direct mail has limitations, such as a low response rate [11], it is still a highly popular resource [7] that can increase the volume of customers [12].

Incorporation of new-media marketing tools such as social-media has made it possible for businesses to communicate and engage directly with current and potential customers while building relationships [13–15]. Establishing a direct line of back-and-forth communication allows consumers to feel their feedback is valued and recognized, thereby increasing the probability of customers engaging in word-of-mouth (WOM) marketing via the digital sphere and physical circles [13]. Ultimately, WOM relies upon community engagement, and in today's digital age it is vital that garden centers create an interactive web presence that can be accessed across multiple platforms in order to facilitate consumer demands and promote WOM [16].

Many businesses are transitioning away from single-channel and passive marketing campaigns and have adopted more interactive strategies that encompass a wider variety of marketing channels [17]. Multiple-channel marketing (MCM) allows businesses to use specific media to market directly to a target audience [18]. Companies must recognize the wide array of channels that can influence consumers, including television, radio, magazines, and online sources. Organizations are starting to focus more on the possibilities of new-media marketing [19].

Businesses that decide to participate in MCM strategies must carefully consider the most efficient and effective channels [18]. Efficiency focuses on the cost per impression or the ability of a channel to reach consumers as economically as possible. In order to do so, marketers must have a clear and full understanding of its unique customer base. Multiple channel marketing must also be effective and yield high sales and positive brand image [18]. Modern businesses are using multiple traditional and new-media channels to market to consumers. Ultimately, the decisions on which channel to use are often the result of organizational tradition and "gut feeling" rather than statistical proof [20].

Marketing campaigns via new-media are free or low cost, and if used correctly, could lead to further promotion [21]. Properly integrating social networking tools can have a positive impact on sales, powerfully establish a company's brand, increase the salience of the business, position the company positively within the community, and reduce advertising costs [22]. However, sufficient and effective measurement practices must be implemented to determine if social-media marketing is successful and yielding a positive return on investment (ROI) [23,24]. Such measurement programs should focus on a social-media marketing campaign, and its ability to raise brand awareness, generate sales, produce customer advocacy, or encourage word-of-mouth marketing [25].

The purpose of this study was to explore the experiences of garden center stakeholders in the Great Plains region of the USA as they use social-media to market their business. Semi-structured, in-depth interviews of Kansas stakeholders explored the following research questions.

Q1: What are garden center stakeholder's perceptions and attitudes towards new-media as it relates to the marketing of their businesses?

Q2: What barriers do stakeholders encounter when using new-media to market their businesses?

This qualitative study is informed by Grunig's [26] Public Relations Theory. Grunig [26] categorizes four models of communication that businesses and public relations (PR) practitioners rely upon: (1) press agentry; (2) public information; (3) two-way asymmetrical; and (4) two-way symmetrical communication. Model one, press agentry, is the least desirable and model four, two-way symmetrical, is the most desirable form of communication. Grunig offers these models to help classify how a business or organization approaches and practices PR.

Press agentry is narrow in focus. Practitioners of this form of communication are primarily concerned with disseminating information on the company's products and increasing brand awareness [26]. Companies that practice press agentry are not bound by truth, and all communication is asymmetrical and focused on a one-way transfer of information. There is no desire for feedback or understanding the customer through strategic research. The public information model evolved from the press agentry in that it focuses on the release and distribution of truthful information [26]. However, the flow of information is still one-way from the organization to the consumer. Unlike press agentry, there is some effort given toward understanding the receiver of information through items like surveys [26].

Model three and four are considered the more desirable models of PR [27]. Model three is the two-way asymmetrical approach. While this form of PR evaluates feedback from a company's target audience, the goal of communication is strictly focused on persuasion and convincing the public to either accept a specific point of view or coerce the consumer to purchase a particular product [26].

The final model is two-way symmetrical communication, and "research shows this model is the most ethical ... and effective approach to public relations" [26] (p. 308). Two-way symmetrical communication establishes constant communication between the business and all stakeholders to mitigate conflict. Businesses do this by understanding the needs and wants of stakeholders to "improve understanding and build relationships with publics" [26] (p. 39). Additionally, small-scale operations are more likely to use two-way communication practices [26]. In the digital sphere, two-way symmetrical communication can help organizations because listening to consumers via social-media allows a company to improve its products and more effectively target potential customers [28].

2. Materials and Methods

This case study used six in-depth interviews with participants from four garden centers. The six participants (Table 1) were two more than the minimum number needed for a qualitative study as identified by Creswell [29]. The participants at each garden center (Table 2) included the owner and/or the employee most responsible for social-media marketing content. All subjects gave their informed consent for inclusion before they participated in the study. The study was conducted in accordance with the Declaration of Helsinki, and the protocol was approved by the Committee for Research Involving Human Subjects/Institutional Review Board for Kansas State University (project #7183) on 19 May 2014.

Table 1. Characteristics of owners and employees at four garden centers in Kansas that were engaged in social-media marketing for their business.

Participant Description	Store
Employee A works at garden center A. She graduated from Kansas State University with a degree in landscape design and took a class in general business marketing. She is the sole landscape designer for the garden center and is also the marketing manager. She uses Facebook and Pinterest for her personal social-media.	A
Owner A owns garden center A. He spent the majority of his career farming. However, when faced with the difficulty of finding a way for the farm to support his children and his retirement, he decided to build a garden center. He does not use social-media in his personal life.	A
Manager B is the general manager of garden center B, and he oversees all of the marketing. Manager B does not use social-media for personal use.	B
President C is the fourth-generation manager of garden center C and received a master's degree in business administration. His current role is president of the garden center. He oversees the operations and marketing of the garden center. He uses Facebook in his personal life.	C
CEO C is the third-generation manager and is the current CEO of garden center C. He identified his primary responsibilities as helping with daily operations, preparing new-media content, and taking pictures for marketing purposes. He operates two blogs for the garden center and has a personal blog.	C
Owner D, of garden center D, works alongside her husband. Her primary responsibilities are with customer service and education. She is also the sole manager of the Facebook page and is in charge of television and radio advertisements. She uses Facebook in her personal life.	D

Table 2. Characteristics of and marketing channels used by four garden centers in Kansas.

Store	Description	New-Media	Traditional Media	Facebook Stats
A	Garden center A is located in Northwest Kansas. There are two other satellite garden center business locations in Nebraska. In addition to offering retail plant material to customers, the garden center also offers landscape design and construction services and does approximately 20% of its sales online through eBay or Amazon. The center is owned by one individual.	B,E,F,G,H,P,T	Radio Billboards Newspaper Direct mail	916 likes 0.07% engagement rate
B	Garden center B is located in Eastern Kansas, and was established in the 1950s. It has gone through several ownership changes. The primary revenue source for the garden center is in retail sales of plant material and gardening supplies such as fertilizer and weed killer.	E, F	Radio Newspaper Direct mail	818 likes 1.3% engagement rate
C	Garden center C is located in Southcentral Kansas, and is in its fourth generation of ownership. The primary focus of this garden center is in retail sales split across two locations in Wichita. In addition to retail plant supplies, the garden center also runs a gift store and a microbrewery store.	B, E, F, I, P, T	Radio Television Newspaper Direct mail	5440 likes 0.14% engagement rate
D	Garden center D is located in Western Kansas and is currently in its first generation. The store focuses on retail plant supplies while a year-round gift shop is also a significant aspect of the business.	F	Radio Television Newspaper Direct Mail	844 likes 1.09% engagement rate

Note: B = blog, E = e-newsletter, F = Facebook, G = Google Plus, H = Houzz, I = Instagram, P = Pinterest, & T = Twitter; engagement rate was calculated on 24 October 2014.

A purposively-selected list of 23 garden centers was generated by a state university Cooperative Extension horticultural specialist with expert knowledge of existing Kansas garden centers. To be included in the list, the garden centers had to be located in Kansas, have exceptional products, good business practices, great customer service, and a presence on Facebook. Since qualitative studies focus on validity and generating a large amount of data from a few participants, the original list of 23 garden centers was scaled down to four garden centers. Two garden centers were selected for a high engagement rate on Facebook and two garden centers were selected that had poor engagement rates. The level of engagement was determined by using Simply Measured's [30] engagement metric which is defined as: engagement rate = (comments + likes + shares)/total number of fans. Simply Measured's [30] engagement rate allows accurate comparisons between Facebook pages. Each of the 23 garden center's previous 60 days' worth of posts were averaged and garden centers were ranked from highest to lowest engagement rate.

Participants were immediately debriefed by the researcher at the end of the interview. Interviews were transcribed by the researcher and a professor's assistant and were entered into NVivo10 (QSR International Pty Ltd., Doncaster, Victoria, Australia) for coding and analysis to determine common linkages and themes. Glaser's [31] constant comparative method assisted the researcher in categorizing participant responses into relevant major themes. Credibility, reliability, and transferability are essential components and concerns of a qualitative study, and the onus is on the researcher to demonstrate the findings result from data and not subjectivities [32]. Shenton [32] also indicates that compromising internal validity is a critical error in qualitative research. In order to mitigate any errors that could decrease credibility, all data was collected and analyzed verbatim with audio recordings and transcribed by the primary author and an assistant. Additionally, after concluding the interview sessions, all participants were debriefed by a researcher to maximize accuracy of the written data as synonymous with participant perception. The research team conducted face-validity analysis of the interview questions to increase validity of the results. External validity in qualitative research is in the eye of the beholder, and it is up to the reader to determine if the information can be generalized to his or her own socially constructed experiences [33].

Although in-depth interviews can yield rich and meaningful data in exploring the experiences of participants, caution should be used in generalizing the findings beyond the specific units of analysis under the specific situations in which they were observed [34]. However, qualitative results may be transferable to other like businesses in similar situations.

3. Results

3.1. *Q1: Stakeholder Perceptions and Attitudes towards New-Media Marketing*

When asked to describe how garden centers market to the public, participant responses yielded two themes: (1) Stakeholders prefer to focus on traditional marketing strategies; (2) Although stakeholders see some positives to social-media marketing, they are skeptical of its ability to positively impact sales.

3.1.1. Stakeholders Prefer to Focus on Traditional Marketing Strategies

Garden center owners and employees indicated a preference for traditional forms of advertising which included television, radio, newspaper, and direct-mail campaigns. Owner D (Table 1), who owns garden center D (Table 2), said, "garden centers are used to being in the regular media." She continued, "[the] newspaper is timely . . . If I advertise in the newspaper I can get them in here; they will bring the coupon in. No one brings their iPhone in and says this is what I want." Manager B, general manager of garden center B, mentioned, "we do a lot of radio advertising . . . we can run radio advertisements, and I can quantify how much I've spend on it because I have the bills to show for it."

The vast majority of strategic planning for garden center marketing also focused on traditional media. President C, of garden center C, talked about his advertising calendar:

> [it has] the number, date, the Monday through Sunday, how we would run our dates, and then at the top of all these we have what we want to promote and seminars. It's really kind of like our Bible. It's got what our spot radio's gonna run. If we're going to run a newspaper that week, if direct mail needs to go out.

Manager B also discussed an in-depth level of planning for advertising:

> [I will] plan out my marketing for next year. The majority of the marketing will get planned out for next year. [It will include] when I'm going to run ads, when we're going to do this, when we're going to do that.

All participants had some form of presence on one or more social-media platforms, with the most popular being Facebook. This is most likely due to the sampling procedures used in this study that drew upon garden centers with an active Facebook page. Other networks used, although to a varying degree, were Twitter, Pinterest, Instagram, Google Plus, blogs, and Houzz.

Participants at three of the four garden centers identified the preferred method for Web 2.0 marketing was through an e-newsletter. Employee A said, "we send out a newsletter every week to all of our local customers. I like to do the newsletter Friday evening, so I can put the new blog on the newsletter." Describing his newsletter, Manager B mentioned, "the e-newsletter is something we've been doing for several years. That gets [the most] attention. We do that every two weeks year round."

President C talked about the weekly newsletter and said, "it goes out weekly and [CEO C] writes those articles . . . He's a good story teller. It's not just a here-we-are company yelling buy our stuff. He'll write a story that's interesting and maybe try to tie a product in with it. It's about a 350-word read." The newsletter has a subscription of approximately 15,000 people and is delivered through Constant Contact, Inc. (Waltham, Massachusetts), which is an e-newsletter program.

Participants varied in the degree to which they used social-media and all were skeptical regarding the ability of social-media to generate a return on investment (ROI). However, participants mentioned the ability for Facebook to facilitate WOM marketing. Discussing why his garden center uses Facebook, Owner A mentioned:

We're too rural. We don't have enough people who could possibly drive two hours here ... I think enough people will come here from enough distance. When they go home they're going to tell their friends about it on social-media. They'll buy from you online because they won't drive that distance ... It's extremely important to [rural garden centers]. I feel it should be more important to us than people in the middle of the city, because we don't have enough demographics. The population isn't here to support how we want to live ... To support that business we have to attract people from a greater distance. Social-media is one way to attract people from the urban area.

Owner D also spoke of the ability of Facebook to generate WOM marketing and offered the following unprompted response, "there's no difference between WOM, us talking, and social-media ... It's the same thing. You're just missing the verbal and non-verbal cues." When prompted, President C also identified social-media could be viewed through the lens of WOM marketing and said, "we could do a better job of building that piece. I think if we were to do that, it would bring some value." Participants indicated a passive strategy for facilitating WOM marketing for their customers, and none of the owners or employees mentioned fostering interaction on social-media to create highly engaged customers.

3.1.2. Stakeholders Were Skeptical of New-Media Marketing Return on Investment (ROI)

Although Kansas garden centers are currently using social-media to some degree and believe it could help facilitate WOM marketing, all participants were highly skeptical of its ability to generate a ROI. When asked how her social-media presence affects the profits of the garden center, Owner D replied:

To be able to tell you it has made me one single dime, I can't. I don't have any way to track it ... [Facebook] has just not been the big boom that I need for me to go spend money on it ... Social-media sometimes is not a help. It doesn't get me stuff sold because the customer is still outside my store ... I'm spending a lot of time on [Facebook], and I cannot justify the amount of time being spent on it for the sales [that are being generated].

Other participants had similar viewpoints. When asked how social-media impacts the garden center, Employee A replied, "there's not often direct sales from [social-media]. If there are, they are really hard to track. It's just generating awareness. [The financial impact] is not much, and it is not direct." Regarding social-media being profitable to his business, Manager B mentioned if you post on "Facebook and you don't sell anymore this week than you did the week prior, then obviously it didn't strike a chord with anybody."

3.2. Q2: What Barriers Do Participants Encounter when Using New-Media to Market Their Business?

Participants were asked questions related to the challenges they face and what materials would help them improve new-media marketing of their business. Participant responses yielded the following themes: (1) Stakeholders lack time and training; (2) Stakeholders desire high-touch channels of education from experienced professionals.

3.2.1. Stakeholders Lack Time and Training

All participants identified the primary barrier to using social-media marketing was a lack of time. Specifically, stakeholders mentioned other job priorities related to the daily operations of the garden center and the large amount of time educating customers as areas that consume the most amount of time. When asked about her role in the garden center, Employee A stated:

I'm in charge of all the marketing and the advertisements. Other than that, my main role is a landscape designer, which works more with the landscape contractor side of the business. It's all under one head, but it's two very separate branches. We all have other jobs, so marketing just isn't . . . it's more my job than anybody else's, but it's not my only job nor is it my most important job.

Even though Manager B identified that his role as general manager of the garden center is to oversee and supervise all advertising, he stated, "[my other responsibilities] are 110% everything [but marketing]." When asked how much time he believed social-media marketing would take, he responded, "lots of time . . . and we just don't have a lot of time with it." When prompted to give a quantitative assessment on the time required to effectively market with social-media, Manager B identified "probably five to ten minutes every day."

Participants at three out of the four garden centers felt they were hindered by the amount of time spent educating potential and existing customers. Manager B mentioned helping customers with questions through the phone or via email "sometimes makes up 10% of the day, or 20% sometimes . . . if I kept track it would probably scare me." Owner A offered similar experiences to those of Manager B. "[Educating the consumer] is what I do all day long. It's my job, my biggest role. It's full time. I do more of that than anything else."

All participants identified a feeling of being lost in an ever-changing world of social-media and felt they did not have the necessary tools or training to keep up. Employee A mentioned her confusion with Facebook advertising and posts not being seen by every follower:

They're pushing more and more in a direction where you're going to have to pay for people to see your post . . . It seemed like it costs a lot of money, and we were confused and weren't understanding how it was being used or why we were getting charged . . . it didn't seem to correlate. It was confusing.

Owner D also identified feeling confused when it comes to Facebook updates. She mentioned, "[Getting up to speed] is the biggest problem I have with social-media. I still have a slide phone. When it comes to paid marketing, is that where I want to go?"

When asked about their desired learning method for new-media marketing training, all stakeholders mentioned a desire for hands-on, high-touch channels of education. Describing what the ideal coaching situation would look like, Employee A added: "Maybe a weekly phone call . . . First [call] would probably be a long one to discuss the overall plan and then like the weekly communication on, what have you done this week, what are you working on, and should maybe try this or that. Just someone to kind of [give you] feedback and keep accountability with." When asked to describe his ideal workshop, CEO C explained it would be a workshop where participants would, "take your laptop to the class and sit down. Actually go through the steps and build a website or whatever you're doing. The [goal would be] a finished blog or website at the end of the course".

One common characteristic participants desired with regards to learning about social-media was to seek out advice from people who, as President C mentioned, are "fighting the same fight" within the garden center industry. Manager B identified that he preferred to learn from events at trade shows or industry meetings, saying, "I attend trade shows, meetings, and hear what other garden centers do . . . If I heard something at a conference, colleagues that are doing something similar . . . I would probably connect with that more than anything" President C echoed this sentiment:

I guess there's that sense of trust . . . it's people that are fighting the same fight that we are. That we're able to learn from what they're doing . . . I don't hold a whole lot of credence for those that call themselves a social-media expert just because it's . . . you can't quantify it. I could go out and say that I'm a social-media expert, read a couple books and probably sound like I know what I'm talking about. The people that have actually been there and done that I think to me have more credibility.

4. Discussion

Participants identified a preference and confidence for traditional marketing channels that included radio, newspaper, television, and print media. This proclivity towards older methods of advertising is in agreement with the findings of Behe et al. [6] and Stone [35]. The preference for older forms of mass communication could demonstrate that garden center stakeholders are contrasting the recommendations of Behe et al. [16] in adopting digital marketing trends to reach the upcoming generation, and marketing strategies have remained the same for nearly 20 years. This could also lend additional support to the findings of Doctorow et al. [20], who identified that decisions for MCM campaigns are often the result of tradition.

Garden center employees and owners were also concerned about the lack of ROI in regards to the time spent marketing on social-media. However, stakeholders were measuring the success of their social-media campaigns by looking at a direct and immediate increase in sales after content was posted online. Since they do not see immediate or direct financial impacts, stakeholders indicated that they do not believe social-media can impact sales. This contrasts the recommendations of Paine [28] who states companies that are the most active on social-media are more profitable than their contemporaries which are not using social-media. Although social-media can have an impact on sales, the greatest impact results from encouraging interaction and developing meaningful and symbiotic relationships [25]. Stakeholders of this study were not focusing on, or measuring, the quality of relationships, level of interaction, or the satisfaction of customers online, which is contrary to the advice and findings of Ledingham [36]. This common perception may indicate that stakeholders are practicing PR through press agentry or public information models [26] and not the two-way symmetrical approach recommended. Since the relationship and awareness benefits can lead to profits that are not directly measurable [37], garden centers most likely are measuring the wrong forms of profit or revenue streams and becoming frustrated with the marketing efforts via new-media.

Garden center stakeholders also demonstrated a lack of understanding for traditional media and were not aware of the potential benefits and analytics of new-media marketing. For example, Owner D stated that advertising in the newspaper was "timely". Furthermore, Manager B had mentioned his preference for radio advertising because he could quantify his advertising reach by determining how much he spent on radio advertising and how it affected the sales for the week. However, new-media marketing is much more rapid in its delivery and response than newspaper, and quantifying the dollars spent on a radio campaign cannot guarantee a consumer has noticed a message. New-media marketing offers advanced analytics that extend beyond simple message reach to include multiple forms of engagement along the online consumer pathway. Furthermore, stakeholders focused on what Keller [18] defined as the efficiency of the advertising message and were not actively tracking the effectiveness of such advertising campaigns. Measurement focused specifically on the short-term sales increase and not the long-term brand awareness or relationship.

Employees and owners were also confused about how to track sales to determine advertising effectiveness. None of the participants indicated asking customers where they heard about sales or promotions or giving any type of survey to determine relevant marketing channels or WOM marketing referrals. This could be especially problematic in tracking the effectiveness and efficiency of social-media advertising and the WOM that comes with it. By not implementing such tracking measures, the participants may never know how effective their social-media marketing efforts are nor how to identify profitable marketing channels to efficiently reach market segments. Although small businesses are more apt to practice two-way symmetrical communication [26], the participants in this study believed social-media should be approached from a public information or two-way asymmetrical communication viewpoint.

The employees who had responsibilities related to social-media had, at best, a split role that involved other garden center duties. These responsibilities quickly overshadowed the marketing responsibilities of the employee. Since "success on social-media is contingent on considerable resources being allocated to the proper use and evaluation" [38] (p. 4), it is possible to conclude stakeholders are seeing little ROI on new-media because they have not fully committed the resources vital to success.

Garden centers are approaching new-media marketing from the same lens as mass communications advertising. The stakeholders identified that they were taking a "broad net" approach to new-media marketing where they send a message out to numerous receivers and hope that results in a purchase. However, this approach of treating new-media like mass communications is in violation of Warshauer and Grimes' [39] findings, which state that social-media should be used for fostering individualized communication and interaction.

Employees and owners stated the majority of their time is spent educating customers through e-mail, phone calls, or in-person conversations. This level of personal interaction could indicate that garden center employees and owners are practicing two-way symmetrical communication offline as an organization. According to employees, customers appreciated a high level of service. However, that level of service also prevented participants from effectively marketing the store because educating customers represented a considerable portion of their time. The stakeholders within this study also had a lack of understanding regarding scheduling and publishing tools for new-media marketing. Only one participant mentioned Hootsuite (Vancouver, Canada) or the scheduled posts feature on Facebook, and she did not use these features. Participants were not actively seeking new information but were not opposed to learning about new-media marketing. If they are going to learn, they expressed a desire for high-touch channels of education from seasoned industry professionals.

5. Conclusions

This study offers several theoretical implications for Excellence in Public Relations theory and how garden centers approach PR in the digital sphere. Grunig [26] identified a two-way symmetrical model of communication as the most effective means of communication between stakeholders. Since social-media is an effective avenue for conducting research and communicating to customers [24,28], this study adds to the body of literature and theory by suggesting that engagement and interaction on social-media could diminish when businesses are not actively participating in two-way symmetrical communication online and do not understand the value it offers beyond direct sales. New-media marketing could garner additional business over time by building a loyal customer base.

Garden center owners and employees should consider implementing principles of two-way symmetrical communication in new-media marketing, and approach it not as a sales tool but, as Constandinindes and Fountain [13] describe, a medium for communicating and engaging directly with potential customers in order to build relationships. In doing so, stakeholders may harness the power of new-media to generate deep involvement with customers. Because customer interaction on social-media can be profitable [40] and WOM can reach an enhanced volume of potential customers for minimal costs [21], using new-media channels could help garden centers that are hindered by resources or geography to reach new target audiences.

Participants also identified using MCM, which included new-media, to reach their target audience. However, the bulk of their efforts focused on traditional marketing that included radio, television, newspapers, and direct mail. Although new-media marketing was used, it was often an afterthought. The popular response for why the stakeholders emphasized traditional media was a mixture of tradition and feeling like they could quantify traditional media. However, stakeholders were not using any form of analysis to determine the effectiveness or efficiency of their marketing efforts. Although stakeholders may be reaching a large number of their target audience via direct mail, radio, and television campaigns; they could be neglecting a very important demographic by ignoring the potential of new-media marketing, which is becoming more vital as traditional forms of media become increasingly segmented. Therefore, this paper recommends that garden center owners and employees implement measurement programs to determine the effectiveness and efficiency of marketing efforts and not rely on traditional or intra-organizational culture to make marketing decisions. Communicators should work to reach this market of garden centers to educate stakeholders on the value of new-media marketing.

This study recommends that future research focus on consumers' perceptions and preferences toward new-media marketing. Since educational and relevant content is paramount to consumers, we recommend identifying content that garden center customers desire as well as which aspects of relationship marketing resonate most. Future research should also identify which new-media platforms are yielding the greatest ROI in regards to increased sales, increased reputation, and increased relationships. Lastly, studies should focus on strategies that are being implemented by garden center stakeholders, how customers perceive those strategies, and how such activities can improve customer loyalty and foster meaningful relationships.

Acknowledgments: This research was supported by the United States Department of Agriculture—Agricultural Marketing Service—Federal State Marketing Improvement Program (number 11402984), James L. Whitten Building 1400 Independence Ave., S.W. Washington, DC 20250, USA. Contribution no. 17-198-J from the Kansas Agricultural Experiment Station. The authors wish to thank Janis Crow (Department of Marketing, Kansas State University, Manhattan, KS, USA) for her guidance and contribution to the graduate committee.

Author Contributions: Scott Stebner planned, executed, and analyzed the study, which involved coordinating with garden center stakeholders, traveling to conduct interviews, transcribing interviews, and determining themes within analysis software and writing. Cheryl Boyer, Lauri Baker, and Hikaru Peterson obtained funding, helped design the study, gave guidance on analysis, and assisted with manuscript writing.

Conflicts of Interest: The authors declare no conflict of interest.

References

1. USDA. 2012 Census of Agriculture. 2014. Available online: https://www.agcensus.usda.gov/Publications/2012/Full_Report/Volume_1,_Chapter_1_US/usv1.pdf (accessed on 1 May 2014).

2. Hodges, A.W.; Hall, C.R.; Palma, M.A. Economic contributions of the green industry in the United States in 2007–08. *HortTechnology* **2011**, *21*, 628–638.

3. Garber, M.P.; Bondari, K. Retail garden outlets: Business characteristics and factors affecting industry performance. *J. Environ. Hort.* **1998**, *16*, 15–19.

4. Hodges, A.W.; Khachatryan, H.; Hall, C.R.; Palma, M.A. *Production and Marketing Practices and Trade Flows in the United States Green Industry, 2013*; University of Florida Agricultural Experiment Station: Gainesville, FL, USA, 2015.

5. Safley, C.D.; Wohlgenant, M.K. Factors influencing consumers' selection of garden centers. *J. Agribus.* **1995**, *13*, 33–50.

6. Behe, B.K.; Dennis, J.H.; Hall, C.R.; Hodges, A.W.; Brumfield, R.G. Regional marketing practices in U.S. nursery production. *HortScience* **2008**, *43*, 2070–2075.

7. Cui, G.; Wong, M.L.; Wan, X. Targeting high value customers while under resource constraint: Partial order constrained optimization with genetic algorithm. *J. Interact. Mark.* **2015**, *29*, 27–37. [CrossRef]

8. Low, S.A.; Vogel, S.J. *Direct and Intermediated Marketing of Local Foods in the United States*; Economic Research Report Number 128; United States Department of Agriculture, Economic Research Service: Washington, DC, USA, 2011.

9. Govindasamy, R.; Hossain, F.; Adelaja, A. Income of farmers who use direct marketing. *Agr. Resour. Econ. Rev.* **1999**, *28*, 76–83. [CrossRef]

10. Gönül, F.; Shi, M.Z. Optimal mailing of catalogs: A new methodology using estimable structural dynamic programming models. *Manag. Sci.* **1998**, *44*, 1249–1262. [CrossRef]

11. Mulhern, F.J. Customer profitability analysis: Measurement, concentration, and research directions. *J. Interact. Mark.* **1999**, *13*, 25–40. [CrossRef]

12. Ishiguro, H.; Amasaka, K. Establishment of a strategic total direct mail model to bring customers into auto dealerships. *J. Bus. Econ. Res.* **2012**, *10*, 493–500. [CrossRef]

13. Constantinides, E.; Fountain, S.J. Web 2.0: Conceptual foundations and marketing issues. *J. Direct Data Digit. Mark. Pract.* **2008**, *9*, 231–244. [CrossRef]

14. Mersey, R.; Malthouse, E.C.; Calder, B.J. Engagement with online media. *J. Med. Bus. Stud.* **2010**, *7*, 39–56. [CrossRef]

15. Verma, V.; Sharma, D.; Sheth, J. Does relationship marketing matter in online retailing? A meta-analytic approach. *J. Acad. Mark. Sci.* **2016**, *44*, 206–217. [CrossRef]

16. Behe, B.K.; Campbell, B.L.; Hall, C.R.; Khachatryan, H.; Dennis, J.H.; Yue, C. Consumer preferences for local and sustainable plant production characteristics. *HortScience* **2013**, *48*, 200–208.

17. Vargo, S.L.; Lusch, R.F. Evolving to a new dominant logic for marketing. *J. Mark.* **2004**, *68*, 1–17. [CrossRef]

18. Keller, K.L. Brand equity management in a multichannel, multimedia retail environment. *J. Interact. Mark.* **2010**, *24*, 58–70. [CrossRef]

19. Winer, R.S. New communications approaches in marketing: Issues and research directions. *J. Interact. Mark.* **2009**, *23*, 108–117. [CrossRef]

20. Doctorow, D.; Hoblit, R.; Sekhar, A. Measuring marketing: Mckinsey global survey results. *McKinsey Q.* **2009**, *2009*, 1–3.

21. Meyers, C.; Irlbeck, E.; Graybill-Leonard, M.; Doerfert, D. Advocacy in agricultural social movements: Exploring Facebook as a public relations communication tool. *J. Appl. Commun.* **2011**, *95*, 68–81.

22. Bolotaeva, V.; Cata, T. Marketing opportunities with social networks. *J. Int. Soc. Netw. Virtual Commun.* **2011**, *2011*, 1–8. [CrossRef]

23. Blanchard, O. *Social-Media ROI: Managing and Measuring Social-Media Efforts in Your Organization*, 1st ed.; Pearson Education: Boston, MA, USA, 2011.

24. Young, A. *Brand Media Strategy: Integrated Communications Planning in the Digital Era*, 2nd ed.; Palgrave Macmillan: New York, NY, USA, 2014.

25. Castronovo, C.; Huang, L. Social-media in an alternative marketing communication model. *J. Mark. Dev. Compet.* **2012**, *6*, 117–134.

26. Grunig, J.E. (Ed.) *Excellence in Public Relations and Communication Management*, 1st ed.; Lawrence Erlbaum Associates, Inc.: Hillsdale, NJ, USA, 1992.

27. Fearn-Banks, K. *Crisis Communications: A Casebook Approach*, 4th ed.; Routledge: New York, NY, USA, 2010.

28. Paine, K. *Measure What Matters: Online Tools For Understanding Customers, Social-Media, Engagement, and Key Relationships*, 1st ed.; Wiley & Sons: Hoboken, NJ, USA, 2011.

29. Creswell, J.W. *Qualitative Inquiry and Research Design: Choosing Among Five Approaches*, 1st ed.; Sage Publications: Chicago, IL, USA, 2007.

30. Simply Measured. Available online: http://simplymeasured.com/blog/2013/08/14/facebook-metrics-defined-engagement-rate (accessed on 21 December 2016).

31. Glaser, B. The constant comparative method of qualitative analysis. *Soc. Probl.* **1965**, *12*, 436–445. [CrossRef]

32. Shenton, A.K. Strategies for ensuring trustworthiness in qualitative research projects. *Educ. Inf.* **2004**, *22*, 63–75. [CrossRef]

33. Crotty, M. *The Foundations of Social Research: Meaning and Perspective in the Research Process*, 1st ed.; Sage Publications: Thousand Oaks, CA, USA, 1998.

34. Flick, U. *An Introduction to Qualitative Research*, 4th ed.; Sage Publications: Thousand Oaks, CA, USA, 2009.

35. Stone, K.E. Impact of the Wal-Mart phenomenon on rural communities. In Proceedings of the Increasing Understanding of Public Problems and Policies Conference, Chicago, IL, USA, 21–24 September 1997.

36. Ledingham, J.A. Explicating relationship management as a general theory of public relations. *J. Public Relat. Res.* **2003**, *15*, 181–198. [CrossRef]

37. Yue, C.; Dennis, J.H.; Behe, B.K.; Hall, C.R.; Campbell, B.L.; Lopez, R.G. Investigating consumer preference for organic, local, or sustainable plants. *HortScience* **2011**, *46*, 610–615.

38. Miller, R.; Lammas, N. Social-media and its implications for viral marketing. *Asia Pac. Public Relat. J.* **2010**, *11*, 1–9.

39. Warschauer, M.; Grimes, D. Audience, authorship, and artifact: The emergent semiotics of Web 2.0. *Annu. Rev. Appl. Linguist.* **2007**, *27*, 1–23. [CrossRef]

40. Jones, N.; Borgman, R.; Ulusoy, E. Impact of social-media on small businesses. *J. Small Bus. Enterp. Dev.* **2015**, *22*, 611–632. [CrossRef]

21

Economic Cost-Analysis of the Impact of Container Size on Transplanted Tree Value

Lauren M. Garcia Chance *, Michael A. Arnold, Charles R. Hall and Sean T. Carver

Department of Horticultural Sciences, Texas A&M University, College Station, Texas 77843-2133, TX, USA; ma-arnold@tamu.edu (M.A.A.); charliehall@tamu.edu (C.R.H.); borrichia@gmail.com (S.T.C.)
* Correspondence: lmgarcia06@gmail.com

Abstract: The benefits and costs of varying container sizes have yet to be fully evaluated to determine which container size affords the most advantageous opportunity for consumers. To determine value of the tree following transplant, clonal replicates of *Vitex agnus-castus* L. [Chaste Tree], *Acer rubrum* L. var. *drummondii* (Hook. & Arn. *ex* Nutt.) Sarg. [Drummond Red Maple], and *Taxodium distichum* (L.) Rich. [Baldcypress] were grown under common conditions in each of five container sizes 3.5, 11.7, 23.3, 97.8 or 175.0 L, respectively (#1, 3, 7, 25 or 45). In June 2013, six trees of each container size and species were transplanted to a sandy clay loam field in College Station, Texas. To determine the increase in value over a two-year post-transplant period, height and caliper measurements were taken at the end of nursery production and again at the end of the second growing season in the field, October 2014. Utilizing industry standards, initial costs of materials and labor were then compared with the size of trees after two years. Replacement cost analysis after two growing seasons indicated a greater increase in value for 11.7 and 23.3 L trees compared to losses in value for some 175.0 L trees. In comparison with trees from larger containers, trees from smaller size containers experienced shorter establishment times and increased growth rates, thus creating a quicker return on investment for trees transplanted from the smaller container sizes.

Keywords: *Acer rubrum*; *Taxodium distichum*; *Vitex agnus-castus*; gain; loss; landscape establishment; tree establishment

1. Introduction

Nurseries over the years have produced trees in increasingly larger container sizes [1,2]. Retail garden centers and even large box stores, such as Walmart®, Lowe's®, and Home Depot®, now sell trees in up to 378.5 L (#100) containers. While debate continues over the relative merits of different container sizes [2], this could in part be due to the appreciation that commercial and residential customers have for the instant impact large trees can provide, such as greater aesthetic value of larger trees [3,4], greater biomass present to withstand environmental anomalies [5], less potential for accidental or malicious mechanical damage [6], instant shade [3,4], and increase in property value [7]. However, these larger trees cost more to grow and occupy a greater amount of nursery space per tree over longer time frames than smaller trees resulting in higher costs of production for growers and higher prices for consumers [6]. Smaller container sizes are ultimately less expensive for consumers as nurseries expend less materials, maintenance costs, and allocate less square footage to produce smaller trees. Also, smaller container sizes, once transplanted to the field, have been reported to experience reduced transplant shock [2], are in a phase of growth more closely aligned with the exponential growth rate of young seedlings [8], have been in containers for shorter times and transplanted to larger container sizes fewer times potentially reducing the chances of circling root development [9], and their smaller size makes for easier handling and staking [6]. The economic benefits and costs of varying container sizes have yet to be fully evaluated to determine which container size affords the most advantageous opportunity for consumers.

The value of a tree, defined as its monetary worth, is based on people's perception of the tree [10]. Arborists use several methods to develop a fair and reasonable estimate of the value of individual trees [11,12]. The cost approach is widely used today and assumes that value equals the cost of production [13]. It assumes that benefits inherent in a tree can be reproduced by replacing the tree and, therefore, replacement cost is an indication of value [10]. Replacement cost is depreciated to reflect differences in the benefits that flow from an "idealized" replacement compared with an older and imperfectly appraised tree. The depreciated replacement cost method uses tree size, species, condition, and location factors to determine tree value [14].

The income approach measures value as the future use of a tree such as in fruit or nut production [15]. In the absence of such products, the income approach could be based on the monetary benefits of the future economic, environmental, and health well-being value of the tree [11]. For example, benefits have been shown to improve the value of the tree, including energy savings [16], atmospheric carbon dioxide reductions [17], storm water runoff reductions [18], and aesthetics [19]. Quantifying and totaling these benefits (ecosystems services) over time can provide an idea of a tree's projected value, but require data outside the scope of this project, thus a derivation of the replacement cost method was utilized within this study.

The objective of the current research was to determine the initial cost and replacement cost value of five different container sizes in three tree species at transplant and after two growing seasons in the landscape.

2. Materials and Methods

In analyzing the impact container size has on the value of the tree, the establishment cost of the tree was calculated and then compared to the replacement price of the tree after two growing seasons. Using the difference, it was then possible to see the net change in value for each container size tree over time. For the purposes of this study, price is the selling price paid by the customer buying the product, cost is the cost of care incurred by the homeowner in maintaining the product, and value is the bundle of attributes important to a homeowner in determining the product's overall worth. The three taxa utilized were selected to represent different niches of the landscape industry. Selections of *Vitex agnus-castus* L. (Chaste Tree), *Acer rubrum* L. var. *drummondii* (Hook. & Arn. *ex* Nutt.) Sarg. (Drummond Red Maple), and *Taxodium distichum* (L.) Rich. (Baldcypress) were chosen due to their widespread use in the southern USA nursery trade and their representation of a variety of classes of landscape trees. Additionally, five container sizes, 3.5 L (#1), 11.7 L (#3), 23.3 L (#7), 97.8 L (#25), and 175.0 L (#45), were selected as demonstrative of a range of typical container sizes purchased in the landscape trade. Clonal selections of these trees grown using as similar inputs as possible [20,21] were transplanted and monitored over the course of two growing seasons in a sandy clay loam (66% sand, 8% silt, 26% clay, 6.0 pH) field in College Station, TX (lat. 30°37′45″ N, long. 96°20′3″ W) beginning June 2013. All replicates of the 3.5 L *Acer rubrum var. drummondii* died within the first season due to deer grazing and pathogens and, therefore, are excluded from the cost analysis. Trunk diameters of all three species were within ANSI (American National Standards Institute) Z60.1-2004 specifications [22] for their respective container sizes [20].

2.1. Initial Costs

In order to analyze the value of the various sizes of the containerized trees, data were collected from 185 different nurseries located across 21 states. Nurseries were contacted and requested to provide wholesale prices of all container sizes available in *Acer rubrum* "Summer Red" or "Red Sunset", *Taxodium distichum*, and *Vitex agnus-castus* "Shoals Creek". Although not all nurseries carried all sizes of each species, data from a minimum of twelve nurseries were acquired for each species and container size combination.

Labor and installation costs are included in analyzing the initial value of a tree. RSMeans is the industry standard source for accurate and expert information on materials, labor, and construction costs [23]. Thus, labor and materials costs were determined utilizing this information. Labor and installation costs, both by hand and using machinery, were compiled for each container size from the RSMeans data. Additionally, twelve companies that produced each container size were contacted and asked to contribute their installation costs to corroborate the data from RSMeans benchmarks.

Finally, maintenance costs were determined by using maintenance records during the two growing seasons for each container size and species. These records were then compared to RSMeans for projected maintenance costs per container size over time. Maintenance included such practices as fertilizing, weeding, pest control, pruning and watering.

2.2. Equivalent Costs

To determine the equivalent value for replacement of the planted trees at the end of two growing seasons, data were collected from the locally-grown trees. Final height and trunk diameter of the trees in the field in October 2014 were utilized to determine ANSIZ60.1 [22] container size approximations. Utilizing these ending container size equivalents, prices were designated according to the mean prices obtained from wholesale growers. Additionally, costs of installation and maintenance were derived for the ending container size of each tree. By subtracting the ending container size costs from the beginning container size costs, the net gain or loss in value over the two post-transplant growing seasons were calculated for each tree.

Data were analyzed using analysis of variance (ANOVA) with JMP 2009 and SAS 9.3 (SAS Institute Inc., Cary, NC, USA) to determine the significance of interactions and main effects for each variable. The overall model was 3 species × 5 sizes with 6 replicates (observations) per treatment combination (Table 1). Means for container size, wholesale cost, installation, maintenance, and total value for each tree were analyzed as the change between the beginning and end of the experiment. Where interactions were significant, Student's t-test (Fisher's Least Significant Difference) was used to compare means among the treatment combinations. When significant main effects were found, a paired t-test comparison was used to indicate values that are significantly different ($p \leq 0.05$).

Table 1. Means and Analysis of Variance of the effects of tree species and initial container size on changes in size, price, costs, and value of trees after transplanting to the landscape and growing for two seasons.

Species	Initial Container Size (L)	Change in Container Size (L)	Change in Wholesale Price ($)	Change in Installation Cost ($)	Change in Maintenance Cost ($)	Gain/Loss in Value ($)
Acer rubrum	11.7	46.5 ± 12.1 [a,b]	45.2 ± 9.2 [a,x,y]	52.4 ± 4.6 [a]	3.8 ± 1.4 [a]	121.4 ± 15.3 [a]
	23.3	49.2 ± 8.3 [a]	38.5 ± 7.2 [a]	20.2 ± 3.9 [b]	5.6 ± 1.2 [a]	94.0 ± 12.4 [a,b]
	97.8	12.4 ± 12.4 [b]	10.1 ± 10.1 [b]	4.9 ± 4.9 [c]	2.0 ± 2.0 [a]	17.1 ± 17.1 [b]
	175	12.4 ± 12.4 [b]	18.0 ± 18.0 [a,b]	4.8 ± 4.8 [c]	9.7 ± 9.7 [a]	0.0 ± 32.5 [b]
Taxodium distichum	3.5	1.8 ± 1.8 [c]	2.0 ± 1.3 [c]	6.9 ± 5.0 [c]	0.2 ± 0.1 [b]	−38.4 ± 6.5 [b]
	11.7	29.5 ± 5.6 [b]	26.0 ± 4.9 [b]	42.5 ± 6.0 [a]	1.8 ± 0.3 [b]	67.3 ± 11.0 [a]
	23.3	55.2 ± 7.9 [a]	46.2 ± 6.6 [a]	23.1 ± 3.6 [b]	6.6 ± 1.2 [a]	68.0 ± 11.5 [a]
	97.8	12.4 ± 12.4 [b,c]	11.5 ± 11.5 [b,c]	4.9 ± 4.9 [c]	2.0 ± 2.0 [b]	−6.6 ± 18.4 [a,b]
	175	0.0 ± 0.0 [c]	0.0 ± 0.0 [c]	0.0 ± 0.0 [c]	0.0 ± 0.0 [b]	−45.0 ± 0.0 [b]
Vitex agnus-castus	3.5	65.4 ± 11.3 [b]	53.8 ± 10.0 [b]	74.4 ± 3.9 [a]	6.1 ± 1.3 [a]	132.9 ± 15.2 [b]
	11.7	127.1 ± 20.4 [a]	138.5 ± 27.0 [a]	82.7 ± 8.3 [a]	15.5 ± 3.2 [a]	235.8 ± 38.6 [a]
	23.3	80.6 ± 12.4 [a,b]	77.1 ± 18.4 [a,b]	33.9 ± 4.9 [b]	10.5 ± 2.0 [a]	120.3 ± 25.4 [b]
	97.8	50.3 ± 15.8 [b]	73.8 ± 23.3 [a,b]	19.6 ± 6.2 [b]	8.1 ± 2.5 [a]	101.6 ± 32.1 [b]
	175	14.0 ± 14.0 [c]	12.4 ± 12.4 [c]	4.8 ± 4.8 [c]	9.7 ± 9.7 [a]	−22.6 ± 28.5 [c]
Species		***	*** z	***	***	***
Container Size		***	***	***	***	***
Species * Container Size		*	n.s.	*	n.s.	*

[x] Standard errors, with different letters ([a,b,c]) indicate significant differences using Students t-test at $p \leq 0.05$ within each species; [y] Values within a column represent the mean of six observations ± standard errors; [z] *, *** Indicate significance of the main effect or interaction at $p \leq 0.05$, 0.001, respectively, or not significant (n.s.).

3. Results and Discussion

3.1. Initial Costs

Prices for a range of sizes of commercial container stock were obtained. Similar price trends existed for all three species (Figure 1). They were lowest for the 3.5 L trees and then slowly increased in price until the 56.8 L trees. Trees greater than 56.8 L tree stage were increasingly expensive compared to the smaller trees. While *V. agnus-castus* was slightly less expensive in the smaller container-grown trees, it became much more expensive in the larger container-grown trees than with the other two species. Higher prices associated with trees greater than 56.8 L would indicate the price point at which nursery growers must increase the prices to a higher rate to offset extra supplies, labor, and inventory carrying costs required to maintain larger container sizes.

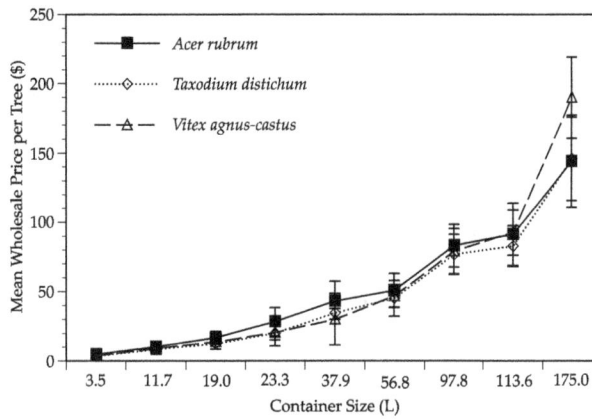

Figure 1. Mean (±standard error) wholesale price [US$] by container size for three tree species (*A. rubrum*, *T. distichum*, and *V. agnus-castus*) in 2013 where n ≥ 12.

Similar trends were observed with the costs to transplant each container-grown tree (Figure 2). The cost to transplant increased gradually with each container size. The 56.8 L container size trees indicated another break point as the cost to transplant by hand was more cost-efficient than by machinery until this point. With 97.8 L and 175.0 L trees, machinery would be necessary to efficiently transplant these trees. Additionally, the 175.0 L trees were eight times more expensive to transplant than 3.5 L trees.

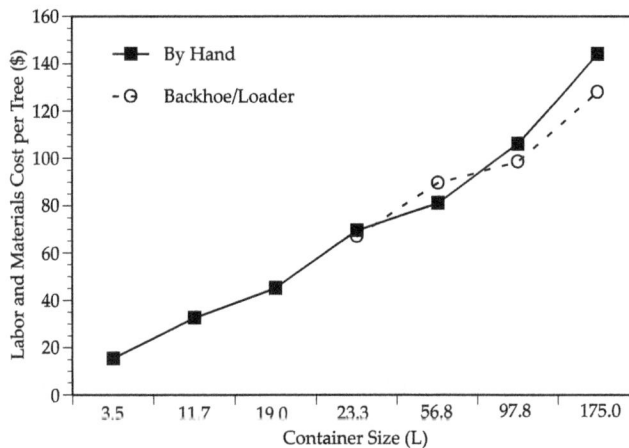

Figure 2. Labor and materials cost [US$] per tree for transplant by hand or machinery of various container size trees in 2013 (excluding wholesale cost of tree) as determined from RSMeans [23].

The maintenance costs for each container size were determined using general practices tree owners would implement during a typical year. This included fertilization, pest control, weeding, pruning, and watering. Cost of fertilization, pest control, and weeding remained nearly constant across all container size trees (Figure 3). However, the cost of pruning increased beginning at container sizes greater than 56.8 L with trees from 175.0 L containers requiring the most pruning labor. Finally, watering costs were relatively similar across all container sizes; however, a slight increase was found for the watering costs of larger container sizes. Despite more water being applied to larger container-grown trees, the current low cost of water mitigates the impact of this differential input. If in future years the cost of water increases, more substantial differences in cost of watering different container-grown trees could become apparent. Regional variation in water costs may also impact this estimate.

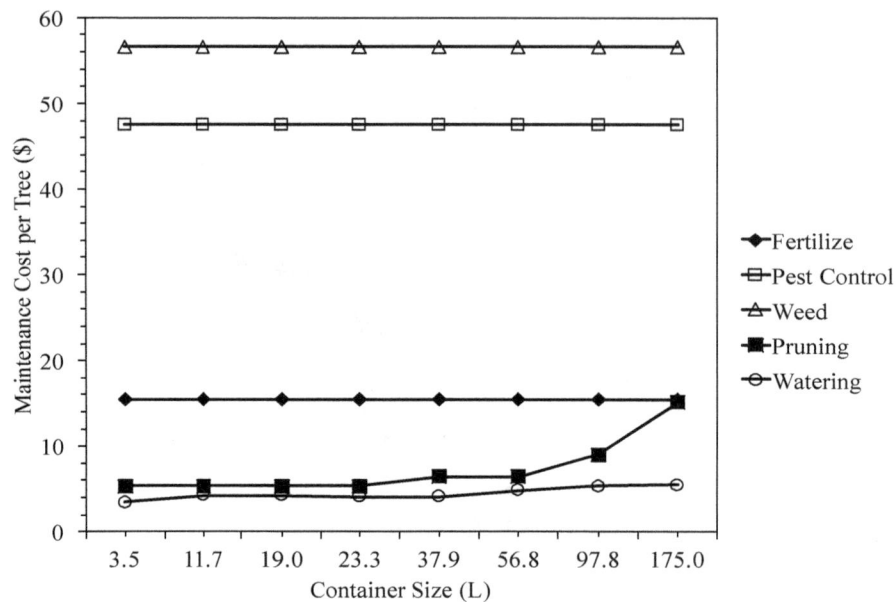

Figure 3. Maintenance costs [US$] per tree for fertilization, pest control, weeding, pruning, and watering of various container sizes summed over a two-year period of growth as determined by RSMeans [23].

3.2. Equivalent Costs

In order to predict the future value of each tree, height and trunk diameter at the end of the second growing season were compared to ANSIZ60.1 [22] to determine equivalent size container-grown trees. Given the different growth rates of the three species of tested trees, the value varies depending on species [20]. Growth and value may also differ among planting sites; however, data from first-year establishment of these species in contrasting environments in Texas and Mississippi indicated similar growth trends [21].

The main effects of species and container size were highly significant for all variables and the interaction between species and container size was significant for changes in installation costs, changes in container sizes, and net gain/loss (Table 1). Therefore, results are presented by species.

The greatest container size changes for *A. rubrum* occurred with the 11.7 L and 23.3 L trees which ended the second growing season at mean sizes of 56.8 L and 75.7 L, respectively (Figure 4A; Table 1). In contrast, 97.8 L and 175.0 L trees ended with very little change from their initial container sizes. Both 97.8 L and 175.0 L *A. rubrum* ended the second season with only one of the six replications increasing their equivalent container size (data not shown).

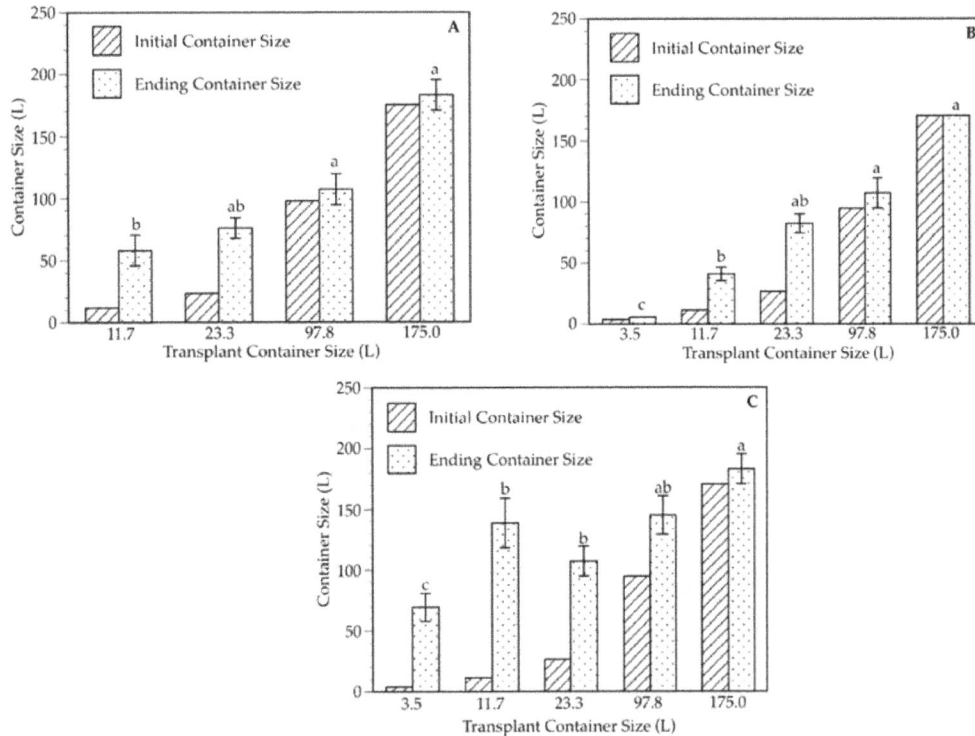

Figure 4. Mean (±standard error) of initial and ending container size of *Acer rubrum* (**A**); *Taxodium distichum* (**B**); or *Vitex agnus-castus* (**C**) trees from transplant (diagonal hatching) to the end of the second growing season (stippled hatching). Initial sizes were 3.6, 11.7, 23.3, 97.8 and 175.0 L; n = 6 or *T. distichum* and *V. agnus-castus* and 11.7, 23.3, 97.8 and 175.0 L; n = 6 for *A. rubrum*. Means of ending container sizes with the same letter are not significantly different at $p \leq 0.05$ using Student's *t*-test.

To predict the gain or loss in value over two growing seasons, the wholesale price of the tree at planting is shown with the wholesale price equivalent of the tree at the end of the second growing season (Figure 5A). The 11.7 L and 23.3 L trees had the greatest increase in replacement price, while the 97.8 L and 175.0 L barely increased (Table 1). Analyzing the cost to install the initial container size versus the cost to install the ending container size after two growing seasons also indicated that costs were lower for 11.7 L and 23.3 L container sizes, but with the greatest increase in installation costs of equivalent trees after two seasons (Figure 5B). Finally, maintenance costs remained steady for the two growing seasons with no differences between container size trees (Figure 5C).

This information allowed analysis of the overall value of the tree. The value of the tree increased the most for the 11.7 L trees of *A. rubrum*, yet the ending value was still not equal to the value of the 175.0 L trees (Figure 5D; Table 1). Therefore, while overall gains were largest for 11.7 L and 23.3 L trees (Figure 5E), 175.0 L trees still maintained the greatest overall value after two growing seasons (Figure 5D). Trends over longer time frames are unknown but suggest trees from smaller sizes may catch up to those from larger size containers if the same growth trends continue.

The stress and initial growth rates of *A. rubrum* greatly influenced final container sizes at the end of the two growing seasons of this study. The increased container sizes ultimately increased the wholesale cost of the equivalent tree, the cost of labor, and the cost of maintenance. Therefore, overall value of the tree was increased, although the final value of the smaller container sizes did not catch up to or surpass that of the larger container sizes for *A. rubrum* during the first two growing seasons. However, the gain or loss estimates for trees from each container size helps to present the overall trends. Smaller container-grown *A. rubrum* produced a greater gain for homeowners over the two growing seasons after transplanting to the landscape than did trees from larger container sizes (Figure 5E; Table 1).

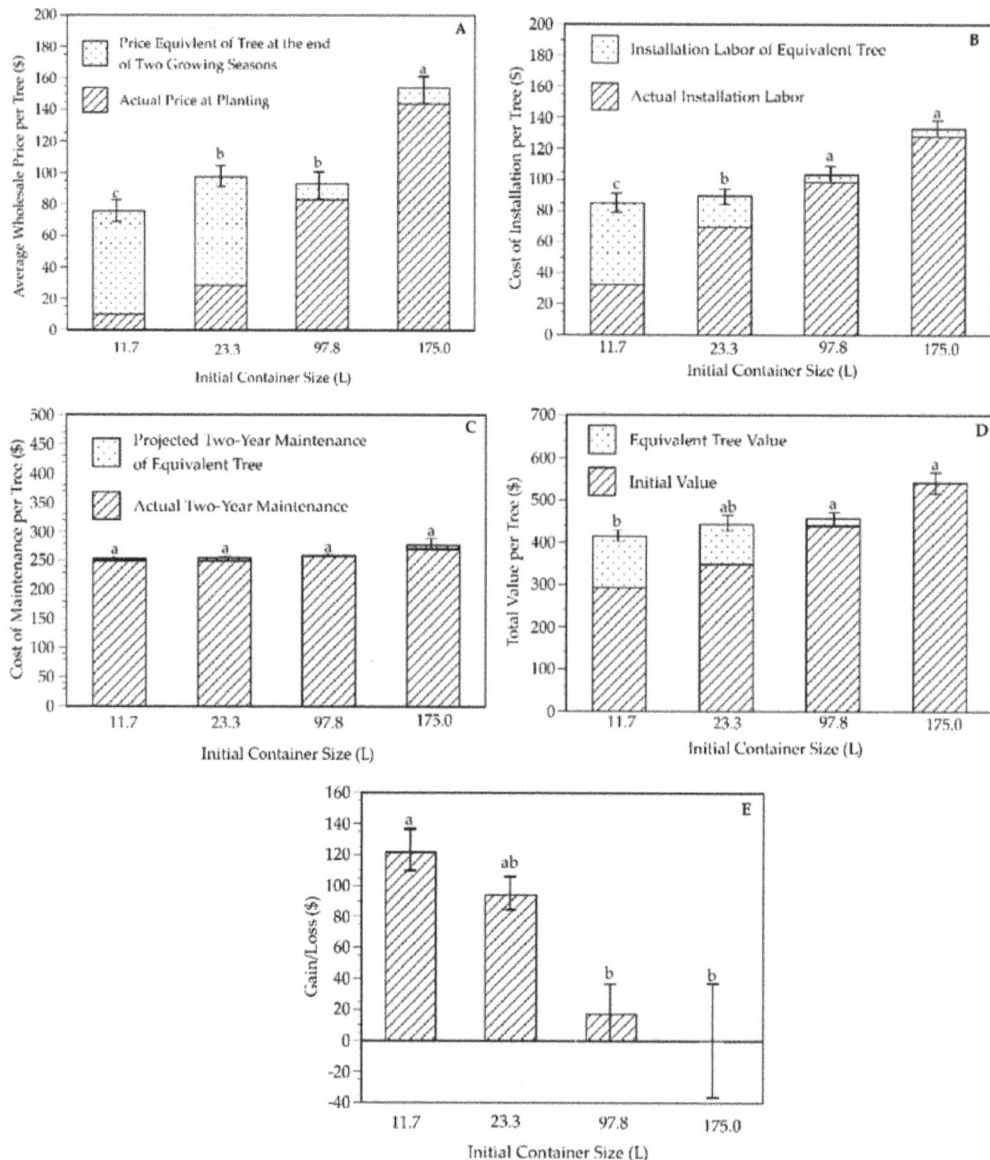

Figure 5. Mean (±standard error) wholesale cost (**A**), installation (**B**), maintenance cost (**C**), value (**D**), and gain or loss in dollars [US$] (**E**) of *Acer rubrum* trees from transplant (diagonal hatching) to the end of the second growing season (stippled hatching) for initial container sizes of 11.7, 23.3, 97.8 and 175.0 L trees. Means of final values after two growing seasons for initial container sizes with the same letter are not significantly different at $p \leq 0.05$ using Student's *t*-test.

For *T. distichum*, the greatest container size change occurred with the 23.3 L trees which ended the second growing season at a mean equivalent size of 83.3 L (Figure 4B). In contrast, the 11.7 L and 97.8 L trees changed less and the 3.5 L and 175.0 L *T. distichum* trees ended with very little change from their initial container sizes. The 97.8 L *T. distichum* trees ended the second season with only one of the six replicates increasing its equivalent container size and 175.0 L trees did not have any increase in container size equivalents (data not shown). One of the six 3.5 L trees died during the two years, which was calculated as a 0.0 L container tree, thus decreasing the mean equivalent of the remaining container sizes. Mortality was greater in the 3.5 L trees most likely due to their small size, which exposed them to more drift of salinity in the irrigation water from the mini-spray-stakes used during irrigation, greater predation by white-tailed deer (*Odocoileus virginianus*) and provided a small biomass with which to withstand environmental variation.

The wholesale price of the tree at planting was compared to the wholesale price equivalent of the tree at the end of the second growing season. The 23.3 L trees had the greatest increase in wholesale price, followed by the 11.7 L and 97.8 L trees, while the 3.5 L trees barely increased and 175.0 L trees had no increase above the actual price at planting (Figure 6A; Table 1). The 175.0 L trees were the costliest to purchase initially, but retained the greatest wholesale price equivalent at the end of the two growing seasons despite no increase in size equivalent. Analyzing the cost to install the initial container size versus the cost to install the ending container size after two growing seasons also indicated that while the costs were low for the smaller container sizes, it was also more cost-efficient to plant the smaller container sizes as greatest savings on transplant costs occurred with the 11.7 L and 23.3 L trees (Figure 6B; Table 1). Maintenance costs remained steady for the two growing seasons with no differences between container size trees (Figure 6C).

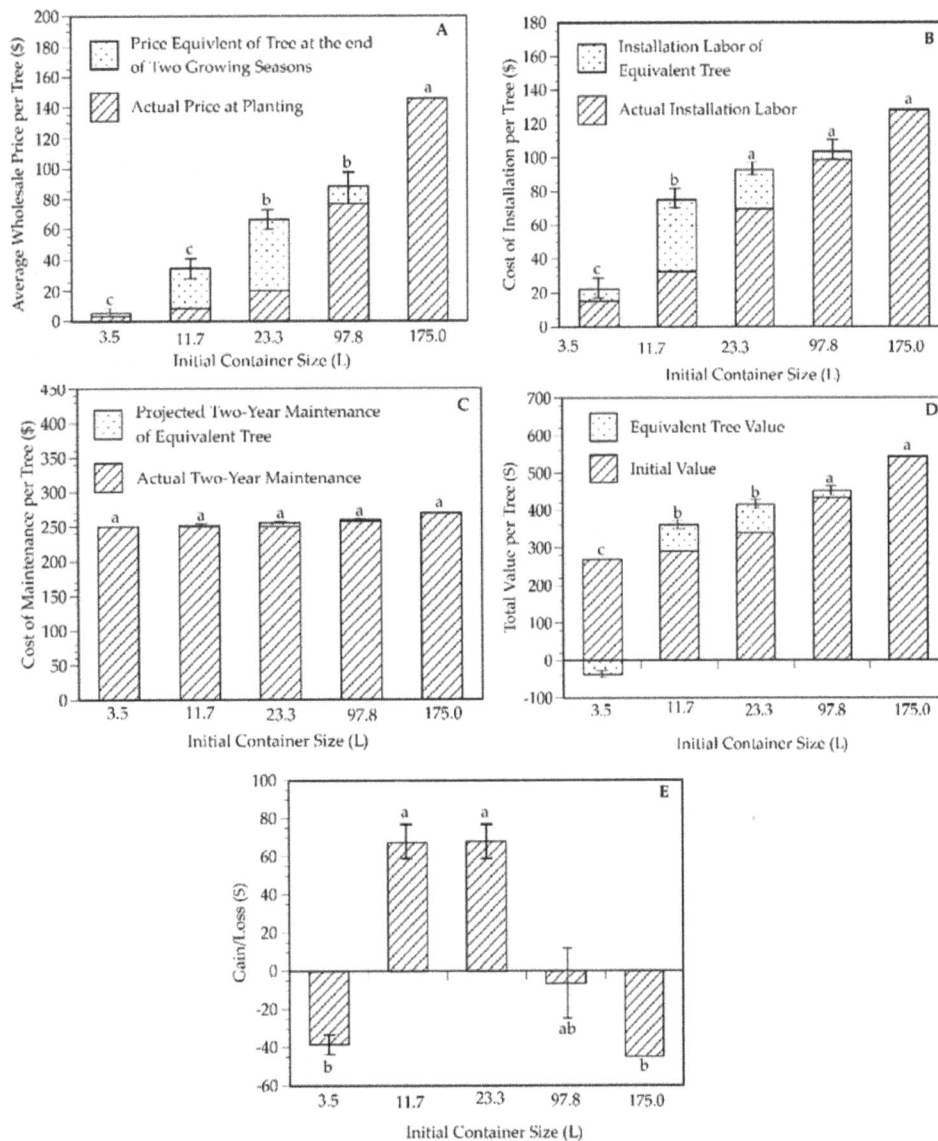

Figure 6. Mean (±standard error) wholesale cost (**A**), installation (**B**), maintenance cost (**C**), value (**D**), and gain or loss in dollars [US$] (**E**) of *Taxodium distichum* trees from transplant (diagonal hatching) to the end of the second growing season (stippled hatching) for initial container sizes of 3.5, 11.7, 23.3, 97.8 and 175.0 L trees. Means of final values after two growing seasons for initial container sizes with the same letter are not significantly different at $p \leq 0.05$ using Student's *t*-test.

The summation of this information allowed analysis of the overall value of the tree. The value of the tree increased the most for 11.7 L and 23.3 L container sizes for *T. distichum* (Figure 6D,E). However, the ending value of both sizes was still not equal to the value of the larger trees transplanted from 175.0 L containers. Therefore, while overall gains were largest in *T. distichum* from 11.7 L and 23.3 L containers (Table 1; Figure 6E), initially transplanted 175.0 L trees still maintained the greatest overall value after two growing seasons (Figure 6D). However, because the 175.0 L trees did not increase in size, money put into maintenance over the two years was considered a loss, as it did not generate an output in increased growth (Figure 6E). Losses were also seen with the 3.5 L and 97.8 L trees (Table 1; Figure 6E).

Slow growth ultimately impacted the economic cost analysis for *T. distichum*. Ending container size equivalents of *T. distichum* were similar to initial size for all container sizes, except 11.7 L and 23.3 L containers (Figure 4B). While the greatest changes occurred with 11.7 L and 23.3 L trees, only the 23.3 L trees increased enough in size so as to not statistically differ from the 97.8 L or 175.0 L trees after two growing seasons (Figure 4B). As a result, the total value and the gain in value were the greatest for 11.7 L and 23.3 L trees, and losses in net value occurred for the remaining container sizes (Figure 6D,E; Table 1).

The greatest container size changes for *V. agnus-castus* occurred with the 11.7 L and 23.3 L trees (Figure 4C; Table 1). The initial 11.7 L and 23.3 L trees ended as 136.3 L and 106.0 L container size trees, respectively. The 3.5 L and 97.8 L container-grown trees ended with similar increases from their initial sizes, and 175.0 L trees increased the least. Ending container sizes were not significantly different among the 11.7, 23.3 and 97.8 L trees, and the 97.8 L trees did not differ from 175.0 L trees (Figure 4C).

The *V. agnus-castus* trees from 11.7 L containers had the greatest increase in wholesale price, while the 3.5, 23.3 and 97.8 L trees had similar increases to one another (Figure 7A; Table 1). The 11.7 L trees would save homeowners the most money after transplant given the higher initial purchasing and planting costs of the 97.8 L container trees. The 175.0 L trees had no increase in value. Analyzing the cost to install the initial container size versus the cost to install the ending container size after two growing seasons also indicated that while the initial installation costs of trees were low for 3.5 and 11.7 L container-grown trees, it was also more cost-efficient to plant the smaller container sizes in relation to installation costs after two seasons (Figure 7B, Table 1). Maintenance costs did not differ across container sizes for the two growing years (Figure 7C).

The overall value of the trees increased the most for the 11.7 L container sizes of *V. agnus-castus*, with an ending value equal to that of 97.8 L trees. (Figure 7D; Table 1). The total value of the 23.3 L trees exceeded that of the initial value of the 97.8 L trees. A slight decrease in total value of the 175.0 L trees occurred after two growing seasons. Gains in total value were greatest for the 11.7 L trees, were similar among the 3.5, 23.3 and 97.8 L trees, and showed a slight loss for 175.0 L trees after two growing seasons in the landscape (Table 1; Figure 7E).

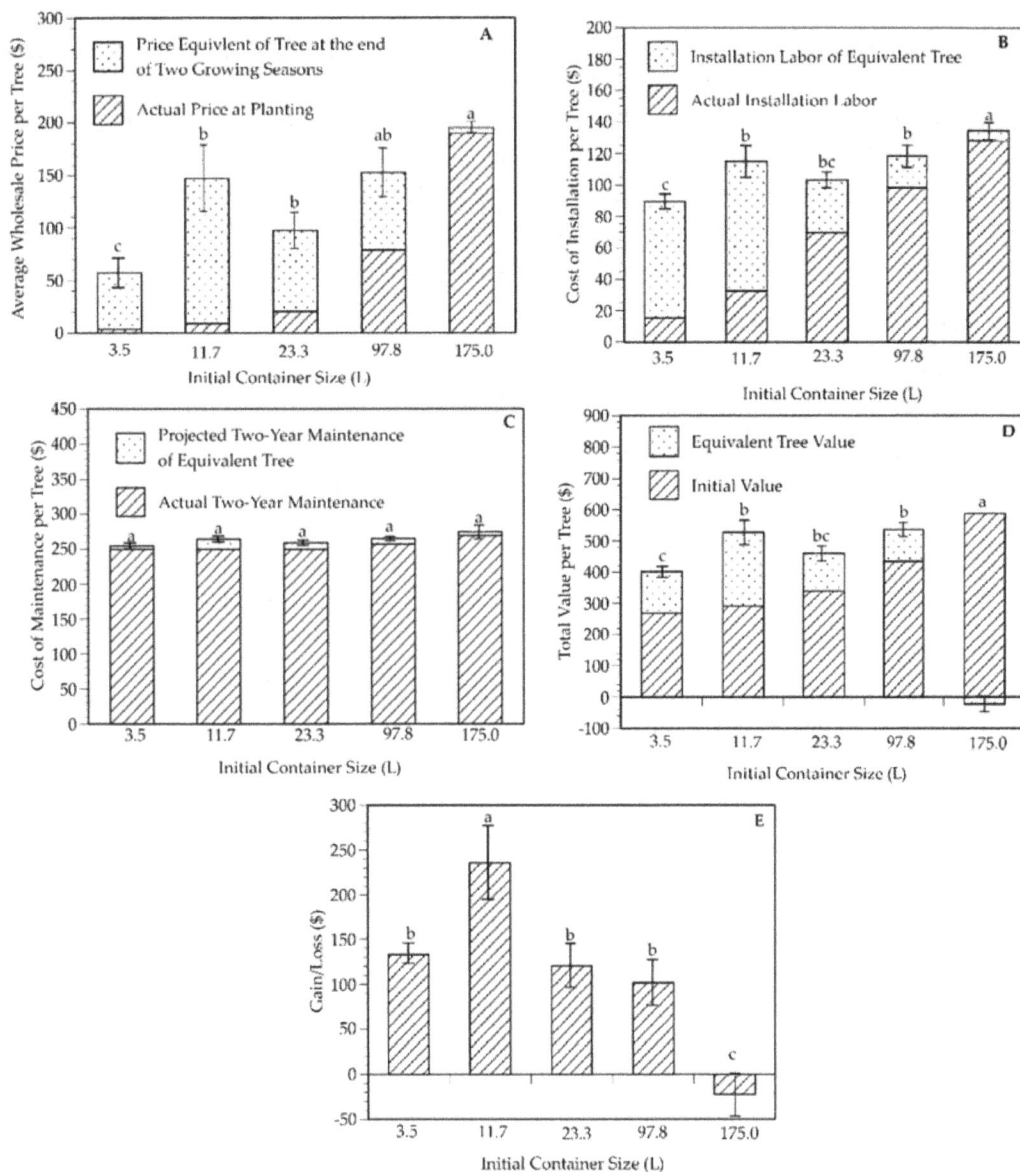

Figure 7. Mean (±standard error) wholesale cost (**A**), installation (**B**), maintenance cost (**C**), value (**D**), and gain or loss in dollars [US$] (**E**) of *Vitex agnus-castus* trees from transplant (diagonal hatching) to the end of the second growing season (stippled hatching) for initial container sizes of 3.5, 11.7, 23.3, 97.8 and 175.0 L trees. Means of container sizes topped by the same letter are not significantly different at $p \leq 0.05$ using Student's *t*-test.

4. Conclusions

Previous research has looked at assigning trees a value for real estate, insurance, production, and other uses [10,14]. However, a lack of research in the value of transplanted trees of various sizes persists. While research can be used to demonstrate that smaller or larger container-grown trees perform better in the landscape [24–26], oftentimes finances are of greater concern to the consumer. By corroborating evidence that smaller container sizes establish quicker in the landscape [8,21,24–27] with results indicating that 11.7 L and 23.3 L trees generally produce a greater profit (net value increase) than larger container-grown trees, steps are being taken to create a complete picture to present to consumers. Continued research should look at cost analysis after a 5-year, 10-year, etc. period or develop projection curves to determine if current findings persist over time. The present results were based on selected species and location (Table 1). However, experiments conducted simultaneously in a different growing environment produced similar results [21]. Additional determination of value trends across growing environments and the time value of money during longer growing periods should be considered. Furthermore, research should analyze the impacts on growers if a shift back toward smaller container-grown trees occurred. Finally, as water shortages become a very real problem [28], future studies should monitor the impacts of irrigation costs on the overall cost of transplanting and growing trees. The current study also does not address the aesthetic value of the "instant landscape" provided by larger size stock immediately after installation, nor the potentially greater ecosystem services of larger stock sizes, which may still be justification for planting larger-sized container plants.

Acknowledgments: This work was supported in part by hatch funds from Texas A&M AgriLife Research provided by the National Institute of Food and Agriculture (NIFA) and funding from the Tree Research and Education Endowment (TREE) Fund. Mention of a trademark, proprietary product, or vendor does not constitute a guarantee or warranty of the product by the authors, Texas A&M University, or Texas A&M AgriLife Research and does not imply its approval to the exclusion of other products or vendors that also may be suitable. Special thanks to Leo Lombardini, Todd Watson, and Andrew King for their assistance during the nursery production and field transplant portions of this experiment which permitted this economic analysis to be conducted.

Author Contributions: Michael A. Arnold and Lauren M. Garcia Chance conceived and designed the experiments; Lauren M. Garcia Chance performed the experiments; Lauren M. Garcia Chance and Sean T. Carver analyzed the data; Charles R. Hall contributed economic assistance and references; Lauren M. Garcia Chance wrote the initial paper with edits from the coauthors.

Conflicts of Interest: The authors declare no conflict of interest.

References

1. Arnold, M.A. Challenges and benefits of transplanting large trees: An introduction to the workshop. *HortTechnology* **2004**, *15*, 115–117.
2. Watson, W.T. Influence of tree size on transplant establishment and growth. *HortTechnology* **2004**, *15*, 118–122.
3. Kalmbach, K.L.; Kielbaso, J.J. Residents' attitudes toward selected characteristics of street tree plantings. *J. Arboric.* **1979**, *5*, 124–129.
4. Schroeder, H.; Flannigan, J.; Coles, R. Residents' attitudes toward street trees in the UK and U.S. communities. *Arboric. Urban For.* **2006**, *32*, 236–246.
5. Nowak, D.J.; Hoehn, R.; Crane, D. Oxygen production by urban trees in the United States. *Arboric. Urban For.* **2007**, *33*, 220–226.
6. Watson, G.W.; Himelick, E.B. *The Practical Science of Planting Trees*; International Society of Arboriculture: Champaign, IL, USA, 2013.
7. Maco, S.E.; McPherson, E.G. A practical approach to assessing structure, function, and value of street tree populations in small communities. *J. Arboric.* **2003**, *29*, 84–97.
8. Gilman, E.F.; Black, R.J.; Dehgan, B. Irrigation volume and frequency and tree size affect establishment rate. *J. Arboric.* **1998**, *24*, 1–9.
9. Gilman, E.F.; Kane, M.E. Root growth of red maple following planting from containers. *HortScience* **1990**, *25*, 527–528.
10. Cullen, S. Tree appraisal: What is the trunk formula method. *Arboric. Consul.* **2000**, *30*, 3.
11. Council of Landscape & Tree Appraisers. *Guide for Plant Appraisal*, 9th ed.; International Society of Arboriculture: Champaign, IL, USA, 2000.
12. Cullen, S. Tree appraisal: Chronology of North American industry guidance. *J. Arboric.* **2005**, *31*, 157–162.
13. Cullen, S. Tree appraisal: Can depreciation factors be rated greater than 100%? *J. Arboric.* **2002**, *28*, 153–158.
14. McPherson, E.G. Benefit-based tree valuation. *Arboric. Urban For.* **2007**, *33*, 1–11.
15. The Appraisal Institute. The appraisal of rural property. *Apprais. J.* **2000**, *68*, 20.
16. Mcpherson, E.G.; Kendall, A.; Albers, S. Life cycle assessment of carbon dioxide for different arboricultural practices in Los Angeles, CA. *Urban For. Urban Green.* **2015**, *14*, 388–397. [CrossRef]

17. McPherson, E.G. *Northern Mountain and Prairie Community Tree Guide: Benefits, Costs and Strategic Planting*; Center for Urban Forest Research, USDA Forest Service, Pacific Southwest Research Station: Davis, CA, USA, 2003.

18. Xiao, Q.; McPherson, E.G.; Ustin, S.L.; Grismer, M.E. A new approach to modeling tree rainfall interception. *J. Geophys. Res.* **2000**, *105*, 29173–29188. [CrossRef]

19. Anderson, L.M.; Cordell, H.K. Residential property values improve by landscaping with trees. *South. J. Appl. For.* **1988**, *9*, 162–166.

20. Garcia, L.M. Post-Transplant Establishment and Economic Value of Three Tree Species from Five Container Sizes. Master's Thesis, Texas A&M University, College Station, TX, USA, 2015.

21. Garcia, L.M.; Arnold, M.A.; Denny, G.C.; Carver, S.T.; King, A.R. Differential environments influence initial transplant establishment among tree species produced in five container sizes. *Arboric. Urban For.* **2016**, *42*, 170–180.

22. American Association of Nurserymen. ANSI Z60.1-2004. In *American Standards for Nursery Stock*; American Association of Nurserymen: Washington, DC, USA, 2004.

23. Mewis, B. *RSMeans Residential Cost Data 2014*, 33rd ed.; RSMeans Company: Norcross, GA, USA, 2014.

24. Gilman, E.F.; Masters, F. Effect of tree size, root pruning and production method on root growth and lateral stability of *Quercus virginiana*. *Arboric. Urban For.* **2010**, *36*, 281–291.

25. Lambert, B.B.; Harper, S.J.; Robinson, S.D. Effect of container size at time of planting on tree growth rates for baldcypress (*Taxodium distichum* (L.) Rich), red maple (*Acer rubrum* L.) and longleaf pine (*Pinus palustris* Mill.). *Arboric. Urban For.* **2010**, *36*, 93–99.

26. Struve, D.K. Tree establishment: A review of some of the factors affecting transplant survival and establishment. *Arboric. Urban For.* **2009**, *35*, 10–13.

27. Gilman, E.F.; Harchick, C.; Paz, M. Effect of tree size, root pruning, and production method on establishment of *Quercus virginiana*. *Arboric. Urban For.* **2010**, *36*, 183–190.

28. USGS. Estimated Use of Water in the United States County-Level Data for 2005. Available online: http://water.usgs.gov/watuse/data/2005 (accessed on 6 November 2013).

22

A Preliminary Comparison of Antioxidants of Tomato Fruit Grown Under Organic and Conventional Systems

Apiradee Uthairatanakij [1,*], Sukanya Aiamla-or [1], Pongphen Jitareerat [1] and Ashariya Maneenoi [2]

[1] Division of Postharvest Technology, King Mongkut's University of Technology Thonburi, Bangkok 10150, Thailand; sukanyamlr@gmail.com (S.A.); pongphen.jit@kmutt.ac.th (P.J.)

[2] Lake Rajada Office, Adams Enterprises Ltd., Bangkok 10110, Thailand; ashariya@yahoo.com

* Correspondence: apiradee.uth@kmutt.ac.th

Abstract: Organic farming is rapidly growing due to its perceived potential for producing higher nutritional quality. However, studies of organically- and conventionally-grown crops have not always shown differences between the systems. The objective of this research was to compare the antioxidant activities of organically-grown tomato to those from a conventional production system during postharvest cold storage. "Tub Tim Dang" tomato (*Solanum lycopersicum* L.) fruit were harvested at the breaker stage of maturity from both organic and conventional farms. Fruit were cold-stored at 10 °C for 20 days, and samples were collected at intervals to measure the activities of superoxide dismutase (SOD), catalase (CAT) and ascorbate peroxidase (APX) activities, and total antioxidant activity by the 2,2'-diphenyl-1-picrylhydrazyl (DPPH) and ferric reducing antioxidant power (FRAP) assays. The activities of SOD, CAT and APX of organic tomato fruit did not differ from those of conventional fruit during cold storage. In addition, there was no effect of production system on FRAP activity. In contrast, DPPH activity of organic tomato fruit was lower than conventional fruit through 10 days of cold storage, but it was higher at 15 and 20 days. These results indicated that organic production did not have a significant effects on these antioxidant traits of tomato.

Keywords: tomato; antioxidant enzyme; DPPH; FRAP

1. Introduction

Currently, many consumers are health conscious and aware of food quality, food safety and environmental protection, leading to an increasing demand for organic fresh products. Numerous studies on organic versus conventional crops have shown that organic fresh produce has significantly less chemical residue and greater nutritional content, including ascorbic acid, vitamin E, β-carotene and phenolics [1–5]. However, the differences in chemical composition between organic and conventional produce is dependent on cultural practices and environmental factors, such as time of harvest, fertilizer, water supply and soil properties [6–8]. Tomato (*Solanum lycopersicum* L.) fruit have high economic value and popularity in the consumer's diet, and are an important nutritional source of antioxidants [9]. Antioxidant content may change during the ripening of tomato, and this is reflected in changes in the antioxidant activity [10].

Reactive oxygen species (ROS) are generated during normal metabolism and also under stress conditions, and are eliminated by the antioxidant system [11]. Antioxidant enzymes, including superoxide dismutase (SOD), catalase (CAT) and ascorbate peroxidase (APX), and free radical scavengers provide defense mechanisms against ROS [12]. SOD is found in the chloroplast and cytosol, and is a key antioxidant enzyme that catalyzes conversion of superoxide radicals into dioxygen and hydrogen peroxide [13,14]. Afterwards, hydrogen peroxide is eliminated by CAT or APX. CAT is located in the peroxisome and removes hydrogen peroxide without reducing agents, while APX is located in the chloroplast and cytosol and requires ascorbate as an electron donor to eliminate hydrogen peroxide in the plant cell [15,16]. SOD, CAT and APX are therefore efficient antioxidant enzymes for detoxifying ROS because they are located in all compartments of the cell.

Some antioxidant levels are believed to be higher in organic produce that are activated by natural mechanisms of the plant defense systems against pests and diseases or other stress factors [17]. Therefore, the aim of this study was to compare total antioxidant activity and activity of the antioxidant enzymes SOD, CAT, and APX of organically- and conventionally-grown tomato fruit.

2. Experimental Section

2.1. Plant Preparation

"Tub Tim Dang" tomato seeds were planted in January 2013 under both organic and conventional management systems. Both farms were located in Phakthongchai district, Nakonratchasima province, Thailand ($14°35'–15°00'$ latitude, $101°45'–102°15'$ longitude) approximately 10 km apart. The organic tomato plants were grown according to the standard protocol used by Adam Enterprises, Ltd. (certified by the United States Department of Agriculture, Bangkok, Thailand). Before transplanting the organic tomato seedlings, $40 \ MT·ha^{-1}$ of composted manure was applied to the soil. Then, fertilizer from fermented fish ($100 \ mL·plant^{-1}$) was applied three times a week until harvest. Conventional tomato fruit were grown on farms that followed Good Agricultural Practices. Before transplanting, $187.5 \ kg·ha^{-1}$ of NPK starter fertilizer (15-15-15) was applied to the soil. The plants were side-dressed with fertilizer 10 days after transplanting with $62.5 \ kg·hectare^{-1}$ of 46-0-0, with $187.5 \ kg·ha^{-1}$ of 15-15-15 NPK at 25 days, and with $187.5 \ kg·ha^{-1}$ of 13-13-21 NPK at 40 days. Irrigation was provided when it was necessary to avoid drought stress. Approximately 20 kg of fruit was harvested from both farms at the breaker stage, 55 days after transplanting. Fruit were washed with tap water, and placed in baskets before storage at 10 °C and 90% RH, where they started to ripen as indicated by fruit color change. At 0, 5, 10, 15 and 20 days of cold storage, about 30 fruit were randomly selected for a total approximate weight of 500 g and divided into three replicates. On the sampling day, tissues were frozen in liquid nitrogen and stored at −20 °C until analysis. From each replicate, 3 g of frozen sample was allocated for 2,2'-diphenyl-1-picrylhydrazyl (DPPH) and ferric reducing antioxidant power (FRAP) assays. For each analysis of enzyme activity, 5 g of frozen sample was used.

2.2. Total Antioxidant Activity

DPPH and FRAP assays followed the method of Brand-Williams et al. [18] and Benzie and Strain [19], respectively, using a spectrophotometer. A standard curve between 25 and 800 μM Trolox was linear, and results are expressed in Trolox equivalent antioxidant activity $(TEAC)·g^{-1}·FW$.

2.3. Enzyme Assay

For enzyme assays, 5 g of fruit tissue was homogenized in 10 mL cold extraction buffer (0.05 M sodium phosphate buffer, pH 7.8). The homogenate was filtered through a muslin cloth and centrifuged at 15,000 rpm for 20 min at 4 °C. The supernatant was used as a crude extract for SOD activity assays [20]. For the CAT and APX assays, 5 g of fruit tissue was homogenized in 10 mL of cold 0.1 M phosphate buffer at pH 7.0. The homogenate was filtered through a muslin cloth and centrifuged at 15,000 rpm at 4 °C for 20 min. The supernatant was used as a crude extract for CAT and APX activities [21].

The SOD activity was determined based on its capacity to inhibit the reduction of nitro-blue tetrazolium (NBT) by superoxide radicals generated by xanthine oxidase. One SOD unit was expressed as the amount of extract with percent inhibition of NBT reduction [22]. CAT activity was measured at

240 nm for decomposition of H_2O_2 (mM)·min^{-1} at 25 °C [23]. APX activity measured the decrease of ascorbic acid at 290 nm, and the result was expressed as mM ascorbic acid·min^{-1} at 28 °C [24]. Three replicates were performed for each main treatment, organic and conventional, and each reaction was repeated 3 times.

2.4. Statistical Analysis

Microsoft Excel (Redmond, WA, USA) was used to perform statistical analyses with a paired *t*-test to compare organic versus conventional means.

3. Results

3.1. 2,2'-Diphenyl-L-Picrylhydrazyl (DPPH)

The DPPH assay indicated a significant difference between organic and conventional tomato fruit, with generally higher levels from the conventional system during the first 10 days of storage, while organic tomato fruit showed significantly higher levels than conventional fruit on days 15 and 20 (Figure 1A).

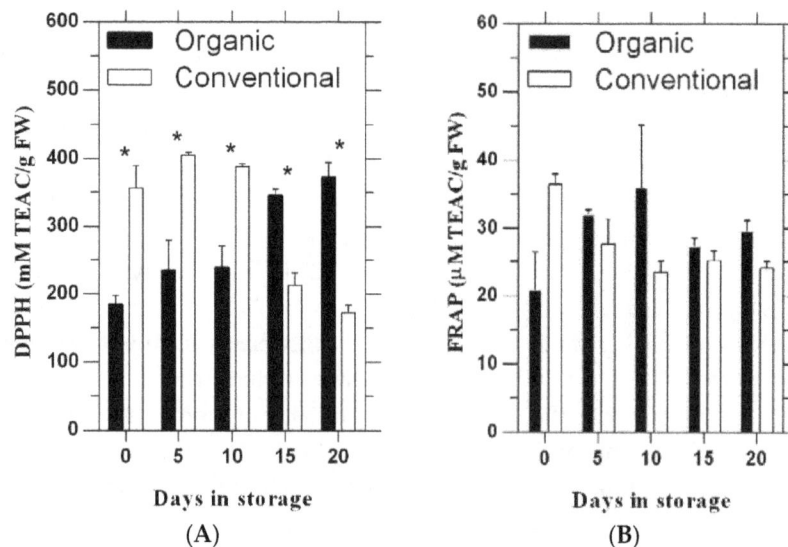

Figure 1. Changes in antioxidant activities by the DPPH (**A**) and FRAP (**B**) assays of tomato fruit during cold storage following production under organic and conventional production systems. Bars represent standard errors. * indicate the significant difference ($p \leq 0.05$).

3.2. Ferric Reducing Antioxidant Power (FRAP)

FRAP values from both systems varied from 21 to 36 μM·TEAC·g^{-1}·FW across storage days (Figure 1B), but the systems did not significantly differ.

3.3. The Activity of SOD, CAT and APX

SOD activity did not statistically differ between the two production systems across days of storage (Figure 2A). CAT activity did not differ between systems except at 15 days of storage when the organic system value was significantly greater than the conventional system (Figure 2B). No significant differences between systems were found for APX activity except that the organic value was significantly lower at 5 and higher at 10 days of storage than the conventional value (Figure 2C).

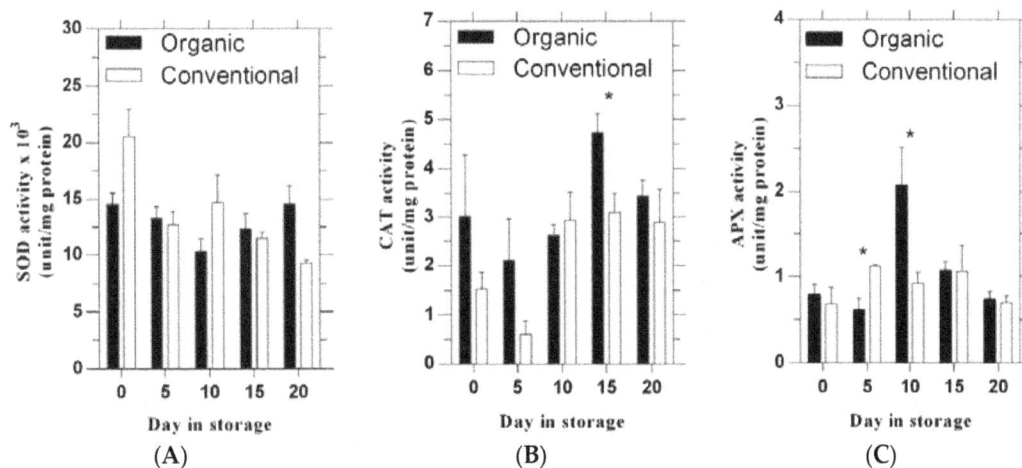

Figure 2. The activity of SOD (**A**), CAT (**B**) and APX (**C**) of tomato fruit during cold storage following production under organic and conventional production systems. Bars represent standard errors. * indicate the significant difference ($p \leq 0.05$).

4. Discussion

Tomato fruit are a major source of phytochemicals and antioxidants such as lycopene, phenolics, flavonoids and vitamin C [24]. In this study, the antioxidant capacity of organic and conventional tomato during cold storage was compared using DPPH and FRAP radical scavenging capacity assays. These methods are based on electron transfer reactions which measure antioxidant reducing capacity [25]. Although DPPH activity was initially higher in conventional fruit through 10 days, it was lower than organic fruit at 15 and 20 days. In contrast, FRAP antioxidant activity did not differ between organic and conventional tomato fruit during storage. These two assays may have different sensitivities to the major antioxidants found in the fruit, as well as different reaction kinetics leading to the contrasting results [25]. Moreover, antioxidant effectiveness may depend on other factors such as polarity and solubility [26]. Some studies have reported significant correlations between the level of antioxidant activity and the phytochemical content of fruit [24,27]. Faller and Fialho [28] reported that organic tomato contained higher phytochemical contents than conventional tomato, but the assays in the present study gave conflicting results so it cannot be concluded that fruit from the two systems differed. No differences in DPPH assays were observed in other comparisons of organic versus conventional apple (*Malus x domestica* Borkh.), mango (*Mangifera indica* L.) or orange (*Citrus sinensis* L.) [28,29]. The antioxidant capacity in organic or conventional fruits can be affected by cultural practices and environmental factors [4,5,9]. Riahi and Hdider [30] reported that organic fertilizer affected the antioxidant capacity in organic tomato. Similarly, Stracke et al. [31] demonstrated that the antioxidant capacity (FRAP, ORAC and TEAC assay) was not significant between organic and conventional apple.

SOD is a primary enzyme that eliminates superoxide radicles, converting them into dioxygen and hydrogen peroxide. Afterwards, hydrogen peroxide is catalyzed by CAT [12,13]. APX utilizes ascorbate as the electron donor reducing H_2O_2 to water, and prevents the accumulation of H_2O_2 to toxic levels under stress conditions [32]. In this study, the SOD activity did not significantly differ between fruit grown under the two cultural systems. In contrast, Oliveira et al. [33] reported that the activities of SOD and CAT were induced by organic production. In addition, no clear patterns in APX activity in organic or conventional tomato were evident.

5. Conclusions

The present study did not provide evidence that would indicate the superiority of nutritional composition of organically-grown tomato fruit in terms of antioxidant capacity by either the DPPH or

FRAP assays, and also by antioxidant enzyme activities. Further work will be required to assess how differing maturities and harvest dates from each system affect these traits.

Acknowledgments: The authors would like to thanks Adams Enterprises, Ltd. for providing organic tomatoes and Postharvest Innovation Center, Commission on higher Education, Bangkok, Thailand for scientific instruments.

Author Contributions: Apiradee Uthairatanakij conceived and designed the experiments and also wrote the paper. Sukanya Aiamla-or performed the experiments and analyzed the data. Pongphen Jitareerat performed the experiments and review manuscript. AshariyaManeenoi contributed plant material production.

Conflicts of Interest: The authors declare no conflict of interest.

References

1. Zhao, X.; Carey, E.E.; Wang, W.; Rajashekar, C.B. Does organic production enhance phytochemical content of fruits and vegetables? Current knowledge and prospects for research. *Horttechnology* **2006**, *16*, 449–456.
2. Lombardi-Boccia, G.; Lucarini, M.; Lanzi, S.; Aguzzi, A.; Cappelloni, M. Nutrients and antioxidant molecules in yellow plums (*Prunus domestica* L.) from conventional and organic productions: A comparative study. *J. Agric. Food Chem.* **2004**, *52*, 90–94. [CrossRef] [PubMed]
3. Parka, Y.S.; Imb, M.H.; Choic, J.H.; Yimc, S.H.; Leontowiczd, H.; Leontowiczd, M.; Suhaje, M.; Gorinstein, S. The effects of ethylene treatment on the bioactivity of conventional and organic growing "Hayward" kiwi fruits. *Sci. Hortic.* **2013**, *164*, 589–595. [CrossRef]
4. Parka, Y.S.; Ham, K.S.; Kang, S.G.; Park, Y.K.; Namiesnik, J.; Leontowicz, H.; Leontowicz, M.; Ezra, A.; Trakhtenberg, S.; Gorinstein, S. Organic and conventional kiwi fruits, myths versus reality: Antioxidant, antiproliferative, and health effects. *J. Agric. Food Chem.* **2012**, *60*, 6984–6993. [CrossRef] [PubMed]
5. Worthington, V. Nutritional quality of organic versus conventional fruits, vegetables, and grains. *J. Altern. Complement. Med.* **2001**, *7*, 161–173. [CrossRef] [PubMed]
6. Winter, C.; Davis, S. Organic foods. *J. Food Sci.* **2006**, *71*, 117–124. [CrossRef]
7. Mehdizadeh, M.; Darbandi, E.I.; Naseri-Rad, H.; Tobeh, A. Growth and yield of tomato (*Lycopersicon esculentum* Mill.) as influenced by different organic fertilizers. *Int. J. Agron. Plant Prod.* **2013**, *4*, 734–738.
8. Reganold, J.P.; Andrews, P.K.; Reeve, J.R.; Carpenter-Boggs, L.; Schadt, C.W.; Alldredge, J.R.; Ross, C.F.; Davies, N.M.; Zhou, J. Fruits and soil quality of organic and conventional strawberry agro ecosystems. *PLoS ONE* **2010**, *5*, e12346. [CrossRef]
9. Toor, R.K.; Geoffrey, P.S.; Anuschka, H. Influence of different types of fertilizers on the major antioxidant components of tomatoes. *J. Food Compos. Anal.* **2006**, *19*, 20–27. [CrossRef]
10. Cano, A.; Acosta, M.; Arnao, M.B. Hydrophilic and lipophilic antioxidant activity changes during on-vine ripening of tomatoes (*Lycopersicon esculentum* Mill.). *Postharvest Biol. Technol.* **2003**, *28*, 59–65. [CrossRef]
11. Foyer, C.H.; Shigeoka, S. Understanding oxidative stress and antioxidant functions to enhance photosynthesis. *Plant Physiol.* **2011**, *155*, 93–100. [CrossRef] [PubMed]
12. Rani, P.; Unni, M.; Karthikeyan, J. Evaluation of antioxidant properties of berries. *Indian J. Clin. Biochem.* **2004**, *19*, 103–110. [CrossRef] [PubMed]
13. Bowler, C.; van Montagu, M.; Inze, D. Superoxide dismutase and stress tolerance. *Annu. Rev. Plant Biol.* **1992**, *43*, 83–116. [CrossRef]
14. Abreu, I.A.; Cabelli, D.E. Superoxide dismutases—A review of the metal-associated mechanistic variations. *Biochim. Biophys. Acta* **2010**, *1804*, 263–274. [CrossRef] [PubMed]
15. Willekens, H.; Chamnongpol, S.; Davey, M.; Schraudner, M.; Langebartels, C.; van Montagu, M.; Inzé, D.; van Camp, W. Catalase is a sink for H_2O_2 and is indispensable for stress defence in C_3 plants. *EMBO J.* **1997**, *16*, 4806–4816. [CrossRef] [PubMed]
16. Kangasjärvi, S.; Lepistö, A.; Hännikäinen, K.; Piippo, M.; Luomala, E.M.; Aro, E.M.; Rintamäki, E. Diverse roles for chloroplast stromal and thylakoid-bound ascorbate peroxidases in plant stress responses. *Biochem. J.* **2008**, *412*, 275–285. [CrossRef] [PubMed]
17. Brandt, K.; Mølgaard, J.P. Organic agriculture: Does it enhance or reduce the nutritional value of plant foods? *J. Sci. Food Agric.* **2001**, *81*, 924–931. [CrossRef]
18. Brand-Williams, W.; Cuvelier, M.E.; Berset, C. Use of free radical method to evaluate antioxidant activity. *Lebensm. Wiss. Technol.* **1995**, *28*, 25–30. [CrossRef]

19. Benzie, I.F.F.; Strain, J.J. The ferric reducing ability of plasma (FRAP) as a measure of "antioxidant power": The FRAP assay. *Anal. Biochem.* **1996**, *239*, 70–76. [CrossRef] [PubMed]

20. Cakmak, I.; Marschner, H. Magnesium deficiency and high light intensity enhance activities of superoxide dismutase, ascorbate peroxidase, and glutathione reductase in bean leaves. *Plant Physiol.* **1992**, *98*, 1222–1227. [CrossRef] [PubMed]

21. Nakano, Y.; Asada, K. Hydrogen peroxide is scavenged by ascorbate specific peroxidase in spinach chloroplast. *Plant Cell Physiol.* **1981**, *22*, 867–880.

22. Bartoli, C.G.; Simontacchi, M.; Guiamet, J.J.; Montadi, E.; Puntarulo, S. Antioxidant enzymes and lipid peroxidation during aging of *Chrysanthemum morifolium* RAM petals. *Plant Sci.* **1995**, *104*, 161–168. [CrossRef]

23. Kato, M.; Shimizu, S. Chlorophyll metabolism in higher plants, VII. Chlorophyll degradation in senescence tobacco leaves, phenolic-dependent peroxidative degradation. *Can. J. Bot.* **1987**, *65*, 729–735. [CrossRef]

24. Guil-Guerrero, J.L.; Rebolloso-Fuentes, M.M. Nutrient composition and antioxidant activity of eight tomato (*Lycopersicon esculentum*) varieties. *J. Food Compos. Anal.* **2009**, *22*, 123–129. [CrossRef]

25. Huang, D.; Ou, B.; Prior, R.L. The chemistry behind antioxidant capacity assays. *J. Agric. Food Chem.* **2005**, *53*, 1841–1856. [CrossRef] [PubMed]

26. Pokorny, J.; Yanishlieva, N.; Gordon, M. *Antioxidants in Food: Practical Application*; CRC Press: New York, NY, USA, 2001; p. 380.

27. Jin, P.; Wang, S.Y.; Wang, C.Y.; Zheng, Y. Effect of cultural system and storage temperature on antioxidant capacity and phenolic compounds in strawberries. *Food Chem.* **2011**, *124*, 262–270. [CrossRef]

28. Faller, A.L.K.; Fialho, E. Polyphenol content and antioxidant capacity in organic and conventional plant food. *J. Food Compos. Anal.* **2010**, *23*, 561–568. [CrossRef]

29. Bogs, J.; Bunning, M.; Stushnoff, C. Influence of biologically enhanced organic production on antioxidant and sensory qualities of (*Malus x domestica* Borkh. cv. Braeburn) apples. *Organ. Agric.* **2012**, *2*, 117–126. [CrossRef]

30. Riahi, A.; Hdider, C. Bioactive compounds and antioxidant activity of organically grown tomato (*Solanum lycopersicum* L.) cultivars as affected by fertilization. *Sci. Hortic.* **2013**, *151*, 90–96. [CrossRef]

31. Stracke Berenike, A.; Rüfer, C.E.; Bub, A.; Seifert, S.; Weibel, F.P.; Kunz, C.; Watzl, B. No effect of the farming system (organic/conventional) on the bioavailability of apple (*Malus domestica* Bork. cultivar Golden Delicious) polyphenols in healthy men: A comparative study. *Eur. J. Nutr.* **2010**, *49*, 301–310. [CrossRef] [PubMed]

32. Salandanan, K.; Bunning, M.; Stonaker, F.; Kulen, O.; Kendall, P.; Stushnoff, C. Comparative analysis of antioxidant properties and fruit quality attributes of organically and conventionally grown melons (*Cucumis melo* L.). *Hortscience* **2009**, *44*, 1825–1832.

33. Oliveira, A.B.; Moura, C.F.H.; Gomes-Filho, E.; Marco, C.A.; Urban, L.; Miranda, M.R.A. The Impact of organic farming on quality of tomatoes is associated to increased oxidative stress during fruit development. *PLoS ONE* **2013**, *8*, e56354. [CrossRef] [PubMed]

23

Climate Change Impacts on Water Use in Horticulture

Richard L. Snyder

Department of Land, Air and Water Resources, University of California, Davis, CA 95616, USA;
rlsnyder@ucdavis.edu

Abstract: The evidence for anthropogenic global climate change is strong, and the projected climate changes could greatly impact horticultural production. For horticulture, two of the biggest concerns are related to the scarcity of water for crop production and the potential for increased evapotranspiration (ET). While ET is known to increase with air temperature, it is also known to decrease with increasing humidity and atmospheric CO_2 concentration. Considering all of these factors and a plausible climate projection, this paper demonstrates that ET may increase or decrease depending on the magnitude of atmospheric changes including wind speed. On the other hand, the evidence is still strong that water resources will become less reliable in many regions where horticultural crops are grown.

Keywords: evapotranspiration; environmental conditions; CO_2 concentrations; water requirements

1. Introduction

Most scientists agree that global climate change is occurring at an alarming rate and that these changes are likely to impact water use in horticulture, agronomy, and natural ecosystems. In some locations, climate change can potentially increase agricultural production, but it is generally believed that widespread detrimental impacts on agricultural production are more likely in much of the world. Currently, the global climate change trend is for increasing air temperature, mainly at night and during winter, and more near the poles than in lower latitudes. Because of global warming, energy storage in water has increased dramatically (much more than in the air), and higher water temperatures has led to rising sea level due to water expansion and additional heat storage mainly in the oceans. In addition, increasing air and water temperature enhances evaporation, and higher air temperature increases the saturation water vapor pressure, i.e., the amount of water vapor that is held in the air at saturation. These atmospheric changes can impact plant growth, production, and water usage, and this chapter presents ideas on the possible impact of higher temperature, humidity, and CO_2 concentration on the evapotranspiration (ET) of horticultural crops.

At this time, the main cause for global climate change is the increasing concentration of CO_2. According to the "Keeling curve," the CO_2 concentration recorded at the Mauna Loa Observatory in Hawaii (USA) increased from about 315 to 395 ppm from 1958 to 2013, and recently passed 400 ppm. This increase has serious implications for global climate change and its impact on nature and horticulture.

The greenhouse gases (GHGs) that contribute most to global climate change from the worst to least worse are as follows: (1) H_2O—water vapor; (2) CO_2—carbon dioxide; (3) NH_4—methane; (4) N_2O—nitrous oxide; (5) O_3—ozone. While H_2O is actually a more effective GHG than CO_2, the atmospheric concentration is spatially and temporally variable and it is not increasing rapidly at this time. However, as the oceans warm, higher concentrations of atmospheric H_2O are likely, and the higher levels can greatly contribute to warming. Carbon dioxide is a less effective GHG than H_2O, but the more evenly distributed global concentration is increasing steadily, and it is currently the GHG causing the most rapid global temperature rise. Methane is a concern for the future because large amounts of methane are stored in the ocean and in permafrost. Scientists believe that thawing the permafrost and warming the oceans might release this stored NH_4 and cause a rapid temperature increase. Nitrous oxide is used as an aerosol propellant, anesthetic, and an oxidizer in rockets and engine fuel. Ozone is not evenly distributed over the Earth's surface, but man-made O_3 does contribute to atmospheric warming in urban areas.

GHGs mostly reside in the troposphere, i.e., the lowest level of the atmosphere, where they intercept long waveband radiation (LWR) and raise atmospheric temperatures. The stratosphere, which is a stable layer of air above the troposphere, has considerably less greenhouse gas than the troposphere, and, interestingly, this upper atmospheric layer is cooling. If the global warming were caused by increasing solar output or reduced reflection of solar radiation, the troposphere and stratosphere would both likely have increasing temperatures. Consequently, the simultaneous troposphere warming and stratosphere cooling is a good indicator that the warming is caused by anthropogenic increases in GHG additions (mostly CO_2) to the atmosphere. The melting of polar ice packs is a strong indication of global warming, and it will likely compound the warming because less solar radiation will be reflected back to space as the ice melts.

Climate change is likely to cause sea level rise and damage plant and animal life in the oceans and on land. Of course, changes in temperature and humidity could also lead to changes in general circulation of the atmosphere, greater frequency of storms and floods, and changes in the length and severity of drought. In general, climate scientists are projecting more frequent and severe storms and drought due to changes in atmospheric circulation [1]. In terms of horticulture, the authors of [2] and [3] discuss several potential effects of climate change on horticulture. In some cases, increasing CO_2, temperature, humidity, and other greenhouse gases might be beneficial in regions where crop production is limited by cold temperatures due to (1) lower potential for frost damage, (2) faster growth, and (3) lengthened growing seasons. In addition, there is some benefit coming from CO_2 fertilization, which can enhance photosynthesis. On the other hand, climate change could negatively impact agriculture in regions where climate conditions are currently good for production. For example, climate change might (1) decrease chilling and inhibit bloom and fruit set in horticultural crops, (2) lead to high temperature and wind during bloom or ripening that could negatively impact fruit set or fruit quality, (3) increase ET rates that could lead to water deficits, and (4) increase problems with heat stress.

Some possible climate change impacts on agriculture include (1) droughts, (2) floods, (3) faster phenological development, (4) inadequate chilling requirements, (5) pollination affected by rainfall and other extreme events, (6) frost and chill damage, (7) the spread of new insects and diseases, and (8) lower or higher yield and quality due to warming and water relations during summer [2,3]. One example of warming impacts is decreased winter chilling, which leads to bad pollination, staggered bloom, reduced fruit set, and poor fruit quality. In low chill years, apricots and cherries can drop flower buds or not produce a crop when chilling is inadequate.

In some regions, winter fog is an important factor in achieving adequate chilling for some crops. Rainfall during bloom can inhibit bee pollination, and precipitation around harvest time can increase fruit and nut diseases, cause fruit cracking, and destroy a crop. In deciduous orchards, late spring and summer rainfall has a negative effect on fruit and nut quality and production. Clearly, there are a multitude of climate factors that can change to wreak havoc on the production of horticultural crops. For example, climate scientists are projecting more bimodal precipitation in California, with more precipitation in the spring and fall and less in the winter. However, California depends on water storage in the mountain snowpack, and higher snow lines in the mountains, especially in the north where the mountains are shorter, will have less snowpack storage and will result in less water delivery to agricultural land in the summer [4]. Snowpack storage will affect the collection and distribution of irrigation water, but precipitation timing will also affect rainfall damage to crops in the spring and fall as well as changes in fog formation due to winters with higher temperatures and less rainfall.

While all of the aforementioned changes and impacts are important to consider, it is difficult to project changes in general circulation, storms, and drought with a high degree of accuracy in any particular region on Earth. However, the general impact of widespread higher temperatures, humidity, and increased CO_2 on factors affecting crop production is possible. Following a short discussion on how rising CO_2 concentrations cause global warming, this paper will present some ideas on the possible impact of higher temperature, humidity, and CO_2 on ET.

2. Increasing CO_2 and Global Warming

The impact of atmosphere CO_2 concentration on the greenhouse effect and the Earth's surface temperature was first described by Svante August Arrhenius [5]. Although climate science has advanced considerably since Arrhenius first made his estimates of increasing CO_2 concentration, its impact on global temperature is still quite accurate. Equations (1)–(3) are modified versions of the original equations from [5]. Equation (1) provides an estimate of Earth's emission temperature (T_e), which is the mean surface temperature it would have if there were no atmosphere:

$$T_e = \left[\frac{\pi r^2 (1 - \alpha_p) R_{sc}}{4\pi r^2 \sigma} \right]^{\frac{1}{4}} = \left[\frac{(1 - \alpha_p) R_{sc}}{4\sigma} \right]^{\frac{1}{4}} \approx 255 \text{ K} \tag{1}$$

where $R_{SC} = 1631$ W·m^{-2} is the solar constant, $\alpha_p \approx 0.30$ is the albedo (reflection of solar radiation from the surface), $r \approx 6371 \times 10^3$ m is the mean Earth radius, $\pi = 3.1415927$, and $\sigma = 5.67 \times 10^{-8}$ J·s^{-1}·m^{-2}·K^{-4} is the Stefan–Boltzmann constant. Based on Equation (1), if we had no atmosphere, the Earth's mean surface temperature would be approximately $T_e = 255$ K $= -18$ °C.

Adding gases to the atmosphere has a minimal impact on the surface solar radiation balance, but greenhouse gases in the atmosphere are known to intercept upward and downward fluxes of LWR, which raises the atmospheric temperature. Assuming a single layer atmosphere that is opaque to LWR, it can be shown that Equation (2) provides an estimate of the maximum Earth surface temperature (T_s) assuming that all of the LWR is absorbed by the atmosphere.

$$T_s = 2^{0.25} \cdot T_e \approx 1.19 \cdot T_e = 303 \text{ K}. \tag{2}$$

Thus, if the GHG in the atmosphere was 100% efficient at intercepting LWR, the maximum surface temperature would be $T_s = 303$ K $= 30$ °C. Therefore, with current solar radiation, short waveband radiation reflection, and GHGs, the actual surface temperature of Earth falls somewhere between -18 °C and 30 °C.

Assuming a single layer leaky atmosphere, which allows for some loss of LWR back to space, it can be shown that Equation (3) provides an estimate of the Earth surface temperature (T_s) as a function of T_e and the fraction of upward LWR absorbed by the atmosphere (ϵ). Until the GHG concentration began to increase, about 78% of the LWR was absorbed by GHGs in the atmosphere. Therefore, using Equation (3) and $\epsilon = 0.78$ shows that GHG increased the Earth's surface temperature from -18 °C to 15 °C (288 K).

$$T_s = \left(\frac{2}{2 - \epsilon} \right)^{\frac{1}{4}} T_e = \left(\frac{2}{2 - \epsilon} \right)^{\frac{1}{4}} (255) \approx 288 \text{ K}. \tag{3}$$

Equation (3) illustrates that T_s is affected only by T_e and ϵ, and Equation (1) shows that T_e is affected only by the amount of incoming and reflected solar radiation. Thus, only the incoming and reflected short waveband radiation and GHG concentration determine the Earth's global surface temperature. Since 1958, changes in R_{SC} and α_p were insignificant; however, the anthropogenic releases of CO_2 and other GHGs led to a rise of more than 80 ppm in atmospheric CO_2 concentration. The rising CO_2 concentration increased ϵ and it is the most likely cause for most of the corresponding 1 °C increase in global temperature [6]. Based on [1], considerably more global warming is projected as CO_2 levels continue to rise. While global temperature is rising, other climate factors, e.g., humidity, precipitation, wind speed, cloudiness, and precipitation, are also changing, and the assessment of climate change impact on weather and horticultural crops should consider all factors.

3. Impact on Evapotranspiration

Climate change is likely to increase temperature, humidity, and stomatal resistance of plants, and all of those parameters affect ET. It is common for people to associate higher ET with higher temperature, because evaporation rates do increase with higher temperature. However, evaporation is a physical process, whereas ET is both physical and biological. Increasing temperature will affect ET, but radiation, humidity, wind, and CO_2 concentration also affect ET. All of these factors are needed to properly assess the impact of climate change on plant water usage.

4. Estimating Crop Evapotranspiration

For many decades, scientists, engineers, and irrigation managers have used ET and the water balance method to determine agricultural and urban water demand for water resource planning and delivery and for on-farm and urban irrigation management. Because ET is affected by soil, plant, and atmospheric factors, spatial ET variation is common on different scales. In some climates, one can estimate ET using weather data of large areas, e.g., 50–100 km, but in other locations, microclimates can limit weather-based estimates of ET to small areas, e.g., less than 5 km. In general, the most common practice is to estimate "potential" or "energy-limited" crop evapotranspiration (ET_c) as the product of reference evapotranspiration (ET_{ref}) and a crop coefficient (K_c). Reference evapotranspiration is the energy-limited ET rate from a broad expanse of a well-watered vegetated surface, e.g., grass or alfalfa.

There are several methods to determine ET_{ref}, but the most common is to monitor weather over a large grass surface and use an equation to estimate the water use for a selected reference surface. Recently, the authors of [7,8] recommended fixed coefficients to estimate the canopy and aerodynamic resistances as inputs to the Penman–Monteith equation [9] to estimate ET_{ref} for 0.12-meter-tall and 0.50-meter-tall vegetated surfaces and assigned the symbol ET_0 for the short canopy and ET_r for the tall canopy. The equations were called "standardized reference evapotranspiration" because the procedures to compute the ET_0 and ET_r were standardized. Strictly speaking, the ET_0 and ET_r equations provide estimates of ET from a virtual surface with input coefficients to estimate the appropriate canopy and aerodynamic resistances. However, in practice, the ET_0 and ET_r approximate the ET of 0.12-meter-tall cool-season grass and 0.50-meter-tall alfalfa, respectively. Once the ET_{ref} is known, the ET_c is estimated as either $ET_c = ET_0 \times K_c$ or $ET_c = ET_r \times K_{cr}$, where K_{cr} is specific to ET_r. Because ET_0 is more widely used than ET_r, the remaining discussion will only use the ET_0 method. Crop coefficients are generally determined as $K_c = ET_c / ET_0$ by calculating the K_c ratio using measured daily ET_c and ET_0 from weather data. Global climate change could affect either the ET_0 or the K_c, so it is important to assess the impact of climate change on both factors.

5. Reference Evapotranspiration

To assess the possible impact of climate change on ET_0, Snyder et al. [10] used the standardized reference evapotranspiration (ET_0) equation for short canopies [7,8] that uses daily radiation, temperature, humidity, wind speed, and canopy resistance data to calculate ET_0. The following version of the ET_0 equation [11] was used for the analysis:

$$ET_0 = \frac{0.408\Delta(R_n - G) + \frac{900\gamma(e_s - e_a)u_2}{T+273}}{\Delta + \gamma(1 + 0.34u_2)} \qquad \left(\text{mm·d}^{-1}\right). \qquad (4)$$

In Equation (4), Δ (kPa·K^{-1}) is the slope of the saturation vapor pressure curve at the mean daily air temperature, R_n (MJ·m^{-2}·d^{-1}) is the net radiation over well-watered grass, G (MJ·m^{-2}·d^{-1}) is the soil heat flux density, γ (kPa·K^{-1}) is the psychrometric constant, T (°C) is the mean daily air temperature at a 1.5–2.0 m height, u_2 (m·s^{-1}) is the wind speed measured at 2 m above the ground, and e_s and e_a (kPa) are the saturation and actual vapor pressures of the air measured at a 1.5 m height. Information on how to compute R_n, G, Δ, etc. is provided in [7].

The aerodynamic resistance to sensible and latent heat transfer (r_a) occurs indirectly in two locations in Equation (4). The 0.34 in the denominator comes from the following:

$$\frac{r_c}{r_a} = \frac{70}{208/u_2} \approx 0.34u_2. \tag{5}$$

In Equation (4), the right-hand side of the numerator could be written as

$$\frac{900\gamma(e_s - e_a)u_2}{T + 273} = \frac{187200[\gamma(e_s - e_a)/(T + 273)]}{208/u_2} = \frac{187200[\gamma(e_s - e_a)/(T + 273)]}{r_a}. \tag{6}$$

Therefore, the 208 coefficient is also included in the numerator of Equation (4) (within the 900). The r_c = 70 s·m^{-1} was estimated from the typical stomatal resistance r_s = 100 s·m^{-1} (corresponding to stomatal conductance g_s = 0.010 m·s^{-1} = 10 mm·s^{-1}) for the actively transpiring C_3 grass leaf surface, which was estimated as half of the LAI = 2.88. Therefore, the canopy resistance for 0.12-meter-tall C_3 species grass r_c was calculated as

$$r_c = \frac{r_s}{0.5 \cdot LAI} = \frac{100 \text{ s·m}^{-1}}{0.5 \times 2.88} = 69 \approx 70 \text{ s·m}^{-1}. \tag{7}$$

Long et al. [12] studied the effect of CO_2 concentration on stomatal conductance of C_3 species plants, and the relationships they reported were used to estimate the impact of increasing CO_2 concentration on canopy resistance of the standardized reference surface. Assuming that the $r_c \approx 70$ s·m^{-1} applies to a 2004 CO_2 concentration of about 372 ppm, estimating a new r_c value for higher CO_2 concentration provides a method to estimate possible impacts of higher CO_2 on ET_o.

Based on eight climate models, the mean air temperature projection for California is about 2.2 °C by 2050 and 4.0 °C by 2100 [13]. The global CO_2 concentration is projected to reach about 550 ppm by 2050 and more than 700 ppm by 2100 [14]. Long et al. [12] reported that stomatal conductance for many C_3 species plants decreased by about 20% when the CO_2 concentration was increased from 372 to about 550 ppm from about 200 independent measurements. Assuming this is true for the stomatal conductance of 0.12-meter-tall C_3 species grass with a stomatal resistance of 100 s·m^{-1}, the stomatal conductance for C_3 grass should decrease from about 10 mm·s^{-1} to 8 mm·s^{-1}, which corresponds to r_s = 125 s·m^{-1}. Using the same approach used to calculate r_c in the ET_o equation [7], the r_c for 550 ppm is calculated as

$$r_c = \frac{r_s}{0.5 \cdot LAI} = \frac{125 \text{ s·m}^{-1}}{0.5 \cdot 2.88} \approx 87 \text{ s·m}^{-1}. \tag{8}$$

Thus, increasing CO_2 concentration from 372 to 550 ppm should increase canopy resistance of 0.12-meter-tall C_3 grass from 70 to 87 s·m^{-1}.

Roderick and Farquhar [15] observed that the global mean daily maximum temperature (T_x) increased by approximately 0.1 °C per decade and that the daily minimum temperature (T_n) increased by about 0.2 °C per decade, but there was no change in the vapor pressure deficit ($VPD = e_s - e_a$) in the few preceding decades. Cayan et al. [13] projected a 3 °C mean temperature increase by 2050 as a worst-case scenario for California. Therefore, Snyder et al. [10] evaluated the possible impact of climate change on ET_o by assuming that the mean air temperature would increase by 3.0 °C, with T_n and T_x increasing by 4.0 °C and 2.0 °C, respectively. Globally, the temperatures were rising, with no increase in relative humidity, which implies that the vapor pressure was also rising. In much of California, the mean daily dew point temperature (T_d) is often nearly equal to T_n, so it was assumed that the T_d like T_n would increase by 4 °C in 2050. It was also assumed that the CO_2 concentration would increase from 372 to 550 ppm by volume by 2050, which corresponds to increasing the r_c from 70 to 87 s·m^{-1}. It was assumed that aerodynamic resistance, which is dependent on the atmospheric stability, wind speed and plant canopy roughness, would not change.

Using 18 California weather stations from a wide range of climates, daily data from 2003 were used to calculate how the annual ET_o might change with the aforementioned climate projections. The results (Figure 1) indicate that the climate change scenario would have little impact on annual ET_o. The annual ET_o increased slightly where there were mean wind speeds less than 1.7 m·s^{-1}, and it decreased for wind speeds greater than 1.7 m·s^{-1}. However, the absolute magnitude of variation from the current annual ET_o was small for all weather stations. Based on the regression lines in Figure 1, the 4 °C rise in T_d and the 17 s·m^{-1} increase in r_c counteracted the impact of the 3 °C mean temperature increase.

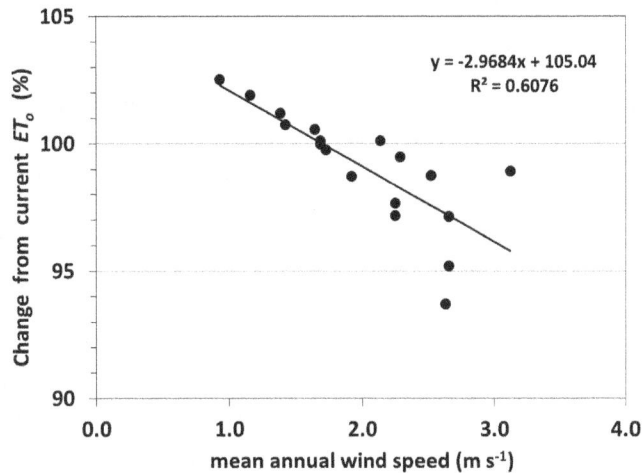

Figure 1. A plot of the change from current annual ET_o (mm·y^{-1}) assuming T_x increases by 2 °C, T_n increases by 4 °C, T_d increases by 4 °C, and r_c increases from 70 to 87 s·m^{-1} versus the mean annual wind speed. Figure 1 is from Snyder et al. [10].

6. Crop Coefficients

Allen et al. [7,8,11] reported a daily time step, standardized reference evapotranspiration equation for short canopies (Equation (7)). While the generated ET_o is actually for a virtual surface, i.e., having characteristics that determine the coefficients in Equation (4), the ET_o approximates the ET_c for a broad expanse of well-watered, cool-season grass. Allen et al. [7,8] also reported a daily time step equation for tall canopies (ET_r) that is expressed as

$$ET_r = \frac{0.408\Delta(R_n - G) + \frac{1600\gamma(e_s - e_a)u_2}{T+273}}{\Delta + \gamma(1 + 0.38u_2)} \quad \left(\text{mm·d}^{-1}\right). \tag{9}$$

Note that Allen et al. [7] used r_c = 45 s·m^{-1} and r_a = 118/u_2 to obtain the 0.38 = 45/118 in Equation (9). The canopy resistance for about 372 ppm CO_2 was derived using Equation (10) with r_s = 66.7 s·m^{-1} and LAI = 2.96.

$$r_c = \frac{r_s}{0.5 \cdot LAI} = \frac{66.7 \text{ s·m}^{-1}}{0.5 \times 2.96} = 45.1 \approx 45 \text{ s·m}^{-1}. \tag{10}$$

Assuming a 20% reduction in alfalfa stomatal conductance when changing from 372 to 550 ppm CO_2 [12], the stomatal conductance changes from 0.015 m·s^{-1} to 0.012 m·s^{-1}, the stomatal resistance changes from 66.7 to 83.3 s·m^{-1}, and the canopy conductance at 550 ppm is estimated as

$$r_c = \frac{r_s}{0.5 \cdot LAI} = \frac{83.3 \text{ s·m}^{-1}}{0.5 \times 2.96} = 56.3 \approx 56 \text{ s·m}^{-1}. \tag{11}$$

Strictly speaking, the ET_r equation estimates the ET_c for a virtual surface with characteristics represented by coefficients used in Equation (9), but the ET_r rates are approximately equal to the ET_c of a broad expanse of well-watered, 0.5-meter-tall alfalfa. Derivations and explanations of the ET_0 and ET_r equations are addressed in [7,8]. To accurately estimate ET_0 and ET_r, the input weather data for Equations (4)–(9) are collected over a broad expanse of well-watered grass. Derivation of the net radiation (R_n) and ground heat flux (G) in $MJ\cdot m^{-2}\cdot d^{-1}$ are presented in [7]. The T (°C) is the mean air temperature measured at a height between 1.5 and 2.0 m, u_2 ($m\cdot s^{-1}$) is the mean daily wind speed monitored at a 2 m height, the saturation vapor pressure (e_s) in kPa is calculated from T, and the actual vapor pressure (e_a) in kPa is measured at the same height as T. The slope of the saturation vapor pressure (Δ) in $kPa\cdot°C^{-1}$ and the psychrometric constant γ in $kPa\cdot°C^{-1}$ are computed as described in [7].

While little change in ET_0 rates is expected due to global climate change, crop coefficient (K_c) values might be affected depending on how climate factors change in the future. To evaluate the effect of changing weather variables on K_c values, an analysis was done using data from 49 California Irrigation Management Information System (CIMIS) weather stations [16]. Equations (4) and (9) were used to compute ET_0 and ET_r using canopy resistances corresponding to both 372 and 550 ppm CO_2. The stomatal conductances g_s = 0.010 $m\cdot s^{-1}$ (grass) and g_s = 0.015 $m\cdot s^{-1}$ (alfalfa) were reduced by 20% to 0.008 $m\cdot s^{-1}$ (grass) and 0.012 $m\cdot s^{-1}$ (alfalfa) at 550 ppm CO_2 based on [12]. The K_c for alfalfa was calculated for 372 ppm CO_2 for all 49 stations using the mean climate data from July 2003. For 550 ppm CO_2, the K_c for alfalfa was calculated using the mean climate data from July 2003, but with the monthly mean daily maximum, minimum, and dew point temperatures increased by 2 °C, 4 °C, and 4 °C, respectively. The wind speed and equations for aerodynamic resistance were not changed.

A plot of the 550 ppm versus 372 ppm K_c values (Figure 2) indicates that the crop coefficients are likely to decrease slightly for a wide range of climates. The biggest K_c decrease was about 0.03 when $K_c \approx 1.10$, and the smallest decrease was about 0.01 for $K_c \approx 1.40$. The differences were approximately 0.01 when K_c values were high. The response of K_c to the projected climate change is most likely related to the alfalfa canopy being more coupled with the environment than the grass canopy, which has a higher aerodynamic resistance. Plant canopies that are more coupled to the environment are more likely to exhibit a reduction in transpiration rate than a canopy that is more controlled by the aerodynamic resistance, e.g., grass. This analysis provides some evidence that coupling with the environment might lead to reductions in crop coefficients due to global climate change.

While changes in the CO_2 concentration can have an effect on the canopy resistance, this analysis showed that the same percentage decrease stomatal conductance can lead to a bigger reduction in transpiration if the aerodynamic resistance is lower and the canopy is more coupled with the environment. There is a lack of information on how crop coefficients of orchard and vine crops might respond to climate change, but the alfalfa example in this paper provides some insight. Since taller rougher canopies, e.g., orchards and vineyards, have considerably lower aerodynamic resistance than alfalfa, it is likely that the K_c values for orchard and vine crops might decrease even more.

For both C_4 and C_3 species, stomatal conductance is reduced by the increasing CO_2 concentration external to leaves [17]. However, small differences in C_4 and C_3 stomatal conductance responses to CO_2 concentration have been reported for grasses with the conductance differences decreasing at higher CO_2 concentrations [18]. On the other hand, grasses are more decoupled from the environment than taller, rougher plant canopies, so the K_c response of grass species due to a projected climate is likely to be smaller than the K_c response of taller rougher plant canopies with lower aerodynamic resistance.

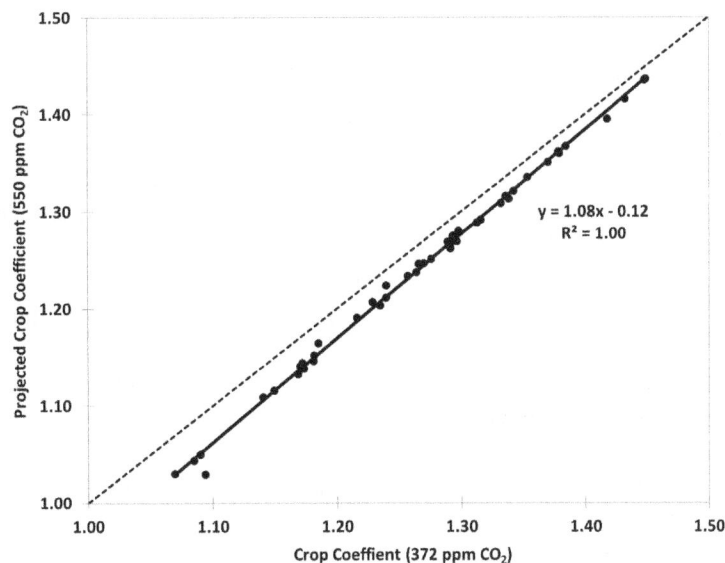

Figure 2. A plot of K_c values for alfalfa calculated from July 2003 mean daily weather data from 49 CIMIS weather stations in California for a climate with 550 ppm versus 372 ppm CO_2. For the projected 550 ppm CO_2 climate, the original daily maximum temperature data were increased by 2 °C, and the minimum and dew point temperatures were increased by 4 °C relative to July 2003 data. The wind speed and solar radiation were not changed from the original data. The alfalfa ET_c was computed using the ET_r equation (Equation (9)), and ET_0 was computed using Equation (4).

7. Conclusions

In summary, the evidence for anthropogenic global climate change due to the excessive release of CO_2 into the atmosphere is strong. While there are many possible impacts of climate change on horticultural crops, the effect of changes on water use of horticultural crops is particularly important. The fact that global climate change is dependent only on global receipt and reflection of solar radiation and GHGs provides strong evidence that anthropogenic global climate change is real and concerning. The global change can impact many weather factors in addition to affecting physical and biological factors, and these weather factors can affect plant growth and agricultural production. The FACE studies showed that increasing atmospheric CO_2 will decrease stomatal conductance, and this will increase canopy resistance of C_3 species plants, which decreases plant transpiration. Additionally, climate change projections indicate that water vapor content of the air will increase as temperature rises, and increased atmospheric H_2O also decreases transpiration. Using the standardized reference evapotranspiration equation for short canopies to calculate ET_0, the impact of projected increases in atmospheric temperature and CO_2 and H_2O concentrations were evaluated, and a large effect on ET_0 rates is unlikely. There was some evidence that ET_0 would increase slightly at low wind speeds and it would decrease as wind speeds increased. The calculation of alfalfa K_c values at 372 and 550 ppm CO_2 with an increase of 2 °C, 4 °C, and 4 °C for maximum, minimum, and dew point temperatures in the higher CO_2 environment, showed that K_c values will probably slightly decrease. This decrease is likely due to the higher coupling of canopy to the environment for the alfalfa canopy. There is little information available about how K_c values might change for tree and vine crops, but trees and vines are even more coupled to the environment, so an even bigger decrease in K_c seems plausible for the taller rougher canopies. A similar K_c response to climate change is expected for both C_3 and C_4 plants. While the evapotranspiration responses to global change seem small, the projected changes in precipitation and water storage in snowpack are large and could have devastating impacts on horticulture in some regions.

Conflicts of Interest: The author declares no conflict of interest.

References

1. IPCC. *Climate Change 2013. The Physical Science Basis Working Group I Contribution to the Fifth Assessment Report of the Intergovernmental Panel on Climate Change*; Cambridge University Press: New York, NY, USA, 2014; pp. 1–1535.

2. Dixon, G.R.; Collier, R.H.; Bhattacharya, I. An assessment of the effects of climate change on horticulture. In *Horticulture: Plants for People and Places*; Dixon, G.R., Aldous, D.E., Eds.; Springer: Dordrecht, The Netherlands, 2014; Volume 2, pp. 817–857.

3. Glenn, D.M.; Kim, S.H.; Ramirez-Villegas, J.; Läderach, P. Response of Perennial Horticultural Crops to Climate Change. In *Horticultural Reviews*, 1st ed.; Janick, J., Ed.; John Wiley & Sons, Inc.: Hoboken, NJ, USA, 2013; Volume 41, pp. 47–129.

4. Anderson, J.; Chung, F.; Anderson, M.; Brekke, L.; Easton, D.; Ejeta, M.; Peterson, R.; Snyder, R.L. Progress on incorporating climate change into management of California's water resources. *Clim. Chang.* **2008**, *87*, S91–S108. [CrossRef]

5. Arrhenius, S.A. On the Influence of Carbonic Acid in the Air Upon the Temperature of the Ground. *Philos. Mag. J. Sci.* **1896**, *41*, 237–276. [CrossRef]

6. NOAA Layers of the Atmosphere. Available online: http://www.srh.noaa.gov/jetstream/atmos/layers.html (accessed on 18 November 2016).

7. Allen, R.G.; Walter, I.A.; Elliott, R.L.; Howell, T.A.; Itenfisu, D.; Jensen, M.E.; Snyder, R.L. *The ASCE Standardized Reference Evapotranspiration Equation*; American Society of Civil Engineers: Reston, VA, USA, 2005; pp. 1–173.

8. Allen, R.G.; Pruitt, W.O.; Wright, J.L.; Howell, T.A.; Ventura, F.; Snyder, R.L.; Itenfisu, D.; Steduto, P.; Berengena, J.; Baselga Yrisarry, J.; et al. A recommendation on standardized surface resistance for hourly calculation of reference ETo by the FAO56 Penman-Monteith method. *Agric. Water Manag.* **2006**, *81*, 1–22. [CrossRef]

9. Monteith, J.L. Evaporation and Environment. *Symp. Soc. Exp. Biol.* **1965**, *19*, 205–234. [PubMed]

10. Snyder, R.L.; Moratiel, R.; Zhenwei, S.; Swelam, A.; Jomaa, I.; Shapland, T. Evapotranspiration Response to Climate Change. *Acta Hortic.* **2011**, *922*, 91–98. [CrossRef]

11. Allen, R.G.; Pereira, L.S.; Raes, D.; Smith, M. *Crop Evapotranspiration: Guidelines for Computing Crop Water Requirements*; Irrigation and Drainage Paper No. 56; FAO of United Nations: Rome, Italy, 1998; pp. 1–300.

12. Long, S.P.; Ainsworth, E.A.; Rogers, A.; Ort, D.R. Rising atmospheric carbon dioxide: plants FACE the future. *Ann. Rev. Plant Biol.* **2004**, *55*, 591–628. [CrossRef] [PubMed]

13. Cayan, D.; Luers, A.L.; Hanemann, M.; Franco, G. *Scenarios of Climate Change in California: An Overview*; CEC-500-2005-186-SF; California Energy Commission: Sacramento, CA, USA, 2006.

14. Prentice, I.C.; Farquhar, G.D.; Fasham, M.J.R.; Goulden, M.L.; Heimann, M.; Jaramillo, V.J.; Kheshgi, H.S.; Le Quéré, C.; Scholes, R.J.; Wallace, D.W.R. The Carbon Cycle and Atmospheric Carbon Dioxide. In *Climate Change 2001: The Scientific Basis. Contribution of Working Group I to the Third Assessment Report of the Intergovernmental Panel on Climate Change*; Cambridge University: Cambridge, UK; New York, NY, USA, 2002; pp. 183–238.

15. Roderick, M.L.; Farquhar, G.D. Changes in New Zealand Pan Evaporation since the 1970s. *Int. J. Climatol.* **2005**, *25*, 2031–2039. [CrossRef]

16. Snyder, R.L.; Pruitt, W.O. *Evapotranspiration Data Management in California*; American Society of Civil Engineers: New York, NY, USA, 1992; pp. 128–133.

17. Morison, J.I.L.; Gifford, R.M. Plant growth and water use with limited water supply in high CO_2 concentrations. I. Leaf area, water use and transpiration. *Aust. J. Plant Physiol.* **1984**, *11*, 361–374.

18. Hager, H.A.; Ryan, G.D.; Kovacs, H.M.; Newman, J.A. Effects of elevated CO_2 on photosynthetic traits of native and invasive C_3 and C_4 grasses. *BMC Ecol.* **2016**, *16*, 28. [CrossRef] [PubMed]

PERMISSIONS

LIST OF CONTRIBUTORS

Xiaoya Cai and Mengmeng Gu
Department of Horticultural Sciences, Texas A&M AgriLife Extension Service, College Station, TX 77843, USA

Gaurav Sharma, Naresh Prasad Sahu and Neeraj Shukla
Department of Floriculture and Landscape Architecture, Indira Gandhi Agricultural University, Krishak Nagar, Raipur, Chhattisgarh 492012, India

Chandran Somasundram, Zuliana Razali and Vicknesha Santhirasegaram
Institute of Biological Sciences & Centre for Research in Biotechnology for Agriculture (CEBAR), Faculty of Science, University of Malaya, Kuala Lumpur 50603, Malaysia

Henning Krause and Ulrike Grote
Institute for Environmental Economics and World Trade, Leibniz Universität Hannover, Königsworther Platz 1, 30167 Hannover, Germany

Rattiya Suddeephong Lippe
Institute of Development and Agricultural Economics, Leibniz Universität Hannover, Königsworther Platz 1, 30167 Hannover, Germany

Uwe Schindler, Lothar Müller and Frank Eulenstein
Leibniz Centre for Agricultural Landscape Research (ZALF), Institute of Landscape Hydrology, Eberswalder St. 84, Müncheberg D15374, Germany

Robert Veberic
Department of Agronomy, Biotechnical Faculty, University of Ljubljana, Jamnikarjeva 101, Ljubljana 1000, Slovenia

Guglielmo Costa, Lorenzo Rocchi, Brian Farneti, Nicola Busatto, Francesco Spinelli and Serena Vidoni
Department of Agricultural Science, Alma Mater Studiorum, University of Bologna, Bologna 40127, Italy

Rodel Maghirang, Maria Emblem Grulla, Gloria Rodulfo, Ivy Jane Madrid and Maria Cielo Paola Bartolome
Institute of Plant Breeding, College of Agriculture, University of the Philippines Los Baños, Laguna 4031, Philippines

Haozhe Gan, Erin Charters, Robert Driscoll and George Srzednicki
School of Chemical Engineering, University of New South Wales, Sydney 2052, Australia

Babul C. Sarker
Principal Scientific Officer, Pomology Division, Horticulture Research Centre, Bangladesh Agricultural Research Institute, Joydebpur, Gazipur 1701, Bangladesh

Mohammad A. Rahim
Department of Horticulture, Bangladesh Agricultural University, Mymensingh 2202, Bangladesh

Douglas D. Archbold
Department of Horticulture, University of Kentucky, Lexington, KY 40546-0091, USA

Francesco Giovanni Ceglie
Organic Agriculture Department, Mediterranean Agronomic Institute of Bari, CIHEAM-IAMB, Via Ceglie, 9, Valenzano, BA 70010, Italy
Department of Science of Agriculture, Food and Environment, University of Foggia, Via Napoli 25, Foggia, FG 71122, Italy

Maria Luisa Amodio and Giancarlo Colelli
Department of Science of Agriculture, Food and Environment, University of Foggia, Via Napoli 25, Foggia, FG 71122, Italy

Elda Esguerra, Dormita Del Carmen, Roxanne Delos Reyes and Ryan Anthony Lualhati
Crop Science Cluster, Postharvest and Seed Sciences Division, Postharvest Horticulture Training and Research Center (PHTRC), College of Agriculture, University of the Philippines Los Baños (UPLB), Laguna 4031, Philippines

Almudena Simón, Carmen García, Fernando Pascual, Leticia Ruiz and Dirk Janssen
Instituto Andaluz de Investigación y Formación Agraria, Pesquera, Alimentaria y de la Producción Ecológica (IFAPA), Centro La Mojonera, Camino de San Nicolas 1, La Mojonera 04745, Spain

Kanchit Thammasiri
Department of Plant Science, Faculty of Science, Mahidol University, Rama VI Road, Phayathai, Bangkok 10400, Thailand

Frank Eulenstein
Leibniz-Centre for Agricultuiral Landscape Research (ZALF) Müncheberg, Eberswalder Straße 84, Müncheber 15374, Germany
Department Agro-Chemistry, Kuban State Agrarian University, Krasnodar 350044, Russia

Marcos Alberto Lana, Marion Tauscke and Axel Behrendt
Leibniz-Centre for Agricultuiral Landscape Research (ZALF) Müncheberg, Eberswalder Straße 84, Müncheber 15374, Germany

Sandro Luis Schlindwein
Departamento de Engenharia Rural, Universidade Federal de Santa Catarina, Florianópolis 88034-000, Brazil

Askhad Khasrethovich Sheudzhen
Department Agro-Chemistry, Kuban State Agrarian University, Krasnodar 350044, Russia

Edgardo Guevara and Santiago Meira
Department of Crop Production INTA — Instituto Nacional de Tecnologia Agropecuaria, Pergamino 2700, Argentina

Carla Benelli
Trees and Timber Institute (IVALSA-Istituto per la Valorizzazione del Legno e delle Specie Arboree), National Research Council, 50019 Florence, Italy

Panida Boonyaritthongchai, Chalida Chimvaree, Mantana Buanong, Apiradee Uthairatanakij and Pongphen Jitareerat
Division of Postharvest Technology, King Mongkut's University of Technology Thonburi, Bangkok 10150, Thailand

Michael D. Cahn
University of California, Cooperative Extension, Monterey County, 1432 Abbott St., Salinas, CA 93901, USA

Lee F. Johnson
NASA ARC-CREST/CSUMB, MS 232-21, Moffett Field, CA 94035, USA

Ruben Madayag Gapasin, Jesusito Laborina Lim, Elvira Lopez Oclarit and Mannylen Coles Alde
Department of Pest Management, College of Agriculture and Food Science, VisayasState University, Visca, Baybay City 6521-A, Philippines

Leslie Toralba Ubaub
University of Southeastern Philippines (USeP Tagum Campus), Davao City 6521-A, Philippines

Scott Stebner
Former Graduate Research Assistant, Department of Communications and Agricultural Education, Kansas State University, 1612 Claflin Rd., Manhattan, KS 66506, USA

Cheryl R. Boyer
Department of Horticulture and Natural Resources, Kansas State University, 1712 Claflin Rd., Manhattan, KS 66506, USA

Lauri M. Baker
Department of Communications and Agricultural Education, Kansas State University, 1612 Claflin Rd., Manhattan, KS 66506, USA

Hikaru H. Peterson
Department of Applied Economics, University of Minnesota, 1994 Buford Ave., St. Paul, MN 55108, USA

Lauren M. Garcia Chance, Michael A. Arnold, Charles R. Hall and Sean T. Carver
Department of Horticultural Sciences, Texas A&M University, College Station, Texas 77843-2133, TX, USA

Apiradee Uthairatanakij, Sukanya Aiamla-or and Pongphen Jitareerat
Division of Postharvest Technology, King Mongkut's University of Technology Thonburi, Bangkok 10150, Thailand

Ashariya Maneenoi
Lake Rajada Office, Adams Enterprises Ltd., Bangkok 10110, Thailand

Richard L. Snyder
Department of Land, Air and Water Resources, University of California, Davis, CA 95616, USA

Index

www.ingramcontent.com/pod-product-compliance
Lightning Source LLC
Chambersburg PA
CBHW080631200326

41458CB00013B/4583